Lecture Notes in Electrical Engineering

Volume 392

About this Series

"Lecture Notes in Electrical Engineering (LNEE)" is a book series which reports the latest research and developments in Electrical Engineering, namely:

- Communication, Networks, and Information Theory
- Computer Engineering
- Signal, Image, Speech and Information Processing
- Circuits and Systems
- Bioengineering

LNEE publishes authored monographs and contributed volumes which present cutting edge research information as well as new perspectives on classical fields, while maintaining Springer's high standards of academic excellence. Also considered for publication are lecture materials, proceedings, and other related materials of exceptionally high quality and interest. The subject matter should be original and timely, reporting the latest research and developments in all areas of electrical engineering.

The audience for the books in LNEE consists of advanced level students, researchers, and industry professionals working at the forefront of their fields. Much like Springer's other Lecture Notes series, LNEE will be distributed through Springer's print and electronic publishing channels.

More information about this series at http://www.springer.com/series/7818

Massimo Conti · Natividad Martínez Madrid
Ralf Seepold · Simone Orcioni
Editors

Mobile Networks for Biometric Data Analysis

Springer

Editors
Massimo Conti
Dipartimento di Ingegneria dell'Informazione
Università Politecnica delle Marche
Ancona
Italy

Natividad Martínez Madrid
Reutlingen University
Reutlingen
Germany

Ralf Seepold
HTWG Konstanz
Konstanz
Germany

Simone Orcioni
Dipartimento di Ingegneria dell'Informazione
Università Politecnica delle Marche
Ancona
Italy

ISSN 1876-1100 ISSN 1876-1119 (electronic)
Lecture Notes in Electrical Engineering
ISBN 978-3-319-39698-9 ISBN 978-3-319-39700-9 (eBook)
DOI 10.1007/978-3-319-39700-9

Library of Congress Control Number: 2016943066

Printed on acid-free paper

This Springer imprint is published by Springer Nature
The registered company is Springer International Publishing AG Switzerland

Preface

Prevention and treatment of common and widespread chronic diseases is a challenge in any modern society and vitally important for health maintenance in aging societies. Capturing biometric data is a cornerstone for any analysis and treatment strategy. Latest advances in sensor technology allow accurate data measurement in a nonintrusive way.

In many cases, it is necessary to provide online monitoring and real-time data capturing to support patients' prevention plans or to allow medical professionals to access the current status. Different communication standards are required to push sensor data and to store and analyze them on different mobile platforms.

The objective of this book is to show new and innovative approaches dedicated to biometric data capture and analysis in a nonintrusive way maintaining mobility. Examples can be found in human centered ambient intelligence or even in methodologies applied in automotive or home automation.

The book is organized in five parts:

- ICT for Health Care
- Embedded Systems for Biometric Data Acquisition
- Biometric Data in Automotive and Home Applications
- Biometric Data Analysis of Cardiac Signals
- Gait Analysis

The book presents a selection of papers presented in the international workshop "Mobile Networks for Biometric Data Analysis" held in Ancona. Italy, during October 30–31, 2014, organized by the Università Politecnica delle Marche (Italy), Hochschule Konstanz (Germany), Hochschule Reutlingen (Germany) approved and financed by DAAD German Academic Exchange Service.

Ancona, Italy — Massimo Conti
Reutlingen, Germany — Natividad Martinez Madrid
Konstanz, Germany — Ralf Seepold
Ancona, Italy — Simone Orcioni

Preface

Contents

Part I
ICT for Health Care

The Role of ICT in the Italian Health Care System

Maurizio Ciaschini, Monica De Angelis, Andrea Monteriù, Rosita Pretaroli, Francesca Severini and Claudio Socci

Abstract The size of the Italian expenditure for the health care on Gross Domestic Product (GDP) and on total public expenditure, is in line with those of the major industrialized countries, and somewhat lower than the European average. The Italian healthcare sector does not simply face a contraction of the expenditure: in effect, it should be highlighted that Italy currently does not spend "a lot" and, above all, invests "little" in the mentioned sector. Nevertheless, the health sector can be a very important flywheel for the economic recovery, provided that a lot of attention is paid, and the best resources are invested. In this work, the authors intend developing an analysis of the desirability of investment on Information and Communication Technologies (ICT) in the health sector, and the consequent compatibility with current expenditure, by identifying at the outset the channels multipliers related to the resources at present allocated to the current health expenditure and capital account. For this analysis, two different simulation scenarios have been considered: in the first, an increase in Health services demand by Government has been studied, while in the second, an increase in final demand of Health services by Households has been considered.

M. Ciaschini · R. Pretaroli · F. Severini · C. Socci
Dipartimento di Economia e Diritto, Università di Macerata, Macerata, Italy

M. De Angelis
Dipartimento di Management, Università Politecnica delle Marche, Ancona, Italy

A. Monteriù (✉)
Dipartimento di Ingegneria dell'Informazione, Università Politecnica delle Marche, Ancona, Italy
e-mail: a.monteriu@univpm.it

M. Conti et al. (eds.), *Mobile Networks for Biometric Data Analysis*,
Lecture Notes in Electrical Engineering 392, DOI 10.1007/978-3-319-39700-9_1

1 Introduction: The Importance of Investing in Health Sector to Protect the Right to Healthcare[1]

Among current problems of the Italian health system, we can include the lack of investment and expenditure restraint. Surely years of crisis and negative expectations about the economic system recovery still make the curbing of public spending a priority. However, in its latest report on local authorities,[2] the Court of Auditors noted that without investments it becomes problematic to maintain the essential levels of care. It is worth noting that since 2010, health care spending has been decreasing about 3 bn, since, between 2010 and 2013, the spending review resulted in a cut of 2.8 % of spending, mainly affecting the pharmaceutical fixed agreement (also for the increase in ticket) and the cost of staff (especially with the block staff turnover).

No contraction came from spending on goods and services, while there are continuous overruns on hospital pharmaceutical. Given these data, the Court calls for better community care and focus on infrastructure where there are fewer services. The Court, in other words, in addition to photographing and analyzing the situation, indicates areas on which to intervene, warning about the future risks.

Unfortunately, the need to restore public finances and to re-balance the weight of the health sector on government spending has often led to consider the health system as a source of potential costs and inefficiencies, rather than as a strategic sector that plays a crucial economic and social importance for the entire country. The statement by the supreme public accounts' auditing body, therefore, seems to herald new perspectives.

Over the last years, this has been triggering the sequence of public control operations mainly oriented to the short term, with a logic of linear spending cuts, which have tended, however, to fundamentally weaken the areas where public health has more difficulty and to fail to distinguish the virtuous areas from the non-virtuous ones as well as to reward the virtuous.

But the issue of health sector is not simply related to the contraction of the expenditure[3]: you must be well aware, in fact, that Italy currently does not spend "a lot" and, above all, invest "little" in the mentioned sector.

---

[1]This paragraph has been autored by M. De Angelis.

[2]Corte dei Conti, Relazione ente territoriali Sanità, 2014 in http://www.quotidianosanita.it/allegati/create_pdf.php?all=7650214.pdf; See also: Rapporto 2013 sul Coordinamento della Finanza Pubblica- 28.05.2013 in http://www.regioni.it/it/show-presentazione_del_rapporto_2013_sul_coordinamento_della_finanza_pubblica_-_28052013_/news.php?id=299026.

[3]In fact the impact of the Italian public health spending on GDP and on total public spending is in line with those of more industrialized countries and somewhat lower than the European average. According to the latest available data from Istat (2012), in Italy public health spending per capita is well below the average of OECD countries. See http://www.sanita.ilsole24ore.com/art/dibattiti-e-idee/2014-02-10/dallassociazione-dossetti-class-action-130644.php?uuid=Ab0MKYMJ.

Perhaps, then, it is necessary to focus attention on issues related to the redistribution (reallocation) of the expenditure rather than on the simple policy of contraction.

The immediate requirement is therefore to address the issue of "how to spend" (what is important is to spend better and focus on quality of expenditure) and at the same time to prepare the system to support future demand [1][4]: these are the elements that make up the ratio of the future direction of policy maker who is responsible for the political accountability of health care spending.

What the Court of Auditors says is clear in the absence of these structural interventions you would risk worsening the economic sustainability of the system of cares and to return to an accounting statement out of control.

Moreover, in the current situation, it is not just a problem of economic sustainability of public finances, but also of competitiveness of the economic system: for instance, with the new rules on cross-border healthcare, the countries with a range of services that do not match the needs of people will be bound to lose competitiveness.

In this respect, and with the opening of the healthcare abroad thanks to the Directive 2011/24/UE the demand for health services that will not be satisfied on the Italian territory will impact hard on the balance of payments precisely because of imports of care services [2].

The imported services will trigger a drain of economic resources, both public and private, giving benefit to other health systems with significant consequences for the Italian economy, in particular for the ability to generate income and employment with unfavorable outcomes on the quality of human capital of national health sector.

Indeed, health sector can be a very important flywheel for the economic recovery, but to make this possible it is necessary to pay a lot of attention and invest the best resources.

Tangible experiences along this direction are rapidly growing and show how the production of health services play the role of key productive sector within a local context.

In this sense, for example, you can highlight the need to create health districts that play the role of very supply chains where the main activity made by healthcare services is able to trigger the production of goods and services from different sectors (chemical, engineering, transport, etc.) with many advantages in terms of employment.[5]

Economic policies in the health sector, moreover, should not only meet the need to find resources in view of the containment of the main indicators related to structural public deficit and debt; rather they should be (and create an occasion to remove structural inefficiencies and prevent the system from being vulnerable in

---

[4]The authors argue that the problem is not the sustainability of the system rather than its survival.

[5]An example is the industrial district wellness spa: http://ideario.formez.it/content/distretto-produttivo-del-benessere-termale-siciliano.

front of protracted crises that could seriously undermine its stability.[6] The Court of Auditors itself, in its mentioned Report, argues that further possible savings from efficiencies, if not reinvested mainly in the areas where the supply of health services is most lacking—such as, for example, in helping local and home care or in technological modernization and infrastructure—could bring out, in the medium term, serious deficits in healthcare, more marked in the southern regions, where these shortcomings are relatively more frequent. In many of these Regions the health spending review—started with the so called Repayment plans [3]—to be effective without compromising the principle of equity in the LEA (basic levels of healthcare), will have to be more selective and reinvest resources in health services which are relatively inadequate to the needs expressed, by taking them from areas where there is still room for inefficiencies to be recovered, as in the case of purchases of goods and services not performed by use of regional central procurement, of non-appropriateness of prescriptions and of services rendered in hospitals. The development of preventive medicine programs too, improving the general health of the population, is an instrument that, in the medium and long term could generate lower healthcare costs and more appropriate use of resources.

Moreover, all the Regions will also have to make a more careful and detailed annual planning of new needs in healthcare in their territories, in order to adjust the supply of services to the new needs, which are also caused by the growing weight of degenerative diseases resulting in the gradual aging of the population, or by the new, and relatively more expensive "biological" drugs with high technological content, which are able to treat more effectively and selectively different categories of diseases.

Unfortunately, due to the difficulty of decision-making and layering regulations, health sector is now at risk of not improving in quality and efficiency. It is true that "the health care system has confirmed the progress already highlighted in recent years in cost containment and absorption of unjustified deficits in the management; but it has now become crucial and urgent to strengthen the tools available to local government to accelerate action of re-adjustment of structures and improvement of the appropriateness of the services provided to citizens, bringing up to speed the review of the mechanisms that govern the operation of the sector".[7]

---

[6]A serious policy of investments in ICT, in fact, may generate significant improvements both from the point of view of the ability of care and with respect to containment of expenditure. For example, one of the cornerstones to a more effective and efficient health care system is digital innovation: according to the data, if the Italian health system realized completely the potential of digital technologies, it could save 6.8 billion a year—much more than the cuts expected in 2015 and 2016 for financing public health sector. See—as an example—the case of the Region "Marche": http://www.ecommunity.marche.it/AgendaDigitale/tabid/174/Default.aspx.

[7]The key elements are: "an evaluation network that allows an effective monitoring capable of making the benchmarking technique a tool to control and effectively manage the system; a redesign of accounting rules that gradually helps to strengthen the reasons for a structural adjustment; a central government that, in comparison to the local government, accompanies the pursuit of cost containment interventions with structural adjustment, with improvements in equipment and with investments in human resources training".

These are the critical factors that can create, from the economic point of view, large diseconomies of scale, of variety and location; diseconomies that must be tackled with the introduction of innovations of both process and product.

Just with regard to the benefits that can bring innovations of process and/or product, in this work we intend to develop an analysis of the desirability of investment in ICT in the health sector and the consequent compatibility with current spending by identifying at the outset the channels multipliers related to the resources at present allocated to the current health expenditure and capital account.

In this way it will be possible to identify the different multiplicative intensity of investment spending compared to current spending.

This first result is the premise to justify and support the choice of policy makers to redistribute resources from the pockets of inefficiency in current spending towards programs for research and innovation in ICT (in other words: it is not beneficial to spend less but you need to spend better).

As a result this not only generate immediate and positive effects on income and employment, but you would have in addition a structural change of the production processes of health care services towards greater efficiency and effectiveness.

2 Main Areas of Digital Innovation for the Italian Health Care System

The Information and Communication Technologies applied to Italian health sector is a possible innovation but especially needed. In effect, in this sector, the ICT can significantly help to improve the offered care services with savings opportunities for the citizens and for the National Health Service, in terms of time and especially money. Numerous studies have been carried out, and by the introduction of ICT in healthcare, it has been proven to save, at European level, billions of euros. Moreover, the quality and efficiency of health services must be improved, and ICT can help in this process. In this section, we analyze the main, significant and priority areas of ICT with high desirability of investment in the Italian health sector.

The Italian national healthcare service (SSN) was created in 1978 to replace a previous system based on a multitude of insurance schemes. The SSN was inspired by the British National Health Service and has two underlying principles. Firstly, every Italian citizen and foreign resident has the right to healthcare and, secondly, the system covers all necessary treatments. Local Health Units (USL) are responsible for the management of all health services in their area and private providers can also operate within the SSN.

The public health expenditure consists of recurrent and capital spending from Italian government (central and local) budgets, external borrowings and grants (including donations from international agencies and nongovernmental organizations), and social (or compulsory) health insurance funds. The Italian healthcare public expenditure is significantly increased during the last years, from 5.1 % of

GDP in 1979 to 7.1 % in 2012 (see Fig. 1 and Table 1), and currently it is over 9 % of GDP.

About the Italian total expenditure on healthcare, in this work it is crucial to know the total amount devoted to the ICT, composed by current expenditure and investments on ICT. From 2010 to 2013, the current expenditure on ICT is not changed, while the total investments on ICT are drastically decreased (see Fig. 2).

In order to understand how the limited ICT resources are allocated on the Italian health care system, it has been characterized the main areas of innovation, analyzing not only current investments associated with them, but also the key importance attributed to the various innovation areas and their development barriers (e.g., complexity of realization, high required investments, resistance to change, etc.). This analysis [4] outlines three main clusters of ICT innovation:

- *Breakthrough*: radical innovation fields with strong strategic benefits that, however, require major changes in the processes and organization. These include Electronic Medical Records (EMRs), document management systems, digital services for the citizen, clinical governance solutions, business intelligence solutions for medication management, business continuity systems, disaster recovery and interoperability solutions integrated with regional and/or national health care digital systems.
- *Must-have*: fields with limited benefits and impacts on the key performance of the health structures, but without barriers to development. Some examples are:

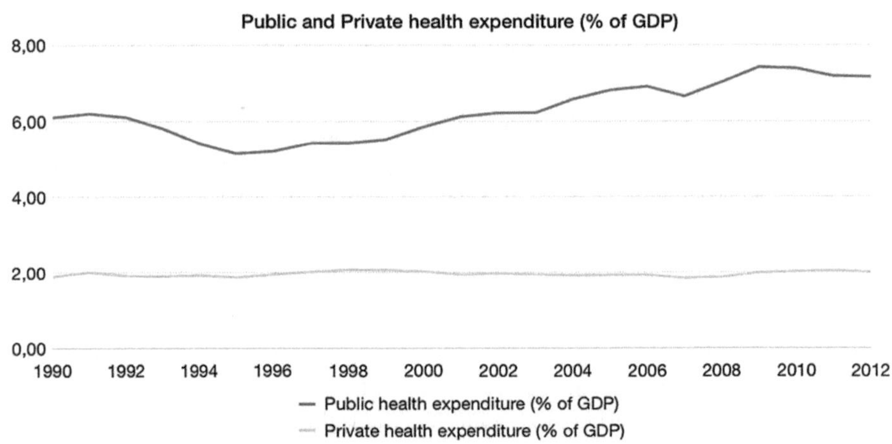

Fig. 1 Public and private health expenditure (% of GDP)

Table 1 Growth rate of public health expenditure (%)

	Average nominal growth rate	Average real growth rate
1978–1986	17.9	3.7
1987–1995	8.1	2.7
1996–2012	6.1	3.9

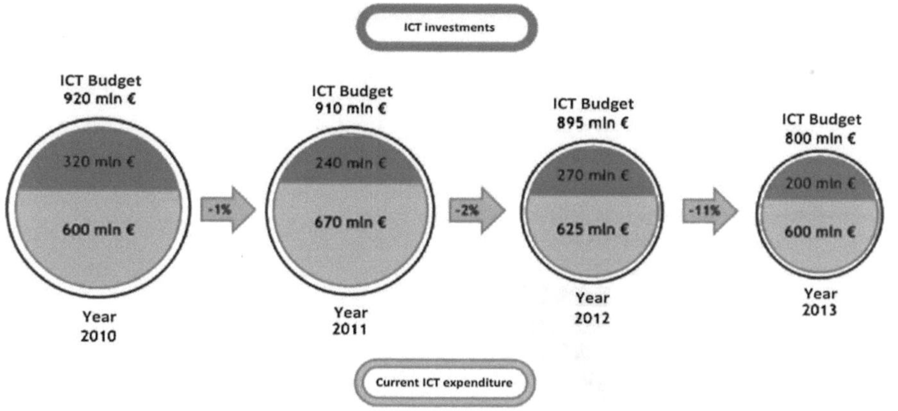

Fig. 2 ICT expenditure in the Italian healthcare system: from 2010 to 2013

the administrative management systems, human resources management systems, front-end systems and ward management systems.

- *Challenge*: fields that appear to be less important to the pursuit of the strategic objectives, and which require significant changes for their implementation, such as overcoming high cultural barriers. Among these fields, there are the mobile health, cloud computing, ICT solutions to support medicine and care (including tele-medicine), ICT solutions for social assistance and electronic social record (that is, however, extremely important for the 94 % medical directors).

Breakthrough is the cluster on which the important is strongly perceived. The *Must-have* cluster is the most consolidated one because referring to available ICT solutions, also coming from other public medicine sectors. Conversely, the *Challenge* cluster contains areas which currently are identified as most problematic, but which can represent the real challenge for the development of the Italian health sector in the near future.

In order to determine the significant and priority fields of ICT with high desirability of investment in the Italian health sector, it is critical to determine how the limited ICT resources are now allocated. The research study carried out by the "ICT in Health Observatory" [4] shows not only the overall ICT spending on health care, but also the budget allocated to the different ICT fields, that is the ICT budget that the Italian health authorities have allocated in 2013 for each field. The results of this analysis are resumed in Fig. 3.

From this study, it is glaring that the Italian investments on ICT are mainly oriented to the development of hospital care solutions, while to the territorial and social developments are devoted only limited resources. This is primarily due to the fact that the decision-makers have limited vision on the challenging ICT solutions based on cloud computing, on mobile health and on innovative ICT digital solutions. In the following, we study and analyze the compatibility with current

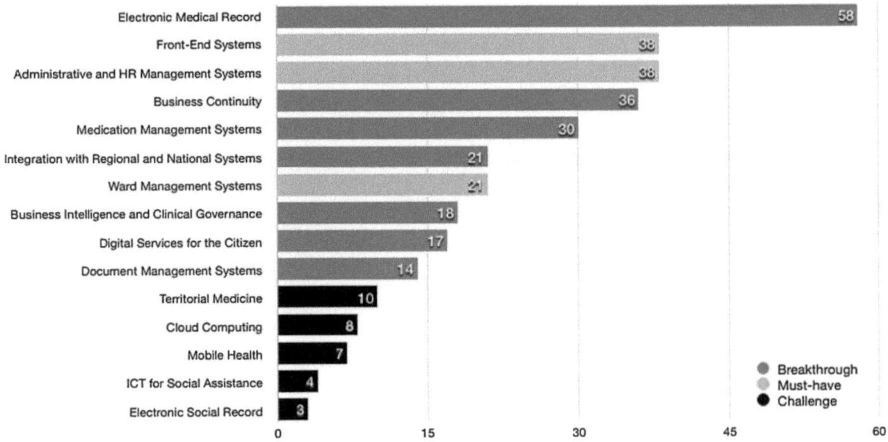

Fig. 3 Budget (Mln euro) on different ICT fields of the Italian health system in 2013

spending to invest on these challenging and crucial ICT fields, by identifying at the outset the channels multipliers related to the resources at present allocated to the current health expenditure and capital account.

3 The Integration Between Human Health Services and ICT Industry

The increase in efficiency and quality of producing health is deeply linked with the possibility to introduce ICT in the process of creation of Health services [5]. As for the other services sectors the industry of Health care services is ruled by innovation and it was shown that a proper policy of investment in ICT for health care procedure has the potential to drive economic growth and innovation [6, 7].

The relevance of Health care services and its relation with the ICTs production can be stressed given a proper data showing the production setting by mean it is possible to measure the incidence of ICT expenditure on total public health expenditure. Actually, public health system spent in ICT about 1.3 billion euro for the year 2011 [4]. This value is affected by the cyclical trend of the economic variables that has substantially been reduced. In per capita terms the total expenditure on ICT in public health services is around 22 euro per inhabitant. Comparing this value to those of the main European countries, it is possible to show a clear difference. In Germany, in fact, the per capita expenditure in ICT in health is about 36 euro, in France rises to 40 euro and Britain reaches a value of 60 euro. The countries of northern Europe and especially Sweden and Denmark, are above these prices with a per capita total expenditure on ICT by 63 and 73 euro.

According to the definition of ICT [8, 9], we can quantify the absorption of ICT in the production of health care services using the available data base represented by the Social Accounting Matrix for Italy, 2009 [10]. The quantification of ICT absorption is made considering the entire production of health services (both public and private) and taking into account that some ICT productions are included in activities that are more traditional. The approximation does not affect the results of the analysis.

In this study, we include in ICT the following sectors: Computer, electronic and optical products (17), Electrical equipment (18), Wholesale trade services, except of motor vehicles and motorcycles (29), Publishing services (37), Motion picture, video and television programme production services, sound recording and music publishing (38), Telecommunications services (39), Computer programming, consultancy and related services; information services.

Using the USE table (the table of intermediate absorption) it is possible to identify the amount of intermediate goods of the ICT typology purchased by health industry according the definition of ICT industry. For the year in which the analysis is performed the production of Health care services is around 107,412 million of euro represented by value added, the 61.53 %, and absorption of intermediate goods, the 38.47 %. Then the value added is represented by compensation of employees (69.7 %), mixed income (16.3 %), gross operating surplus (9.5 %) and finally other taxes less subsidies on domestic production (4.5 %). Within the intermediate consumption the absorption of ICT goods are stressed in Fig. 4.

The intermediate consumption of ICT in order to produce Human Health care services is 2816 million of euro that represents the 2.6 % of the whole health care production. As it can be seen in Fig. 4 ICT goods that play a major role in the

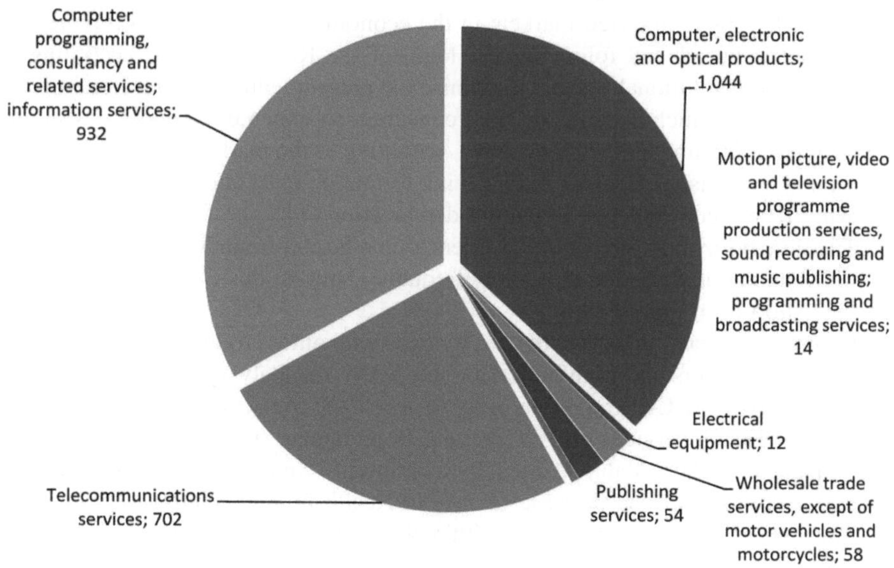

Fig. 4 ICT intermediate expenditures by Human Health care services

production of Human Health care are: Computer programming, consultancy and related services; information services (37.03 %), Computer, electronic and optical products (33.10 %) e Telecommunications services (24.93 %).

4 Assessing the Role of Human Health Services Demand in ICT Industry with a Dynamic CGE Model

In this chapter we aim to quantify the impact on GDP, total output by commodity and income by institutional sectors of an increase of 2 % in Health final demand. In particular, we consider two different scenarios:

- SIM1 we assume an increase in Health services demand by Government of 17,962 millions of Euro;
- SIM2 we consider an increase in final demand of Health services by Households for 2752 millions of Euro.

In both scenarios the fiscal policy is made under budget constraint.

We use the recursive dynamic Computable General Equilibrium (CGE) model where the behaviour of agents depends on adaptive expectations. It is calibrated on the basis of SAM structure and the blocks of the SAM determine the blocks of the equations in order to replicate and provide a fast check of macroeconomic aggregates. The model is characterized by nested production function (CES technology to determine VA, LEONTIEF to aggregate the intermediate goods and VA to determine the domestic production), equalization between supply and demand across all the interconnected markets in the economy. The dynamic component in the model is introduced following the logic of the Ramsey model, according to which all the Institutional Sectors maximise the present value of their intertemporal utility function, which depends on final consumption expenditure and gross saving subject to the lifetime budget constraint. According to the market clearing condition for capital, any change in gross fixed capital formation must affect the capital yearly growth given a constant rate of capital depreciation [11].

The time horizon is finite and considers a time period from 2014 to 2020. Given that ICT and related productions show frequent changes, this does not allow to go beyond short-term time horizons.

Given the interaction of Health care services with other production processes and Institutional Sectors, as highlighted in the SAM for Italy (year 2009), we can calibrate a dynamic CGE model in order to assess the impact of an increase in final demand of Health services on ICT sectors. In particular, it is possible to highlight the impact on GDP, Value Added by commodity and Income by Institutional Sectors resulting from an increase of 2 % in Health final demand.

The use of technological tools within the health sector, is a well-being and development indicator and, at the same time, a potential factor to increase the public and private healthcare expenditure. Indeed, the increase on ICT expenditure is an investment that has two main effects. The first effect is a classic problem of aggregate demand, and the second effect is a problem of income generation. In both cases where the purchase has to be attributed to the public administration or to the private enterprises, this increased demand on capital account is a relevant stimulus for the whole productive and economic system. The multiplier principle connected to the production of required goods, involves direct and indirect effects also on the income generation. Then, the increased demand is accompanied by an increased, earned and available income, whose overall effect, which takes also into account the possible "substitutability" issue, is certainly relevant. The second effect, correlated to the increase of the final demand on capital account, is instead closely connected with the adopted transmission channel. It is certainly not indifferent who and how the Institutional Sectors act in the economic system, whether they are households firms, institutional private sectors and public administration. The purchase of new production means, in particular those of new generation, can generate significant effects on production costs. In detail, for the public administration, besides the potential positive effect described above, this determines a clear situation where the new production mean automatically generate its own demand. In other words, the demand on capital account from the public administration leads to an expenditure increase on current account with potential negative effects on the system in the event that it is necessary to respect the stability regulations and constraints.

About the simulated policies, it is therefore necessary to consider how the different decision-making centers for health expenditure, can potentially exert different economic effects. This study is conducted in case it does prevail the positive part of the increase on income, or in case it does prevail the negative part with respect to the balance on macroeconomic aggregates used to test the performance of the policies. By analyzing the obtained results, namely by the GDP and by its trend over the time, it is possible to observe that the overall effect is always positive (see Fig. 5). Instead, by measuring the deviations of the GDP with respect to the growth path defined by the benchmark (absence of policy), there is a dominance of the second scenario on the first policy, which is mainly due by the effect of the budget constraint on the public administration.

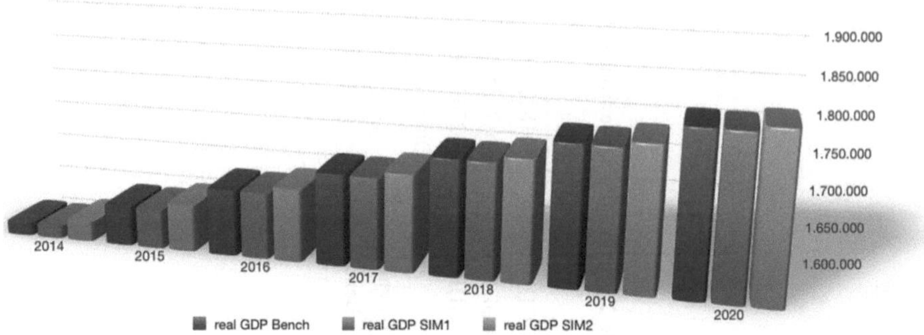

Fig. 5 Real GDP change (millions of Euros) from 2014 to 2020 for Benchmark, SIM1 and SIM2

Table 2 Real GDP change from 2014 to 2020 for the Italian economy (millions of Euro)

	2014	2015	2016	2017	2018	2019	2020
Real GDP Bench	1618227.9	1655274.5	1690308.5	1723783.1	1755994.7	1787049.8	1816823.8
Real GDP Sim1	1617225.3	1654261.6	1689284.3	1722746.8	1754945.3	1785986.3	1815745.4
Real GDP Sim2	1627637.6	1664126.7	1698560.4	1731398.6	1762942.8	1793302.9	1822355.4
	Differences between policy scenarios and benchmark						
Sim1-Bench	−1002.637	−1012.91	−1024.159	−1036.357	−1049.483	−1063.512	−1078.416
Sim2-Bench	9409.7068	8852.2094	8251.9708	7615.4753	6948.0193	6253.0776	5531.6132

In effect, Table 2 shows a negative difference on the GDP trend in case the public administration combine to the increase of the investment expenditure, a reduction of the current expenditure for all other types of goods. The lower growth stems from the fact that other expenditure sectors, of the government consumption, record significant multiplier effects and, in this case, they operate decreasing the overall demand.

From the ICT point of view, it is necessary to highlight how these increments on capital account should generate a positive effect on the economy system both in absolute and relative terms, i.e., as deviations by the benchmark. This effect could be mitigated by the work of the condition of balanced budget to find the necessary resources for institutional sectors.

In Fig. 6 we present the results of ICT impact in scenarios 1 and 2. In the first simulation the changes in total output of ICT commodities are very small, quite equal to zero, but it is possible to observe a different impact on the commodities that react to the increase in health care services demand. In the second scenario all changes are all negative, but still very small.

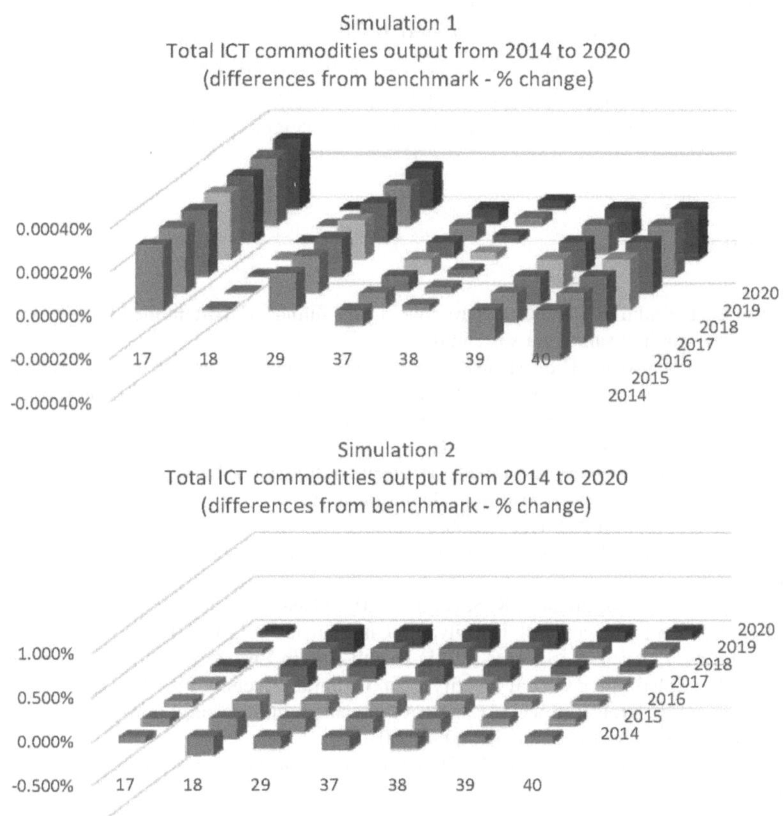

Fig. 6 ICT impact in scenario 1 (SIM1) and 2 (SIM2)

5 Conclusion

This chapter is devoted to the study and analysis of the desirability of investment on Information and Communication Technologies in the Italian health sector. In order to accomplish this analysis and to study the impact of an increase in final demand of health services on ICT sectors, a recursive dynamic Computable General Equilibrium model has been calibrated where the behavior of agents depends on adaptive expectations. In detail, two different scenarios have been considered where the fiscal policy is made under budget constraint: in the first scenario, it has been assumed an increase in health services demand coming from Government, while in the second scenario, it has been considered an increase in final demand of health services coming from Households. By analyzing the obtained results, namely by the GDP and by its trend over the time, the overall effect is always positive. Instead, by measuring the deviations of the GDP with respect to the growth path defined by the benchmark (absence of policy), there is a dominance of the second scenario on the first policy, which is mainly due by the effect of the budget constraint on the public administration. Moreover, this analysis revels that the economic system reacts more when the health services are stimulated by the private demand instead of public demand. Above all that, the value added by ICT commodity has a better performance when the policy is implemented using resources from the Government to finance the Households expenditure in health services.

References

1. Caruso E, Dirindin N Perché tagliare ancora una sanità già sostenibile? http://www.lavoce.info/perche-tagliare-sanita-gia-sostenibile/
2. De Angelis M (2014) The expansion of patients' rights in Europe: a new perspective for healthcare abroad. Bull Transilvania Univ Brasov—Series VII: Social Sci Law 7(56):111–120
3. De Angelis M (2015) Politiche sanitarie e diritti di fronte alla crisi: aspetti generali e un'esperienza. *passim*, in Le Regioni (2)
4. Dipartimento di Ingegneria Gestionale, Politecnico di Milano: Innovazione Digitale in Sanità: l'ICT non basta! Osservatorio ICT in Sanità (May 2014)
5. Ciaschini M, Pretaroli R, Socci C (2009) Number 2/09 in Contenimento della spesa corrente e spesa in conto capitale nel sistema sanitario pubblico. In: Tendenze Nuove. Il Mulino (2009) 115–132
6. Socci C, Ciaschini M, Pretaroli R, Severini F (2010) Health services as a key sector for income and employment change. Bull Transilvania Univ Brasov **3** (2010) 163–172
7. Ciaschini M, Pretaroli R, Socci C (2011) Health-care in Europe as a driving force for output and GDP. Global Forces Local Identity Econ Cultural Diversity 28:106–135
8. Ciaschini M, Pretaroli R, Socci C (2009) A convenient multisectoral policy control for ICT in the US economy. Metroeconomica 60(4):660–685
9. Ciaschini M, Pretaroli R, Socci C (2010) Multisectoral structures and policy design. Int J Control 83(2):281–296

10. Ciaschini M, Socci C (2007) Final demand impact on output: a macro multiplier approach. J Policy Model 29(1):115–132
11. Ahmed I, Ciaschini M, Pretaroli R, Severini F, Socci C (2014) Economic policy effects and financial crisis: a dynamic CGE model for Italy. In: 22nd international conference of input-output, Lisboa, Portugal, 14–18 July 2014

A Non-Intrusive Model for the Collection and Analysis of Data, Feedback and Needs

Lorenza Lupini and Francesca Scocchera

Abstract The present paper focuses on a specific aspect of research with people, and particularly with people with frailty, that is how to prevent the intrusiveness of new solutions, technologies or methodologies and how to foster and promote project innovation among the community.

1 A Non-Intrusive Approach

The internal Department of Research and Training of COOSS has been involved in EU projects since 1993. During its long activity in development and management of project, COOSS participated also into the design of several ICT prototypes and Assistive Technologies with the peculiar mission of providing the users' viewpoint, based on years of analysis about user needs and requirements, acceptability and accessibility issues, ethical implications, field trials and validation protocols. The expertise gained 'on the field' allowed the Department to perfect a method for a 'non-intrusive' management of project activities, in particular for what concerns the collection of confidential data and information.

The non-intrusive model is the core of how to design and manage an ICT project devoted to elderly and real user in general and it grows out of years of analysis on user needs and requirements, acceptability and accessibility issues, ethical implications, field trials and validation protocols.

The methodology starts joining technical validations of many prototypes and contributing to improve them and make them as more user-friendly as possible and focuses on the collection and analysis of data, feedback and needs.

The importance of non-intrusiveness makes necessary a new total approach which starts from the design of the innovation and will accompany it all over its

L. Lupini (✉) · F. Scocchera
COOSS Marche, Ancona, Italy
e-mail: l.lupini@cooss.marche.it

F. Scocchera
e-mail: f.scocchera@cooss.marche.it

M. Conti et al. (eds.), *Mobile Networks for Biometric Data Analysis*,
Lecture Notes in Electrical Engineering 392, DOI 10.1007/978-3-319-39700-9_2

implementation and application. The respect of privacy, ethical issues, and sensitive information must be guarantee also when the solution or device will be on market and out of researcher control [1–8].

1.1 The Project Design

A good solution needs a well designed strategy of implementation. If a device, a methodology or a service is created for a peculiar target of users, his development will be set up on the basis of a real need. But it is important that the survey will perceived by then end users as 'natural' and that their feedback will flow down spontaneously. So specific attention on non-intrusiveness must be paid since the first step of the design of a project.

Step 1 **Identification of users' needs and requirements**: this is the pillar of next steps and it is based on USER CENTRED APPROACH. The usability of a solution, a device or system must be evaluated in a wider framework:

- user as a *central factor*: he/she's the starting point, the drive and motivation of the research, but also the goal.
- user within *home environment*: the innovation will be better accepted if designed according to the routine and daily activities of final user.
- user and *support services*: the integration of support services (persons or devices) with an ICT solution may improve the quality of life and not make it more complicated.
- user within *social and political environment*: to strongly consider the strategies adopted by policy makers, which kind of support the community can offer or not, but also social habits, creed and belief system of the user in order to not to modify his/her life but to empower it.

Step 2 **Scenario Planning and Definition**: identification and forecast of potential scenarios of impact of a given ICT based solution or innovation. Such a methodology is based on the evaluation of several variables and on a multi-disciplinary approach, in particular according to the users' basins, market trends and beneficiaries' acceptability.

The implementation of these two steps will take advantage of instruments wherewith beneficiaries are already in familiarity; for example, to carry on a preliminary surveys among elderly end-users, it can be strategic to use elementary questionnaires in paper form or informal questions and not complex or on-line surveys in order to avoid the sense of inadequacy (the sentence 'I can't do it' is the sign that something in the survey is going wrong). The presence of a researcher helping them in filling it the questionnaire should be another strengthen point.

Furthermore the confidence of the end user with the researcher is very important, it will allow the active participation, the enthusiasm and will prevent the retirement during the pilots.

The acquisition of inputs, viewpoints and key information from primary, secondary and tertiary categories of beneficiaries of a given ICT based solution and the definition, identification and foreseen of scenarios of use and potential scenarios of impact of a given ICT based solution, will allow to:

1. better profile the potential user basin, for the design of a well tailored solution or innovation
2. foreseen of final user acceptability of the ICT solution, allowing a high real user-acceptance
3. give a sense of non intrusiveness, giving a real perception of the empowerment of quality of life

1.2 *The Project Ongoing*

We consider in this section the aspects that must be considered when the project idea started its implementation and the prototype is ready to be tested. The prototype has been designed taking into account early feedback, early hypothesis. Now the innovation have to verify the capacity to meet the final real users and, with them, the possible market.

In following steps the methodology will consider the non-intrusiveness of research activities in all the activities related to the field trials with real users, that is the first contact of the beneficiaries with the innovation itself.

Step 3 **Validation**: ICT based solutions are tested on field trials with real users in real life situations.

It is important to be able to deal with several issues:

- get in touch with beneficiaries from the recruitment up to the end of test
- deal with the possible withdraw from pilot tests
- avoid the digital divide
- manage system, etc.

It is also fundamental to ensure the users that their need will be always the main goal of the research.

For these reasons, the validation must be planned, first of all according to users' demands and after on work plan indications.

Step 4 **Engagement of primary, secondary and tertiary beneficiaries**: The organization of workshops, conferences, round tables should involve users representatives, experts, professional caregivers. Suggestions and feedbacks must be collected, during and after the experimentations, from all actors involved, useful for the design and finalization of prototypes.

The involvement of every target of beneficiaries will guarantee the highest acceptability of the innovation in the system that is represented by the 'community' of the user.

Step 5 **Training of caregivers and users** about the use and the functioning of the system.

The introduction of every kind of innovation, has to be accompanied by a training. A guide cannot be sufficient both for users neither for carers, in the introduction of assistances technologies. Training has to be conducted according the different ability of trainers and their degree of confidence toward innovation.

A solution is represented by the use of the **Mediator of Innovation**, a figure with both social and technological competences able to train the final beneficiaries to use innovative technologies and, above all to generate confidence and acceptability.

The Mediator has to accompany caregivers and users in their training with the innovation. He/she could be:

- an *external figure* who links ICT, user and caregiver

But also he/she might be:

- a *well trained caregiver* who mediate between ICT and user

A not well designed ICT solution can inhibit the users and generate a digital divide, a gap very hard to be mended that can jeopardize the efficacy of the innovation. All these steps aim to guarantee a real acceptability and to create a positive and constructive approach and attitude of the user toward Assistive Technologies.

At the end of the core of the implementation phase, the methodology will allow:

1. collection of continuous real feedbacks, with the aim of a constant improvement of ICT innovation during and after the validation
2. social mediator for a better acceptability. The integration of different typologies of assistance gives the opportunity of a better welfare system
3. a real self serve society as an outcome of innovation

1.3 All the Project Life

The care and the respect for the interest of final users carry on with all the project' tasks; they are also guaranteed thanks to a well designed Dissemination and Exploitation Strategies and, finally, making the Ethical Issues, Privacy and Support a basin for all actions.

Step 7 **Dissemination and Exploitation** activities are another way to introduce and make available and accessible all results of project effort into the market, creating user centred communication and market.

- An innovative ICT solution, integrated with a service, may be the best way to offer an efficient support for users
- To raise awareness with correct communication media about opportunities offered by the solution, may give the ICT solution a better acceptability among beneficiaries

Step 8 **Ethical issues—Support Protocols and Privacy**: dealing with the experimentation, test and validation of prototypes with real users in real life circumstances entails to define and apply detailed and focused procedures aimed to identify possible *Ethical Issues* deriving from such activities.

The aim of taking into account Ethical Issues is to ensure and protect the users' health, privacy and safety according to international and national regulations and legislations.

- *Assistance and Support Protocol*: in order to check the correct managing of all procedures a call centre or assistance centre may be a great solution
- *Privacy*: whenever the research will imply the collection and analysis of personal and, above all, clinical data. The use of Informed Consent approved by an Ethical Committee is recommended

2 Study Cases

Here follows a short list of project carried on by COOSS Marche, that could be considered as best practices for the application of methodology they show.

Users' needs and personalized solution: **AAL 2010 'NaCoDeAL'**, Natural Communication Device for Assisted Living, helping elderly people with personalized solution for temporary memory loss.

The project is implementing an ICT system able to sustain the daily life of elderly people suffering with memory problems, through interactive interfaces that stimulate elderly's cognitive abilities and keep them in contact with the external environment.

- It is an example about how the non-intrusiveness methodology may rich great results in acceptance of ICT among elderly with disability.

Collection and analysis of sensitive data:**CIP-ICT-PSP-2011 'FATE'**, a Fall deTector for the Elderly with chronic health problems, interfacing with a call center and database.

The portable and usable device implemented by the project is validating within 175 individuals with high risk of falling an ICT-based solution that will accurate detect falls, at home and outdoors. The system runs a complex and specific algorithm to detect falls, collects personal and clinical data, carefully analyzes and processes them. A call centre is active 24/24 h to receive alarms and support carers, familiars and users.

- Attention on Ethical Issues, Data Protection, Assistance and Protocols to support all management phases are the strong point of this project.

Some study cases: **IPA Adriatic CBC 'AdriHealthMob'**, a cross border strategy with the aim to develop an Adriatic Model of Sustainable Mobility in the Health & Care Sector, implementing ICT integrated, accessible and shared solutions and databases. **CAALYX**, "Complete Ambient Assisted Living Experiment" developed a wearable light device able to measure specific vital signs of the elderly, to detect falls and to communicate automatically in real time with his/her care provider in case of an emergency, wherever the elderly person happens to be, at home or outside. **CONFIDENCE**, an ICT integrated system to detect falls and abnormal behavior of the elderly in indoor environments; its efficiency, non-intrusiveness and affordability will increase the sense of independence and the quality of life of users, families and carers. **ELDERHOP** project that aims to "provide innovative ICT-based solutions to older persons, which means innovative system addressing identified wishes and needs of the end-users" in order to "prolong the time of the well-being in aging".

References

1. Charness N, Boot WR (2009) Aging and Information Technology Use: Potential and Barriers. Curr Dir Psychol Sci 18(5):253–258
2. De Munari I, Matrella G, Ciampolini P (2012) AAL in ITALIA—Primo libro bianco, foritaal, Parma
3. Eurostat (2010) Social participation and social isolation, Methodologies and Working papers, European Commission
4. Goldberg D, Williams P (1988) A user's guide to the G.H.Q, NFER-Nelson, Windsor
5. Mancini C, De Giacomi G, Carletti P, Dubbini L (2003) Indagine sulla salute nelle terza età. Studio Argento, Italia

6. Meulendijk M, Wijngaert LVD, Brinkkemper S, Leenstra H (2011) Am i in good care? Developing design principles for ambient intelligent domotics for elderly. Info for Health Soc Care. 36(2), 75–88
7. Renau K, Biljon JV (2008) Predicting technology acceptance and adoption by the elderly: a qualitative study. SAICSIT
8. Van Der Heijden K (2006) Scenarios. The art of strategic conversation, West Sussex, England

Analyzing Environmental Conditions and Vital Signs to Increase Healthy Living

Ralf Seepold, Christoph Dermati, Artur Kostka, Lars Pfeil, Ralf Lange, Matthias Hermann and Benedikt Martinez

Abstract Present demographic change and a growing population of elderly people leads to new medical needs. Meeting these with state of the art technology is as a consequence a rapidly growing market. So this work is aimed at taking modern concepts of mobile and sensor technology and putting them in a medical context. By measuring a user's vital signs on sensors which are processed on a Android smartphone, the target system is able to determine the current health state of the user and to visualize gathered information. The system also includes a weather forecasting functionality, which alerts the user on possibly dangerous future meteorological events. All information are collected centrally and distributed to users based on their location. Further, the system can correlate the client-side measurement of vital signs with a server-side weather history. This enables personalized forecasting for each user individually. Finally, a portable and affordable application was developed that continuously monitors the health status by many vital sensors, all united on a common smartphone.

R. Seepold (✉) · C. Dermati · A. Kostka · L. Pfeil · R. Lange · M. Hermann · B. Martinez
HTWG Konstanz, Brauneggerstr 55, 78462 Konstanz, Germany
e-mail: ralf.seepold@htwg-konstanz.de
URL: http://uc-lab.in.htwg-konstanz.de

C. Dermati
e-mail: christoph.dermati@htwg-konstanz.de

A. Kostka
e-mail: artur.kostka@htwg-konstanz.de

L. Pfeil
e-mail: lars.pfeil@htwg-konstanz.de

R. Lange
e-mail: ralf.lange@htwg-konstanz.de

M. Hermann
e-mail: matthias.hermann@htwg-konstanz.de

B. Martinez
e-mail: benedikt.martinez@htwg-konstanz.de

M. Conti et al. (eds.), *Mobile Networks for Biometric Data Analysis*,
Lecture Notes in Electrical Engineering 392, DOI 10.1007/978-3-319-39700-9_3

1 Introduction

With the present demographic change and a growing population of elderly people, meeting new medical needs supported with state of the art technology is a rapidly growing market and surprisingly, not many manufacturers have identified this niche. Meanwhile, the vital and growing quantified-self community shows how to use tracking devices and health monitoring sensors combined with a smartphone to optimize every aspect of a healthy lifestyle. The proposed system SAFESA aims at taking these known concepts and putting them in a medical context. SAFESA stands for *System for alerting and forecasting elder persons with a sensor based architecture* which is developed at HTWG Konstanz. It may be possible to determine the current health state of a user, by measuring his vital signs with sensors and further processing them on a smartphone. This is achieved by applying data mining methods to the gathered data on server side and is illustrated in Fig. 1. If a critical situation arises (e.g. possible upcoming fainting), the system alerts the user about the change in his health status and advises him to take necessary precautions. In case of an unpredictable emergency (e.g. fainting, cardiac arrest, etc.), the system asks the user for manual feedback. If no feedback is given, a group of third persons (professional nursing staff or private persons) is contacted automatically and provides them with the current geo-location of the user. Additionally, the system will include a weather forecasting functionality, which alerts the user on possibly dangerous meteorological events. High humidity or sudden drops in barometric pressure can be of risk to the user. This information is collected centrally and distributed to users based on their location.

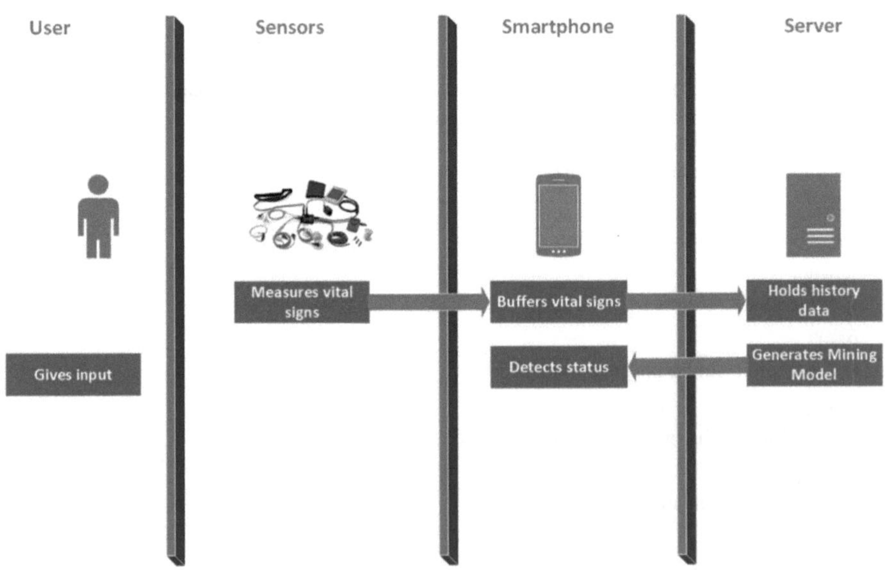

Fig. 1 Logical system architecture and data flow

The system can combine client-side measurement of vital signs with a server-side weather history to reveal unknown correlations in any of these dimensions. This enables personalized forecasting for each user individually. Furthermore, the data gathered can be used to identify area-based clusters of risk groups in the future.

2 State of the Art

It is of great significance, which technical achievements are already on the market and on which state of research it can be built on, during the implementation. The emphasis especially is on measuring body signals with sensors and interprets the signals to detect diseases at an early stage.

2.1 Miniaturized Sensors in Medical Field for Long-Term Monitoring

Former days it was only possible to use sensors for laboratory measurement of blood gases and electrolytes. Nowadays biosensors give us the possibility to continually monitor glucose in diabetes or do a long-term electrocardiogram to identify heart rhythm disorders [1, 2]. Originally these sensors were only used in intensive care units and cost thousands of dollars, but through their reliability, nanofabrication technologies and low production costs, their success in health service is inevitable. The advantage of the inclusion of smartphones is that everyone who already invested in a smartphone doesn't have to buy additional diagnostic devices. For now the integrated optical sensors and a flashlight is already enough for detecting any toxins, proteins, disease-causing bacteria and viruses [3].

2.2 Data Mining in Medical Context

There is quite an amount of research done on data mining, machine learning and artificial intelligence techniques applied in a medical context. In combination to mobile health appliances Alahmadi and Soh [4] proposed a 3-tier architecture. The first tier represents the Body Area Network (BAN), which consists of sensors and a smartphone. The second tier represents the Personal Area Network (PAN), a personal computer of the user. The third tier, which is basically a Wide Area Network (WAN), consists of numerous smart servers. On each tier of the architecture, smart reasoning and decision making is performed on three different levels of process capabilities, meaning the first tier acts as a detector for emergent situations, the second tier enables communication and the third tier includes a full history of the user's medical records, analysis and alerting functionality [4]. An application of a set of data mining techniques to create a decision tree, which predicts the level of heart attack risks, is proposed in [5]. It uses clustering (K-Mean), itemset mining (MAFIA) and decision tree building (ID3) to detect the user's health status.

2.3 Atmospheric Related Syndromes

Atmospheric related syndromes (ARS) or meteor sensitivity is defined as changes of physical, mental or social well-being and incidence of diseases, which are associated with weather dependent factors [6]. Polls show regularly that people between 55 and 70 % call themselves meteor sensitive [7]. But the physical appearance is according to Sönning individually different. Therefore medicine-meteorology distinguishes the following three levels of meteor sensitivity [8]:

1. *Weather reaction* affects unconsciously every organism. But it is only measurable in physiological data like reaction time, blood pressure or heart frequency.
2. *Meteor sensitivity* describes effects on affected people. For instance feeling temporary uncomfortable or triggering individual migraine, depression and tinnitus.
3. *Meteor hypersensitivity* describes effects on a tested collective with a positive rate of 5 to 15 %. This means that a statistically measurable amount of people show an increase in triggering emergencies like heart attacks, epilepsies or strokes.

The so-called medicine-meteorology was shaped in Germany after World War II within the *Deutscher Wetterdienst (DWD)*. Back then, working groups consisting out of biologists, physiologists, doctors and meteorologists tried to investigate the coherence between weather and healthiness. Therefore, typical weather phenomena were classified and they tried to find correlation with the data of the doctors [9]. In 1965 the *Tölzer and Königssteiner working group* defined six weather phases that are identified to impact healthiness [9, 10]. Based on this data many studies tried to find correlations between the phases and different diseases. In 1984 the *Deutscher Wetterdienst (DWD)* summed these studies up and published the so-called *Medicine meteorological meta-analysis* [11].

In 1950 studies showed that there must be an additional factor inside resp. beside the weather phases that is responsible for meteor sensitivity [12]. This bio tropical factor is modeled by the *atmospheric effect-complex* and describes basically the impact of *perceived temperature* and *atmospheric stress* on healthiness [13]. A later study from 1998 seems to corroborate this by showing coherence between perceived temperature and thrombosis [14]. In 1983 Sönning added another factor to the model: the so called *neurotropic factor* [15].

Today the exact bio tropical factor, i.e. the factor that is responsible for meteor sensitivity is still not found [7]. The current state is that there are two factors that could be the causal factor of meteor sensitivity: *Atmospheric pulse radiation* and *air pressure oscillation*. Both tend to be inside the *neurotropic factor* of the model. This assumption is mainly based upon one finding and a study [11]:

- *Atmospheric impulse radiation (AIR)* frequency is very similar to the frequencies occur in the human neuronal cells [16].
- *Air pressure oscillation* caused headache, irritations of airways etc. in a field study [17].

Despite there are stochastic coherences between some weather conditions and selected effects on healthiness, today there are almost none causally clarified effects. This means that in most cases it is not possible to reproduce these effects. Only scientific causally proven factors are Pollutants, UV radiation and Ozone [6].

2.4 Use of Research Results

Based on the stochastic coherences there are bio weather and risk forecast systems since the end of the 1980s. One problem is that these are only based on the trivial weather factors, but not on the modern non-classical medicine-meteorology factors [7]. Another problem is the difficulty in estimating individual risk based on empiric collective risk. This means that the impact on individuals is mostly not foreseeable without the help of an individual examination through a doctor or information about perceived weather factors [6].

As a conclusion, the proposed system will mainly focus on the following areas:

1. Examination of the atmospheric effect-complex and its impact on the individuals with a mobile sensor infrastructure.
2. Matching individuals with their specific risk group based on user feedback or vital signs. When successful a better forecast can be accomplished.
3. Analyzing individual sensor data and user feedback with advanced machine learning techniques. Predict risk levels based on historical data and the knowledge that there is a stochastic coherence between healthiness and weather.

This work focuses last point of the list.

3 Data Mining Methods for Forecasting

The following section describes the theoretical methodic of the SAFESA application. Accordingly, the used input factors, which are relevant, are described as well as how they are used to determine if an emergency is likely to occur. The input data for the creation of a user's individual prediction are distinct in personal input data and external input data, which can give more general forecast information.

- The user generated input data results from sensors, which are worn on the body of a user. The SAFESA application uses the following measurements.
  - Systolic blood pressure, diastolic blood pressure, heartbeat, glucose, body temperature, electrocardiogram, galvanic skin conductance, galvanic skin conductance, galvanic skin response, oxygen saturation.
  - Based on the existing sensors, the used measurements can be extended or reduced.
- The SAFESA application requests also external environmental data like weather information of a predetermined weather station.

  Current temperature, min. temperature, max. temperature, humidity, atmospheric pressure, wind speed, wind direction in degrees, cloudiness in %, weather condition.
- In addition, data from a fall detection system are used, too.

3.1 Data Mining on Server Site

To calculate the prediction of a possible emergency forecast data mining is used as a computational process to discover pattern in user and environmental generated data. The data mining processes will extract information, which is contained in the data and present the information in a useful and understandable way for the user.

Therefore, SAFESA imports the WEKA library, which contains different algorithms for data mining [18]. The data-mining component, which is used in the described application, uses the Simple K-Means algorithm for the implementation of clustering and the J48 algorithm for the implementation of a classifier. The Simple K-Means algorithm uses the measured values, which are saved in the database and creates for each user an individual clustering. Each clustering contains four clusters over all axes, which are generated automatically by the algorithm. The axes represent the different measures. In the case of this mining process the axis describe the different input values like blood pressure and heart beat, as well as the an additional axes for the time. The four cluster representing different daily situations of an individual user like sleeping or physical training. The data-mining component saves the individual user clusters in the database. These clusters are used as an input for the J48 in the next step [19, 20]. The clusters of individual users are used together with the measurement values as input averaged in minutes intervals to create a classifier. Also each time interval contains the information, if an emergency had occurred or not. For this to be possible, each occurring emergency has to be registered and available as input in the database. Each emergency needs a timestamp and the information, which user generated sensor data led to the situation. The SAFESA application connects the input data to the time the emergency

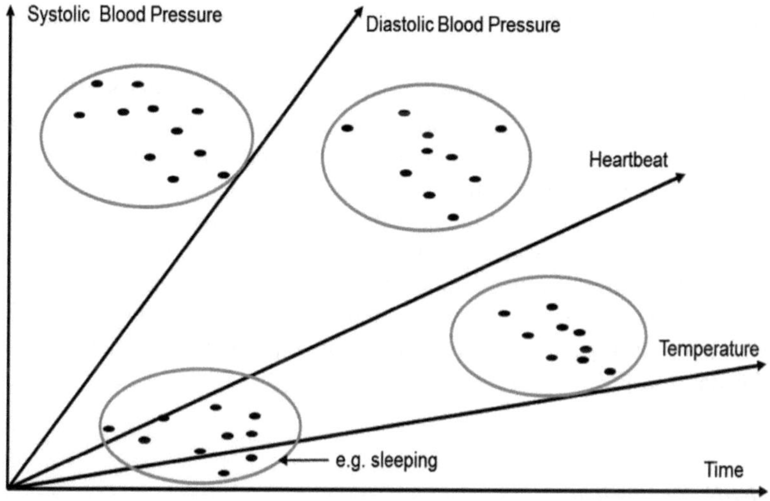

Fig. 2 Simple K-Means cluster

happens as well as the three minutes before, which led to the emergency. The classifier used is the J48, which is a decision tree with automatically splitting. To be able to make predictions a nominal class attribute is needed for all occurring yellow cases. In case of SAFESA system the class attribute is the emergency attribute with values yes and no. The individual trained decision tree is saved in the database [19, 21].

After the implementation the continues incoming data will be clustered by the trained Simple K-Means model as shown in Fig. 2 and processed by the trained decision tree as shown in Fig. 3. The decision tree determines based on the incoming data, if a yellow emergency situation has occurred.

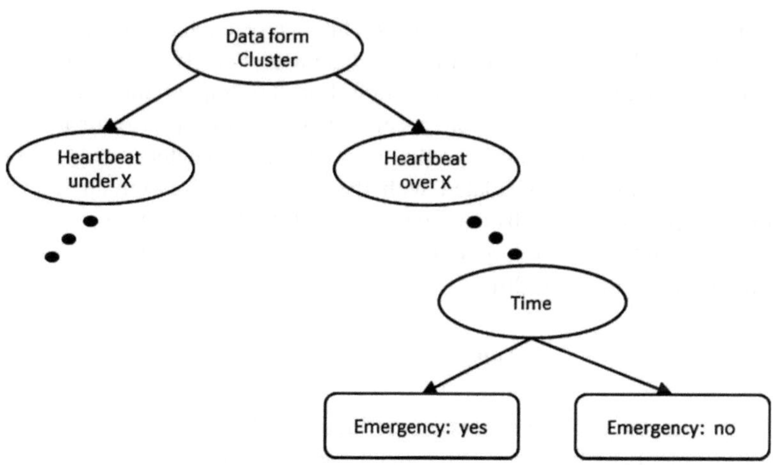

Fig. 3 J48 decision tree

3.2 Weather Service

In addition to the emergency prediction through data mining, the SAFESA system uses the requested weather information from a weather service to give a weather forecast for the next days as well as predictions for generally problematic days, like the occurrence of black ice, which could be dangerous for a user. Similar to a weather report, which calculates the probability for rain based on historical weather data, the SAFESA system warns a user if a day occurs which is problematic for this specific user. Therefore weather data form historical days will be compared with the prediction for the next days. When there are days with similar data, which caused an emergency, the user gets a warning that similar problems could happen on the particular day too [22].

3.3 Fall Detection

Elderly people fall down very often because of tired muscle, poor eyesight or slow reaction rate and this largely affects their health in their old age. Injuries just do not heal as well as at a young age [23]. There are some scientific papers that have dealt with this issue and implemented such an algorithm for falls on mobile devices. Most smartphones have a built-in accelerometer that can be accessed with the native API. In a project, developed at the Institute of Computing Technology Beijing by Zhongtang et al. [24] an activity recognition algorithm was developed which was able interpret the accelerometer signal pattern and to recognize movements or activities (like running, jumping, walking, falling) of a person which was a helpful and an important criterion in the implementation. Because of wrong detected daily activities like running. However, in order to cover this case a maximum velocity can be determined as a threshold. Once this is exceeded, a fall must have happened. A possible threshold value could be a velocity of around 2 m/s. At this speed injuries may not necessarily occur and thus a fall can be excluded [25]. Furthermore the timestamp of the falling body till the acceleration is at 1G (when the body hits the ground) is another indicator of a fall. For the implementation of the fall detection we use the native hardware accelerometer sensor of an Android device. It listens to all acceleration changes of the tri-axial accelerometer sensor as it can be seen. The acceleration is calculated with the use of the Newtonian fundamental equation of mechanics. So falls can be recognized while the sum-vector of all axes exceeds a predefined threshold of a multiple of the force of gravity. This is shown in the following equation [25, 26].

$$Sum_{all} = \sqrt{a_x^2 + a_y^2 + a_z^2}$$

Previously a few different states are defined which make a statement about the current movements of the user. Three main parts determine these states:

- Predefined thresholds for the movements (standing, sitting, walking and falling).
- The calculated sum vector of the tri-axial accelerometer data, which is calculated with the equation shown above.
- An array of the buffered sum vectors.

A pre-recognition of the status with the current sum vector starts and checks whether a fall could have happened. This is of great importance in the following algorithm, because it could be better distinguished whether the user jumped or fell down.

4 Prototype

This chapter will focus on a description of the developed prototype. Therefore it is separated into three parts, which can be regarded as the main components. These components are: Sensors, client application and server side application.

4.1 Sensors

The sensors can be divided into two categories. First we have some build in sensors from our smartphone. Secondly there are the different kinds of sensors from the Arduino Health Kit. The smartphone sensors, which are used, are:

- GPS sensor, which is required for an accurate position in order to send positioning information in a potential emergency SMS and to get the weather information for the current location.
- Accelerometer, which is used for the fall detection as described in the previous chapter.

With regard to the Arduino Health Kit, the decision has been made to implement interfaces in order to connect the *pulse and oxygen saturation sensor* and the *blood pressure sensor* [27]. Both sensors transfer the measured data via Bluetooth LE (low energy), since we have to focus on using less battery, to our connected mobile client application. Thus it is very easy to attach any other sensors, which is available for the Arduino Health Kit platform.

4.2 Client

The client application on the smartphone is developed for Android 4.3, because Bluetooth LE is supported since this version. As already mentioned the measured values are being transferred to the smartphone and stored in a local database

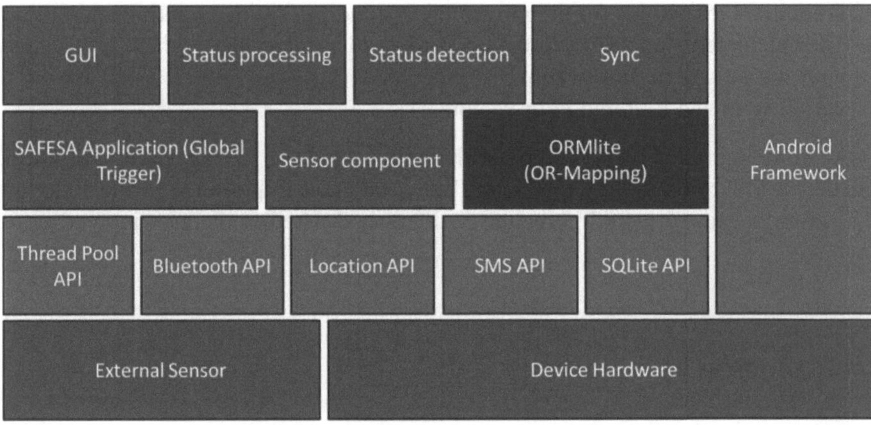

Fig. 4 Client SAFESA API and components stack

(SQLite). The SAFESA client will not directly talk to the SQLite API. Instead it will make use of an OR-Mapping framework called ORM-Lite [28]. These stored measurements are later pushed to the server. This happens with a synchronization process, which is started every 30 min. Though we already have the measured data on the client side, we just have to push them to the server.

Besides new weather information is gathered for our current location from the server during the synchronization process as well as the mining model for our status detecting and processing component. This measured data is user specific and therefor gets tight to a previously defined user. For data transfer JSON API is used. It is easy to use and understand and Android already supports it by default. The layering of the client components is shown in Fig. 4.

4.3 Server

The server architecture for our web service, which is illustrated in Fig. 5 provides weather and synchronization services. It is mainly based on Play Framework in the version 2.2.2 [29]. Main advantage of Play Framework is, that it is designed completely RESTful. Unlike to Jave EE no connection per session is required. In addition the Hibernate Object-relational mapping (OR-mapping) is used, because the provided OR-mapping tool from the Play Framework is not satisfying. Hibernate interprets the Java Persistence API (JPA) annotations in the defined entities. The outcomes of this are database tables and its relationships. Underneath the OR-mapping a MySQL-database is utilized. Just as on the client application JSON is used for client-to-server requests and responses. For scheduling jobs on the server side AKKA is used. It is a default package in Play Framework too and allows building scalable applications. In the context of SAFESA web services it is used to

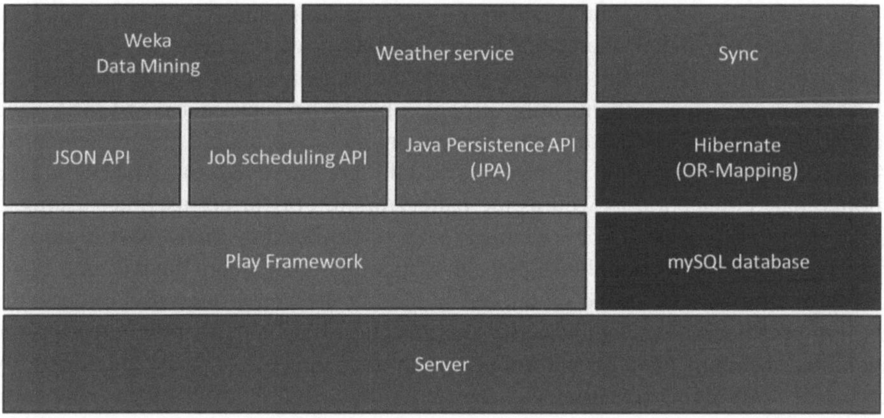

Fig. 5 Server SAFESA API and components stack

trigger current weather updates and forecasts. Therefore HTTP-requests to Open Weather Map [22] are made and the required data from response is being stored in the server side database. In addition another job in AKKA is defined to enable building mining models and decision trees for a user on a weekly basis. This is realized by using Weka Data Mining Software [18], which is built in Java as well. For the generation of the mining model, the user specific measured data is considered as well as the weather information for the attached location of this measurement. To easily access this data, views on MySQL server side database are implemented. The generated mining model and decision tree are stored in the database. However, to run the mining model properly the views on the MySQL server side database must be implemented on the client side as well. SQLite does not provide any ways to create views. Therefore a programmatic way to implement these views has to be made.

5 Conclusion and Outlook

The proposed solution offers a basic implementation of the main planned features. The application architecture is heavily data driven which enables an easy extension of the base application. Sensors, measures and measurements can be added with little effort by implementing the necessary interfaces and adding entries to the according database tables. Furthermore the data mining techniques applied can be easily extended, server-side functionality or communication can be added and possible additional components like web services can be integrated.

Another advantage of the proposed solution is the possibility to configure main features for personalization and fine-tuning. All thresholds for emergency statuses can be changed individually by applying a new configuration.

However, the system could undergo some improvements. The limited database functionality prevented a practicable transformation of the necessary datasets in reasonable time. The client-side application and database lifecycle is handled in a very basic way, the authentication process should be secured by SSL/TSL, and overall IT-Security to protect privacy and avoid fraud should be implemented. Additionally, the location management could be improved to work without a mobile data connection as well as the battery drain. For further usability testing, a field study with test persons wearing sensors, measuring vital signs would be necessary. Such study could take place in a nursing home, where the intended target group of SAFESA could be monitored in a controlled environment. Furthermore, the pre-sent health staff could verify the detection mechanisms by comparing the identified health status of the subject with the respective status measured by common vital sign measurements. To verify the alerting mechanisms, the false positive and false negative rate of all detected and undetected incidences should be deter-mined. A goal for the intended field study could be to reach a false negative rate of fewer than 10 % and a false positive rate of fewer than 20 %.

References

1. Captum Capital Ltd.: Sensors in Medicine 2014, (2014) http://www.sensor100.com/SensMed2014/Conference.html. Accessed 09 Aug 2014
2. Heidelberg Private Clinic: Long-term ECG (long-term electrocardiogram), (2014) http://www.heidelberg-private-clinic.com/services/private-clinic/long-term-ecg/. Accessed 09 Aug 2014
3. Gallegos D, Long KD, Yu H et al (2013) Label-free bio detection using a smartphone. Lab Chip 11:2124–2132
4. Alahmadi A, Soh B, (2011) A smart approach towards a mobile e-health monitoring system architecture. Res Innovation in Inform Sys (ICRIIS), 1–5
5. Hnin Wint Khaing (2011) Data mining based fragmentation and prediction of medical data. Comput Res Dev (ICCRD) 2:480–485
6. Bucher K, Wanka E (2007) Das Problem der Wetterfühligkeit, promet, 33(3/4), 133139http://www.dwd.de/bvbw/generator/DWDWWW/Content/Oeffentlichkeit/KU/KU3/Biowetter/Bioprog/promet,templateId=raw,property=publicationFile.pdf/promet.pdf. Accessed 08 Aug 2014
7. Höppe P, Mackensen S, Nowak D, Piel E (2002) Prävalenz von Wetterfühligkeit in Deutschland. Dtsch Med Wochenschr 127:15–20
8. Sönning W (2007) Wetter und Gesundheit: Die Suche der Medizinmeteorologie nach dem biotropen Wetterfaktor, Umwelt-Medizin-Gesellschaft, 20, 212–218 http://www.diagnosefunk.org/downloads/soenning_umg-307.pdf. Accessed 08 Aug 2014
9. Becker F (1956) Arbeitsgrundlage der medizin-meteorologischen Vorhersage im Königssteiner Arbeitskreis. Medizin-Meteorologische Hefte 11:21–25
10. Brezowsky H (1965) Meteorologische und biologische Analysen nach der Tölzer Arbeitsmethode. Meteorol Rdsch 18:132–143
11. Sönning W (1984) Katalog Meteorotropie. Deutscher Wetterdienst, Medizin-Meteorologie
12. Ungeheuer H, Kügler H (1957) Meteorologie-Biologie-Medizin. Arzneimittel Forschung 7:370–380
13. Hellpach W (1950) Geopsyche. Enke, Stuttgart

14. Durdu N, Chone L, Mueller H, Bucher K, Haas J (1998) Meteorologische Einflüsse und Thromboseentstehung, 79. Deutscher Röntgenkongress
15. Sönning W (1983) Zur biosynoptischen Arbeitshypothese, Z. Phys. Med., Balneologie. u. med. Klimat, 12, 2–4
16. Ranscht-Froemsdor W-R (1962) Beeinflussung der neuralen Information durch niederfrequente Schwankungen von Umweltfaktoren. Zeitschrift für angewandte Bäder und Klimaheilkunde 11:462–477
17. Richner H (1974) Zusammenhänge zwischen raschen atmosphärischen Druckschwankungen, Wetterlagen und subjektivem Empfinden, LAPETH-8. Laboratorium für Atmosphärenphysik, ETH Zürich, pp 97–98
18. The University of Waikato: Weka 3: Data Mining Software in Java, (2014) http://www.cs.waikato.ac.nz/ml/weka. Accessed 21 Aug 2014
19. The University of Waikato: Weka 3: Use WEKA in your Java code, (2014) http://weka.wikispaces.com/Use+WEKA+in+your+Java+code#Clustering-Building a Clusterer. Accessed 21 Aug 2014
20. The University of Waikato: Weka 3: SimpleKMeans, (2014) http://weka.sourceforge.net/doc.dev/weka/clusterers/SimpleKMeans.html. Accessed 21 Aug 2014
21. The University of Waikato: Weka 3: J48, (2014) http://weka.sourceforge.net/doc.dev/weka/classifiers/trees/J48.html. Accessed 21 Aug 2014
22. Extreme Electronics Ltd: OpenWeatherMap—actual and forecast weather (2014) http://openweathermap.org. Accessed 21 Aug 2014
23. Fiß T, Meinke C (2012) Stürze bei Senioren: gefahren erkennen und vermeiden, Pharmazeutische Zeitung, 9, http://www.pharmazeutische-zeitung.de/?id=41065. Accessed 21 Aug 2014
24. Zhongtang Z, Yiqiang C, Junfa L (2010) Fall detecting and alarming based on mobile phone, ubiquitous intelligence & computing. In: 7th International Conference on Autonomic & Trusted Computing (UIC/ATC), 494–497
25. Yabo C, Yujiu Y, Wenhuang L (2012) E-FallD—A fall detection system using android-based smartphone, Fuzzy Sys Knowl Discov (FSKD), 1509–1513
26. Chia-Chi W, Chih-Yen C, Po-Yen L et al (2008) Development of a fall detecting system for the elderly residents, Bioinf Biomed Eng, 1359–1362
27. Libelium Comunicaciones Distribuidas S.L.: e-Health Sensor Platform V2.0 for Arduino and Raspberry Pi [Biometric/ Medical Applications] (2014) http://www.cooking-hacks.com/documentation/tutorials/ehealth-biometric-sensor-platform-arduino-raspberry-pi-medical. Accessed 21 Aug 2014
28. OrmLite.com: OrmLite—Lightweight Object Relational Mapping (ORM) Java Package (2014) http://ormlite.com. Accessed 21 Aug 2014
29. PlayFramework.com: Play Framework—Build Modern & Scalable Web Apps with Java and Scala, The High Velocity Web Framework (2014) https://www.playframework.com. Accessed 20 Aug 2014

Preliminary Study of a Novel Shelving System for Nutrition Habits Measuring

Alessandro Freddi, Sauro Longhi, Andrea Monteriù, Davide Ortenzi and Mariorosario Prist

Abstract Nutrition plays a key role in both the prevention of several critical diseases and lead an healthier life in the modern society. In order to analyse household's eating habits in the Italian scenario, authors propose the design of an innovative system called "Smart Shelve". This system consists in a robotic structure integrable with various appliances, such as refrigerators, which allows to receive and deliver several types of foods in order to control their daily consumption. Smart Shelve is equipped with external optical sensors for food recognition, while inside it is composed of several food compartments in which load cells measure the weight of each type of food. The system is capable to warn the users who have recently conducted an incorrect style. The system is also able to send the acquired food consumption information to the cloud in order to provide statistical services regarding the nutrition style of clusters of users localized in a specific geographical area (e.g. district or region).

1 Introduction

It is scientifically recognized that prevention is better than cure. Among the main aspects of prevention, nutrition plays a key role for keeping people healthy [1]. According to the World Health Organization, about one third of cardiovascular diseases and cancers could be prevented through a healthy balanced diet [2]. Moreover, recent experimental results have shown that a nutrition rich of saturated fats (from animal origin) and lacking of fiber is associated with the risk of developing cancers of the digestive system. Further diseases related to malnutrition are: obesity, diabetes (type two), and osteoporosis [3, 4].

S. Longhi · A. Monteriù · D. Ortenzi (✉) · M. Prist
Dipartimento di Ingegneria dell'informazione, Università Politecnica delle Marche, Via Brecce Bianche, 60131 Ancona, Italy
e-mail: d.ortenzi@univpm.it

A. Freddi
Università degli Studi eCampus, Via Isimbardi 10, 22060 Novedrate (CO), Italy

M. Conti et al. (eds.), *Mobile Networks for Biometric Data Analysis*,
Lecture Notes in Electrical Engineering 392, DOI 10.1007/978-3-319-39700-9_4

In order to classify a diet as healthy, it is necessary that it is customized according to the physical characteristics and conditions of the subject (e.g. age, height, weight, sex, health condition, etc.). During pregnancy, for example, the mother must eat enough to satisfy the basic needs for herself and her foetus, and to avoid certain classes of food which could harm the foetus at his/her early state.

Industry and research are developing several smart systems for nutrition habits measuring and, among monitoring devices at home, smart fridges are the most common. They were introduced as part of smart home concept in the late of '90s [5], and their main purpose consisted in avoiding food waste. In fact thanks to several sensors such as RFID readers and bar code readers installed on the structure, the smart fridge was able to recognize foods and store their information into a database [5, 6]. In this way, the smart fridge was able to warn the users about food close to expiry date. The modern smart fridges are more innovative then early versions, and they posses a much a wider number of features, which can be summarized as follows:

- Monitoring of the user nutrition habits
- Automatic food recognition
- Control of user energetic balance
- Storing family information and medical record
- Operation in multi-user environment
- Non-invasive installation in pre-existing environment
- Multimedia cooking demonstration
- Generating shopping list
- Friendly control panel
- Displaying calories for various foods
- Saving energy consumption

Although numerous designs have been proposed in the last decades, only a few of them have become commercial products [7] such as for example, LG TV refrigerator [8], the Siemens CoolMedia Fridge Freezer [9], Samsung Smart Zipel refrigerator [10] and Electrolux Screen Fridge [11]. The diffusion of cloud computing and of advanced mobile devices such as smartphones and tablets allowed to increase the decentralization of health management systems and applications. By exploiting concept, the research is developing innovative smart fridges, such as, CloudFridge [5], ColudFit [12], Diet4Elders project [13] and SapoFitness [14] which provides a web service that runs on a server in the cloud. In this way, the fridge will be accessible from any internet access point.

This paper presents a concept study for a food shelving system called "Smart Shelve", which belongs to the field of Ambient Assisted Living (AAL), because it integrates assistive functions to those already included in a standard smart fridges. Thanks to its robotic structure, it delivers the suitable food to each user according to special user needs, for example dietary control and nutrition diseases. Moreover, it is based on cloud computing in order to share nutrition habits with medical centres, statistical analysis centre and personal mobile devices.

The article is structured as follow. Section 2 provides a description of the Smart Shelve concept, while Sect. 3 describes its communication architecture. Section 4 presents a critical analysis of the proposed concept, which is finally summarized in the concluding section.

2 Description of the "Smart Shelve" System

Although smart devices capable to monitor the health status of users already exist in the literature [5, 7, 14], these have some lacks about user experience focus, low-intrusion object recognition and automatic food localization detection [5]. Smart Shelve is designed to overcome these limitations thanks to an innovative robotic structure which allows to store and deliver several types of food in order to monitor and control the daily food consumption of the users. Its main components are: the food compartments, the control unit and a friendly user interface. Figure 1 shows the Smart Shelve block diagram which better explains its operating principle.

The system structure has to be easily integrable with various appliances such as refrigerators or kitchen cupboards. The main operation consists in automatically collecting the food loaded by the user, recognizing the food by reading a QR or bar code and finally placing it inside the appropriate food compartment. A food compartment is an intelligent unit that can hold only one category of food, such as dairy products, meat, eggs, sweets, bread, etc. Each compartment is equipped with load cells which allow to determine the weight of individual foods in order to evaluate

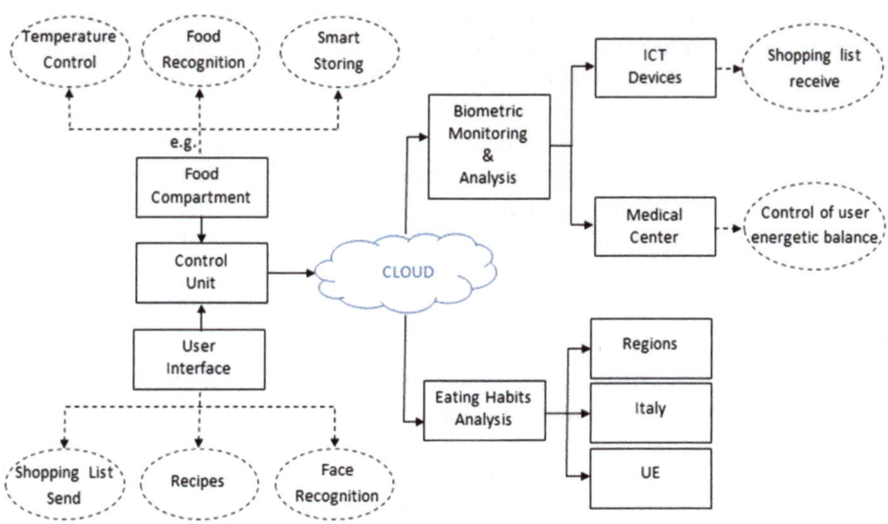

Fig. 1 Block diagram of the novel smart shelving system

their daily consumption. Thanks to the food information achieved from the optical readers, the smart shelving system provides the optimal storage temperature for each food compartment while the delivery of each food depends on their expiry date.

As it is possible to see in Fig. 1, the user can interact with the system through the user interface, equipped with a small RGB camera for user identification, a touch screen for input and output user information, a bar and QR code reader for food recognition. In order to limit privacy issues, the user data captured by the system should contain only generic information about the physical state of the user such as age, sex, pregnancy, predisposition to diabetes mellitus (type two) or celiac disease. These information must be inserted in the system by the user or his/her doctor during the initial configuration of the system.

The data, captured by the interface and the food compartments, allow to obtain a complete information about the type and amount of food assumed by each user. Moreover, in case of an unbalanced diet, the display shows a warning message whenever the user approaches the screen. The warning message is also sent to the medical center and to smart devices.

The concept is designed taking into account the paradigm of the internet of things [15], and thus all the relevant information of the users, such as gender, age, state of health and diet, is stored in the cloud. The use of cloud computing permits to delocalize the storage and computational capabilities of the system, while making the information always available whenever it is needed. Moreover, it makes the system easily interconnected with the user devices and expandable.

The information about nutritional habits can be used in two different ways:

- Biometric Monitoring and Analysis
- Eating Habits Analysis

In the first case, the biometric data can be sent to doctors or to user devices (e.g.: smartphones) in order to warn him/her in presence of an unbalanced diet. In the second case the information about eating habits of people is sent to statistical centres of regional or national level, in order to evaluate the nutritional models. Moreover, the system can be extended to other countries of the European Union in order to provide European nutritional models as well. The knowledge of these models will be useful for researchers to better understand the origin of many diseases that are strongly connected to the food consumption and relative nutrition habits.

Smart Shelve has several innovative functions which can be summarized as follows:

- Smart control of temperature and humidity in each food compartment
- Monitoring of the user nutrition habits
- Automatic food recognition
- Control of user energetic balance
- Transmission of shopping list to user mobile devices

- Suggestions about recipes based on type of stored foods
- Periodic transmission of user nutrition habits to medical centers
- Send of users feeding data to statistical centers for eating statistical analysis
- Nutritional feedback to the users in case of food delivery/consumption
- Operation in multi-user environment by face recognition
- Non-invasive installation in pre-existing ambients
- Conservation of drugs into specific compartment
- Smart graphical user interface

The proposed smart shelving system can be considered as a concept device of the internet of things, and thus its communication architecture, which is based on cloud computing, plays a key role and it is described in the next section.

3 "Smart Shelve" Communication Architecture

The network architecture of Smart Shelve is composed by two parts as depicted in Fig. 2:

1. *Communication* bus based on the IEEE1451 [16]. The serial bus defines a set of communication interfaces to connect transducers (sensors or actuators) to microprocessors, instrumentation systems, and control/field networks. So it is possible to attach different kind of sensors without changing the hardware conditioning of sensor signals, the food compartment firmware and the software for data analysis [17]. The field installation, the system upgrade and the sensors

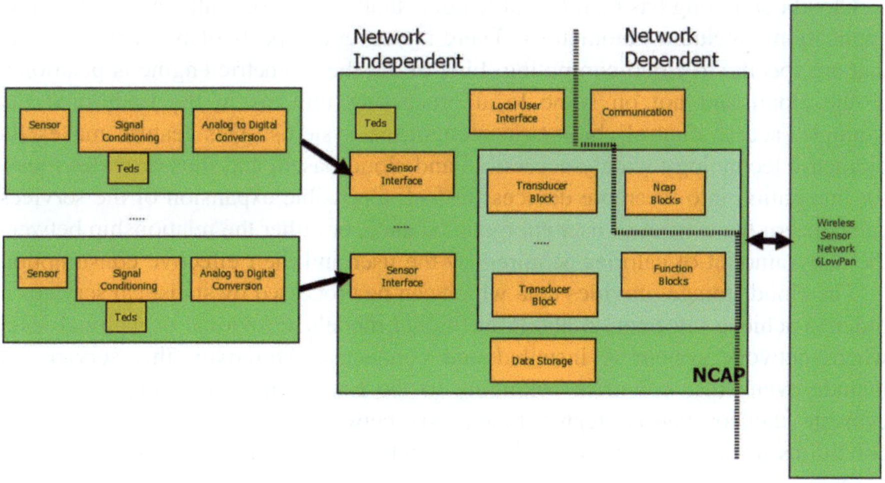

Fig. 2 Block diagram of communication architecture

maintenance are made simple through the "plug and play" of devices to the instruments and the networks [18–21].

2. *Wireless network* based on the IEEE 802.15.4 standards. The Smart Shelve system is fully integrated with the internet of things (IoT). It uses a specific protocol that allows not only to send the data, but also to query each food compartment in the complete system depending on needs. One of the most important protocols developed to achieve full reachability of each element of the Internet of Things, is the 6LoWPAN (IPv6 over Low power Wireless Personal Area Networks) [22].

In particular, the communication bus allows to transfer data from sensors installed in each food compartment to the system control unit, which processes data to transmit them to the gateway through the wireless network.

Regarding the cloud computing, it can be defined as a large-scale distributed computing paradigm that is driven by economies of scale, in which a pool of abstracted, virtualized, dynamically-scalable, managed computing power, storage, platforms, and services are delivered on demand to external customers over internet [23]. Cloud computing provides abstract resources, such as computing power, storage, network and software, and services available through internet [24]. The main characteristics of cloud computing which should be included into the system are [25]:

1. Rapid elasticity
2. Measurement services
3. On-demand self-service
4. Ubiquitous Network Access
5. Pooling of resources

Cloud computing has many great features that make it very attractive for various applications, including biometrics. There are several aspects of biometric systems that are specific to cloud computing. First of all, the biometric engine is positioned in the cloud and not on some local processing unit, as in the traditional (for example, access control) biometric recognition systems. This feature makes the biometric technology widely accessible and cloud-based, and thus provides means for integration into wearable devices. In fact, a possible expansion of the services, can concern the control of user energetic balance, or rather the relationship between the daily amount of calories acquired by the user and their effective consumption.

The cloud storage provides data which can be processed by statistical services in order to achieve information about nutritional models at regional or national level, where network system is installed and connected. Moreover, this service can include even other countries belonging to the European Community in order to evaluate the correlation degree that exists between the distribution of chronic conditions in Europe and the food models relating to each European country.

4 Usability and Acceptability of the "Smart Shelve"

Despite of other Information and Communication Technology (ICT) products for smart homes, such as smart televisions and robotic vacuum, the smart fridge has not obtained a great success on the market yet. Among the several motivations, there are surely a poor analysis of the user acceptance factors and the elevated cost, which both represent very important factors for the success of this technology [26]. About the evaluation of the user acceptance, researchers and experts use the TAM (Technology Acceptance Model) protocol [27], which has been defined the best model to explain the user's attitude towards new technology. The TAM proposes a set of factors that are important to evaluate the user attitude towards smart products. These factors can be grouped into two categories: technological factors and social factors [26]. The first category includes perceived usefulness and perceived ease of use [28], while the social factors indicate the cost of technology [29], technology anxiety [30] and social influence [28].

As previously described, the novel Smart Shelve system belong to the AAL field, because its principal features are oriented towards users with nutrition-related diseases, such as diabetes, hypertension and allergy. For this reason, the only TAM protocol is not sufficient to evaluate the user acceptance towards Smart Shelve system. About this, it should be necessary to add a third category (which we define "health factor"), which permits to evaluate the improvements of user health conditions obtained by using the Smart Shelve. In fact, since this system delivers food only suitable for specific user conditions, the user can not eat any food which could prove dangerous for his/her health. On the other hand, the food constraints imposed by the system could limit the usability by young and healthy people. For these reasons, we are currently defining an acceptance study on sixty potential users composed by thirty questions according to the modified TAM protocol. The testers will be divided into three different ageing groups: under forty, under sixty and over sixty. Moreover, each group will be classified in two subgroups according to the health user condition, "Healthy" and "Sick", in order to evaluate the correlation between the variation of user acceptance and the variation of the quality of life. Another critical point on user acceptance is given by the cost of Smart Shelve. As introduced at the beginning of this section, the high cost of smart fridges is one of the reasons about their low success. In order to project a commercial solution with a competitive price on the market, the system should be composed by low cost hardware, such as low power actuators and low accuracy sensors. Since the processing data is decentralized, the installation of a high performed processor unit is not required and this would lead to a general reduction of the costs. Moreover, it is possible to design different Smart Shelve models with different size, number of food compartments and number of functions. The authors expect the basic version to have a price just higher than a classic fridge while the high performance version could be more expensive than the newest smart fridge models.

Finally we would line to point out that, from a privacy point of view, the smart shelve system has a security protocol capable of protecting the user information stored into the cloud server. The system associates to each user an unique identification number and a password, which are known only to the user and medical center. Moreover, users can choose to transmit own eating nutrition habits to statistical center only after appropriate authorization.

5 Conclusion

Smart Shelve is a system concept which should be able to follow the eating habits of people. Thanks to its innovative structure, it permits to know the daily food consumption of all the people living in the house without invading their privacy. Smart Shelve can be considered a concept device of the internet of things, because it can store user information in the cloud, which makes the system highly scalable, and allows a quick adaptation of the available technology to a growing base of users [31]. The data provided by the system will be very useful to extract a statistical model about eating habits of Italian society, and help the doctors to better follow their own patients. Moreover, if the user has a smartphone, the system is able to calculate the user energetic balance in order to warn him/her in presence of an unbalanced diet.

Future challenges will concern the development of the Smart Shelve system prototype and the study about its user acceptance. In particular, the user acceptance analysis will consist in meetings with some potential users. Each meeting will consist of two steps: prototype presentation and interview. The first will be a short video presentation about smart shelve system prototype and its main features, while in the second step, people will have to answer a questionnaire, which will include thirty questions in according to the modified TAM protocol.

Finally, the Smart Shelve system can have a great social impact, because it may provide an indirect measure of wellness of the modern society, by allowing an effective proactive public health intervention, aimed at reducing both the incidence of nutrition related diseases and the public health care cost.

References

1. Fassino S, Leombruni P, Pierò A, Abbate-Daga G, Rovera GG (2003) Mood, eating attitudes, and anger in obese women with and without binge eating disorder. J Psychosom Res 54 (6):559–566
2. Ministero della Salute della Repubblica Italiana: Alimentazione Corretta Description Available at. http://www.salute.gov.it/portale/salute/p1_5.jsp?lingua=italiano&id=108&area=Vivi_sano
3. World Health Organization: Global Strategy on Diet, Physical Activity and Health Description Available at. http://www.who.int/dietphysicalactivity/publications/trs916/summary/en/
4. Wolf EM, Crowther JH (1983) Personality and eating habit variables as predictors of severity of binge eating and weight. Addict Behav 8(4):335–344

5. Sandholm T et al (2014) Cloudfridge: a testbed for smart fridge interactions. CoRR abs/1401.0585
6. Noutchet AD (2013) Novel user centric RFID fridge design. Comput Info Sci 6(2):151
7. Luo S, Xia H, Gao Y, Jin J, Athauda R (2008) Smart fridges with multimedia capability for better nutrition and health. In: International Symposium on Ubiquitous Multimedia Computing. 39–44
8. LG: TV Refrigeretor. http://www.appliancist.com/refrigerators/lg-tvrefrigerator-lsc27990tt.html
9. Siemens: CoolMedia Fridge Freezer. http://www.appliancist.com/refrigerators/siemenscool media-fridge-freezer.html
10. Samsung: Smart Zipel Refrigerator. http://gizmodo.com/gadgets/homeentertainment/samsung-smart-zipel-refrigerator-156257.php
11. Electrolux: Screen Fridge. http://www.appliancist.com/refrigerators/electroluxscreen-fridge.html
12. Ruiz-Zafra AEA (2014) Cloudfit: a cloud-based mobile wellness platform supported by wearable computing. In: Ambient Intelligence—Software and applications of advances in intelligent systems and computing, Springer International Publishing, 291, 151–159
13. Taweel A, Barakat L, Miles S (2014) A distributed service-based system for homecare self-management. In: On the Move to Meaningful Internet Systems: OTM 2014 Workshops of Lecture Notes in Computer Science, Springer Berlin Heidelberg, 8842, 361–366
14. Lopes IM, Silva BM, Rodrigues JJ, Lloret J, Proenca M (2011) A mobile health monitoring solution for weight control. In: IEEE International Conference on Wireless Communications and Signal Processing, IEEE 1–5
15. Hwang K, Dongarra J, Fox GC (2013) Distributed and cloud computing: from parallel processing to the internet of things. Morgan Kaufmann
16. Baronti P, Pillai P, Chook VW, Chessa S, Gotta A, Hu YF (2007) Wireless sensor networks: a survey on the state of the art and the 802.15.4 and ZigBee standards. Comput Commun 30 (7):1655–1695
17. Jin M et al (2011) Enabling globally unique sensor ID with dual-interface RF tag. In: IEEE Sensors. 1628–1631
18. Prist M, Monteriù A, Longhi S, Giuggioloni F, Freddi A (2015) An integrated simulation environment for wireless sensor networks. In: IEEE International Symposium on a World of Wireless, Mobile and Multimedia Networks (WoWMoM2015), Boston, MA, USA
19. Zhaochun L, Jin H, Yuzhu C (2009) Approach to wireless sensor network development based on IEEE 1451. In: IFITA '09. International Forum on Information Technology and Applications. 1: 443–446
20. Lee K (2000) IEEE 1451: a standard in support of smart transducer networking. In: 17th IEEE Instrumentation and Measurement Technology Conference. 2: 525–528
21. Burch J, Eidson J, Hamilton B (2000) The design of distributed measurement systems based on IEEE 1451 standards and distributed time services. In: 17th IEEE Instrumentation and Measurement Technology Conference. 2: 529–534
22. Naeimi S, Ghafghazi H, Zahedi Y, Ariffin S, Chow CO (2012) Energy evaluation of data aggregation and authentication protocol (daa) in wireless sensor networks in wireless sensor networks. In: IET International Conference on Wireless Communications and Applications. 1–5
23. Fazio M, Celesti A, Puliafito A, Villari M (2014) An integrated system for advanced multi-risk management based on cloud for IoT. In: Advances onto the Internet of Things. Springer, 253–269
24. Dakhane D, Arokar A (2012) Data secuity in cloud computing for biometric application. Int J Sci Eng Res 3(6):1–4
25. Peer P, Bule J, Gros J, Štruc V (2013) Building cloud-based biometric services. Int J Comput Informat 37(1):115–122
26. Alolayan B (2014) Do i really have to accept smart fridges? an empirical study. In: The Seventh International Conference on Advances in Computer-Human Interactions. 186–191

27. Adams DA, Nelson RR, Todd PA (1992) Perceived usefulness, ease of use, and usage of information technology: a replication. MIS quarterly 227–247
28. Moore GC, Benbasat I (1991) Development of an instrument to measure the perceptions of adopting an information technology innovation. Informat Sys Res 2(3):192–222
29. Kuniavsky M (2010) Smart things: ubiquitous computing user experience design, Elsevier
30. Cambre MA, Cook DL (1985) Computer anxiety: definition, measurement, and correlates. J Edu Comput Res 1(1):37–54
31. Kohlwey E, Sussman A, Trost J, Maurer A (2011) Leveraging the cloud for big data biometrics: meeting the performance requirements of the next generation biometric systems. In: IEEE World Congress on Services. 597–601

A Rule-Based Assistant System for Managing the Clothing Cycle

Thomas Walzer, Emre Yay and Natividad Martínez Madrid

Abstract A lot of people need help in their daily life to wash, select and manage their clothing. The goal of this work is to design an assistant system (eKlarA) to support the user by giving recommendations to choose the clothing combinations, to find the clothing and to wash the clothing. The idea behind eKlarA is to generate a system that uses sensors to identify the clothing and their state in the clothing cycle. The clothing cycle consists of the stations: closets, laundry basket and washing machine in one or several places. The system uses the information about the clothing, weather and calendar to support the user in the different steps of the clothing cycle. The first prototype of this system has been developed and tested. The test results are presented in this work.

1 Introduction

Modern living environments are surrounded with sensors and networks to support people's life: lighting, heating or surveillance systems are some examples. However, the domain of clothing is not equipped with sensors although it is daily used and expresses to a high extent the attitude and external appearance of a person. The goal of this work is to design an assistant system (eKlarA) to support the user by giving recommendations to choose the clothing combinations, to find the clothing and to wash the clothing.

Many people can use help in managing their clothing. Especially for functionally impaired people, it is not easy to find the clothing combination for the weather or

T. Walzer (✉) · N. Martínez Madrid
Reutlingen University, Reutlingen, Germany
e-mail: thomas.walzer@reutlingen-university.de

N. Martínez Madrid
e-mail: natividad.martinez@reutlingen-university.de

E. Yay
Daimler AG, HPC T914, 70546 Stuttgart, Germany
e-mail: emre.yay@daimler.com

M. Conti et al. (eds.), *Mobile Networks for Biometric Data Analysis*,
Lecture Notes in Electrical Engineering 392, DOI 10.1007/978-3-319-39700-9_5

events of a day. For example, a blind person has problems to find the clothing items [1]. eKlarA gives the location of clothing in the shelf to the user that only has to sense the content of one shelf to find the item of clothing. A person with autism [2] can use the system as a style recommendation system. The person takes an item of clothing out of the closet and gets a recommendation on which other items of clothing are matching to it. The elderly with forgetfulness can use such a system to get clothing recommendations and to find the location of clothing, so they do not need to ask the nursery staff for help to get clothed and to wash the clothes. The current solutions to these problems often rely on written clothing plans or on the help of other persons. This limits the freedom of these people. The goal of eKLarA is to locate clothing items and to get advises and recommendations at the cycle stations. An example for a recommendation at the closet is the selection of a clothing combination according to a particular style and weather. At the top of system there is a style, which is defined by known styles of the textile industry. Examples for these are casual, business or sportive. The style defines valid clothing combinations using rules. For all styles, the valid combinations will be generated by the system. Additionally, the system gathers the information of calendar and weather. The events and weather of the day are used to select an adequate style. The second area of application of the system is to support the user to locate the items of clothing in the cycle. This can be specially helpful for people with forgetfulness. They can use the system to search for clothing they forget the location. The idea behind eKLarA is that the system works for all people the same, just the way the user interacts with the system changes.

2 Related Work

There are systems that help to manage the clothing at home. The functionality of some of these systems centres on the management of clothing in the closet and available combinations. The app My Fashion Assistant [3] is one example of these. However, such systems are not ideal for functionally impaired people as these apps do not consider the need of impaired people. At the moment, for example, blind persons have to scan their clothing with the hand and connect this feeling with their knowledge, but they miss the color information. Furthermore, persons with autism need help from their parents/relatives to find the right combination. Elderly get help from relatives and nursery staff for that. With eKlarA it is possible to give the functionally impaired people more freedom in their daily life as eKlarA recommends clothing combinations and washing advises considering the dysfunctionality of the people, for example by giving a recommendation to a blind person using a voice output. There are also automatic washing machines developed by different companies; the last prototype of these was shown by NXP Semiconductors

(NXP) in a press announcement[1] in 2012. NXP is not the first company to create a washing machine that uses the Radio-Frequency Identification (RFID) technology for finding the matching washing program. In 2003, the Italian appliance maker Merloni Elettrodomestici (now Indesit Company) launched an RFID washing machine under its Ariston brand, as well as both RFID-equipped refrigerator and oven. The Japanese Dai Nippon Printing company (DNP) developed a similar washing machine, although neither products successfully penetrated the market.[2] At the moment there is no automatic washing machine for the consumer market. eKlarA uses the RFID technology [4] for the tracking/locating of clothing in the cycle. Widely used in the textile industry for logistics [5], the tags are washable and usable in the laundry. These tags are also used for example in hospitals and retirement homes.

3 The eKlarA Assistant System

The name eKlarA stands for a German acronym which is roughly translated as "electronic clothing cycle rule based assistant system". Therefore, the main goal of eKLarA is to observe and control the clothing cycle. An overview over this cycle is shown in Fig. 1.

First of all there is the closet that is the main station and storage for the clothing in the cycle. After the user has taken a item of clothing from the closet and has worn it, he puts it to the laundry room. There the user is adviced to put the item of clothing in the closet-basket that is used for this color-type. The laundry itself is a collection of stations like the washing machine and the closet-basket. From the laundry the clothing goes back to the closet and the cycle starts again. For this function, RFID readers in every cycle station are necessary and every item of clothing has to be equipped with an RFID tag. This tag could be embedded in the clothing or added by the user, for example by seaming it in the washing recommendation label. Alternative actions could be the recommendation for the washing machine to get the ecologic washing program for the clothes in the washing machine. So, there should be an interface for the washing machine to the system, sending the washing programs of the washing machine. Additionally, the washing machine is sending its current load of clothes to the system, and this chooses the matching washing program by analyzing the washing information of the clothes and sending it to the washing machine.

[1]Press announcement of NXP: http://www.marketwired.com/press-release/rfid-nfc-enabled-smart-washing-machine-detects-fabric-supports-remote-maintenance-nasdaq-nxpi-1626149.htm.

[2]Overview of different RFID washing machines: http://www.appliancedesign.com/articles/93065-smart-washer-demo-features-rfid–nfc.

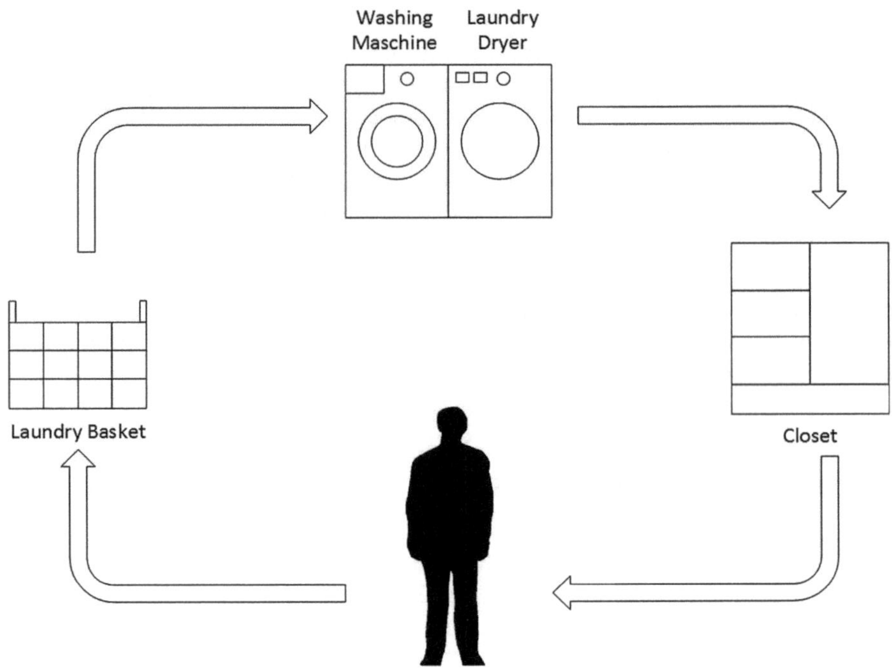

Fig. 1 The clothing components: user, laundry basket, washing machine, dryer and the closet

4 Architecture

In this section the architecture of the hardware and software of the system will be described. In the hardware part there is a description of hardware devices that are used and how they are built. In the software part there is the description of the programming style used and how the system is composed.

4.1 Hardware

The hardware of the system contains two kinds of devices. The main device is the gateway that is placed at home and connected with Internet. The necessary information for the system is collected from the gateway. The gateway is built with a Raspberry Pi.[3] The chosen architecture for the gateway is shown in Fig. 2(1). The hardware layer is the Raspberry Pi, on top of this there is the operating system layer, realized as a Debian for Raspberry Pi. On top of this there is a system service layer that contains a database system realized with MySQL and a Java Virtual Machine

[3]Website of the Raspberry Pi project: http://www.raspberrypi.org/.

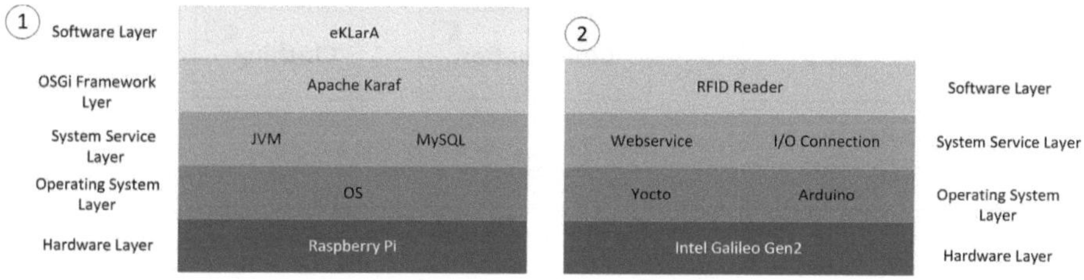

Fig. 2 The gateway stack (*1*) and the control device stack (*2*)

(JVM). In top of this layer there is the OSGi Framework layer realized with an "Apache Karaf" installation. On top of Karaf there is the software of eKLarA.

The second device is the cycle control device shown in Fig. 2(2) which is placed at the clothing cycle stations. It is used to identify the location of clothing. This device contains an RFID reader used with the Intel Galileo Gen 2 Board, which is the hardware layer. On this layer in there are two operating systems. In case of the 'Galileo Gen 2'-Board the first is a Linux called Yocoto, it manages the connection to the network. There above the web-services are located. The second OS is an Arduino compatible system, it is an emulation of such a system. Above this there is the connection to the Input/Output-pins of the board. The both software stacks together are the base for the RFID reader which is the software layer of this component.

4.2 *Software*

The software is programmed in Java. It uses the modular programming concept, by which functions in the same area of the system are grouped in modules. Those modules have interfaces and interface consumers to communicate with the other modules. The concept for modularity [6, 7] in Java is called OSGi[4] which is an extension of Java. Modula ßjava systems use OSGi-frameworks, which is in case of eKLarA Apache Felix.[5] This framework allows to generate modules. These modules need a run time environment likely Apache Karaf[6] [8]. In OSGi, the term bundle is used for a module. It is an OSGi-container based on the Apache Felix framework. In Karaf you can control the system per shell and web services. To work with the database system, there is the Java Persistence Api (JPA). Every module that has to read or write data in the database, has an Entity Manager for the database system.

[4]Website of the OSGi Consortium: http://www.osgi.org/.

[5]Website of the Apache Felix Project: http://felix.apache.org/.

[6]Website of the Apache Karaf Project: http://karaf.apache.org/.

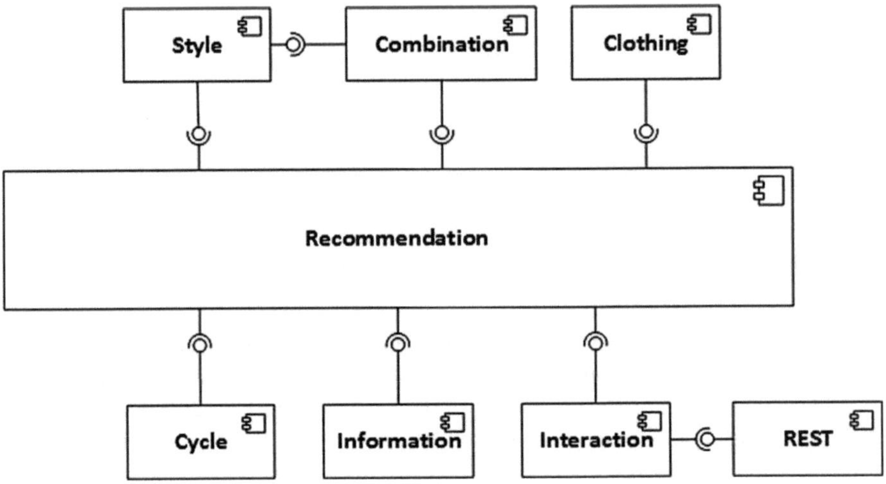

Fig. 3 Modules of the eKLarA clothing system

In Fig. 3 the composition of modules is shown. The main module is the recommendation module that connects all modules and uses their services. The other modules are Style, Combination, Clothing, Cycle, Interaction and Information. They control their subject and contain if necessary the connection to the database tables. The following list gives an overview of the modules and their functionality:

- Style: The Style module manages the different styles. It stores all the styles in an database table called style. The Style module can redefine and personalize the style for the user and the bias of that user. In the style there is a function to generate the combinations with the style rules.
- Combination: The Combination module manages all the clothing combinations in the system. It can create new Combinations if there are new items of clothing in system or if an new style is added. It includes a service to provide combinations that are filtered by style.
- Clothing: The Clothing module manages the clothing items. Therefore, there are functions to update, create, remove and select clothes from the database. An additional function is the recommendation to retire old and not used items of clothing in the closet, this helps the user to clear the closet from clothing.
- Cycle: The Cycle module manages all the features in the cycle. The first feature is the recognition of the items of clothing in each cycle stations. For this, there has to be a list with all the existing cycle stations and a list with all the item of clothing in the cycle. An extended function is to deliver washing advice to devices like the washing machine. With the information of the load of the machine, the cycle module can request the system an advice for the best matching washing program.
- Interaction: The Interaction module provides different kinds of user interaction. For example using a speech interface or a touchscreen in front of the closet. An

integration of mobile apps can be done as well. The interaction module can directly use the function of each module and handing-off this functions to the other modules.
- REST: The REST-Module is an communication module. It offers an REST-Interface for e.g. web clients to the functions of the system by using the functions of the Interaction-Module. The REST-Interface offers the results of the function as JSON-objects. An web-client can consume that objects and parse it to an UI.
- Information: The Information module manages the request and generating information for the recommendation system, therefore the daily information of calendar events and weather forecasts are needed. So the module connects to this data sources and gathers the information. For the weather, there is a parsing method to categorize the weather information to weather data that are known by the system.
- Recommendation: The Recommendation module generates and stores the recommendations. It is possible for a user to create new clothing combination and use the recommendation module to store them. A stored information object contains the information of the weather and the events on the current day and the combinations which are provided by the system and which combination was chosen by the user.

5 Style

The rule-based information source used in eKLarA is the style. Fashion style is an abstract construct that every person recognizes different. The idea to describe style with a data type is to divide the body in four regions and build arrays of clothing items for each region. In the Fig. 4 the regions head, body, legs and feet can be

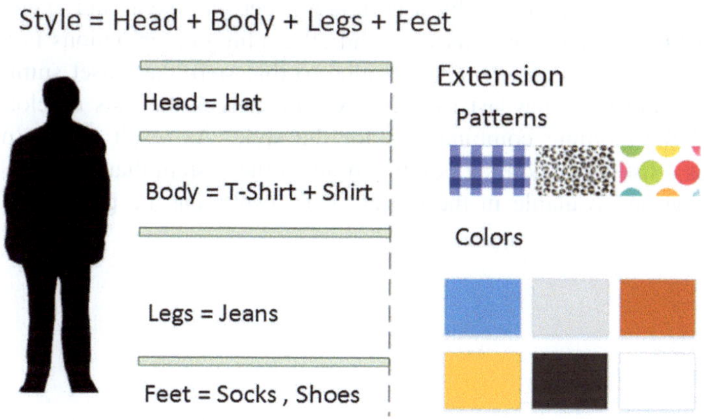

Fig. 4 Definition of clothing style

seen. An extended model could have an region for accessories. A rule is defined by which clothing categories could be combined for that actual style. With this rule the system generates clothing combinations. In Fig. 4 such a combination is shown. In this case it would be a combination for an casual style: he Head with the item hat, the body with the items t-shirt and the shirt, the legs with the jeans and the feet with socks and shoes. The style can be extend with colors and patterns per style.

6 Data Structure

The main data type for the system is the combination, the other data types are clothing, weather, style, cycle station, cycle position, and event categories. For the recommendation, eKlarA needs the information of the combinations [9], the weather and the calendar. The information of the clothing consists of color, size, category, washing recommendations and buying-date, among of others. The weather information is gathered from Internet using the OpenweatherMap[7] API. The data types cycle station and cycle position are defined to locate the items of clothing. The style data type will be defined in the section Fig. 5. The data type combination is an implementation of the style.

7 Recommendation Process

The process of a recommendation illustrated in Fig. 6. It starts by retrieving the information of weather and events for the actual day. To find an adequate style for the events on a day, it is important to prioritize events by their event category. The prioritization depends on their importance, that means an business event is more important than a free-time event. In the recommendation process the system uses the event with the highest priority of the event category. That is used to find the adequate style by matching it by a correlation between style and event category. The result of this matching is used to find all clothing combinations in the system that have this style as an attribute. Parallel to that step, the closet summarizes all containing clothes. In the last step the system merges the lists of clothes in the closet and the clothing combinations for the style. As result of this merge, the system has a list of all clothing combinations in the system that are fit to the actual day and that are available in the closet. This results can be processed by a user interface.

[7]Website of the OpenweatherMap project: http://OpenweatherMap.org/.

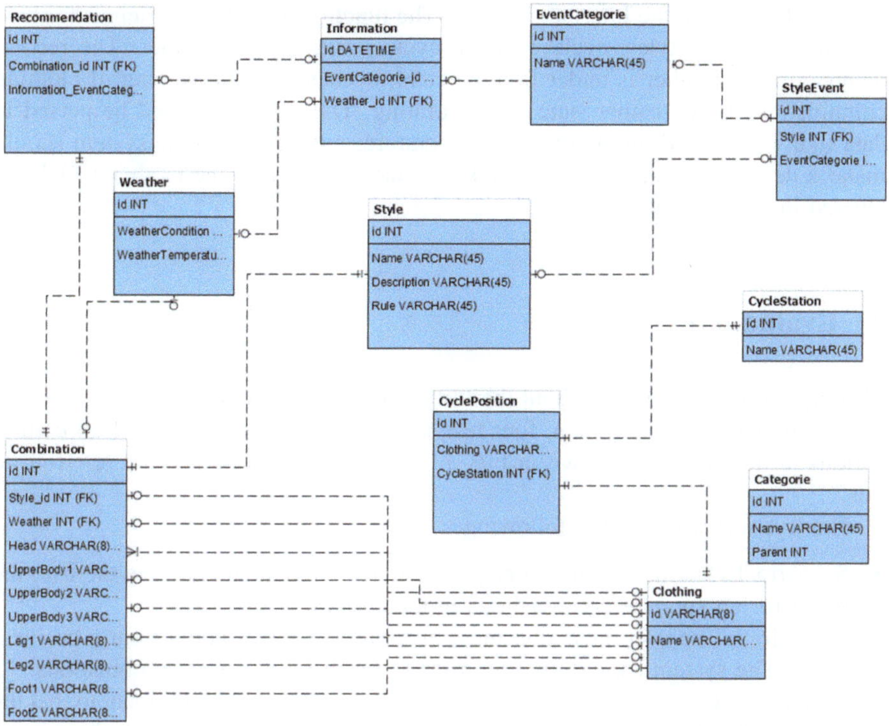

Fig. 5 Datatypes used in eKLarA

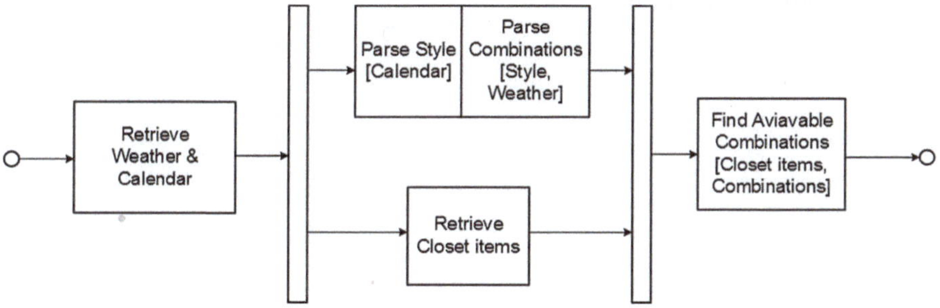

Fig. 6 Clothing recommendation process

8 Prototype

The prototype of the system is installed on a Raspberry Pi on a Karaf installation. The implementation of the modularity functionality in Karaf was done using the blueprint implementation. The client user interface is built by HTML5 and the connection between the eKLarA system and the client is done by a REST-Interface which is built up on the Interface module. The functionality of the system is to show all combinations, styles, cycle stations, cycle positions and clothes. The recommendation by weather is under improvement, because a static parsing of clothing combinations to a weather state is too limiting. The weather should be parsed in data like wind speed, humidity and temperature, with this data the system should make a decision by the material, thickness and structure of the clothes which are layered in an combination.

9 Evaluation

This section is a discussion about the testing and evaluation of the system. To test the functionality of the system, four test scenarios were used. A list of all combinations in the system is shown in Fig. 7. The following scenarios are referring to this.

The test scenarios are the following:

- Scenario 1: The user wants to go jogging. The system should give the user the result for all sportive combinations in the system.
- Scenario 2: The user have a business event at work and should wear a business outfit. The system should give the user as result all business combinations.
- Scenario 3: It is a normal day with event category Normal. Additional to that the trouser HOS0010 is missing. The result are all combinations with the normal and which not containing the trouser HOS0010 is shown. After putting the trouser HOS0010 back in the closet, the system give all combinations with the style normal as result back.

#	id	Style_id	FullBody	UpperBody1	UpperBody2	UpperBody3	Leg1	Leg2
1	1	1	NULL	OBH0003		NULL	HOS0005	NULL
2	2	1	NULL	OBH0002	COA0002	NULL	HOS0007	NULL
3	3	1	NULL	OBH0003	NULL	NULL	HOS0011	NULL
4	4	1	NULL	OBH0002	COA0002	NULL	HOS0011	NULL
5	5	2	NULL	MAO0001	COA0001	NULL	HOS0001	NULL
6	6	2	NULL	MAO0007	MAO0027	NULL	HOS0011	NULL
7	7	2	NULL	MAO0011	NULL	NULL	HOS0006	NULL
8	8	2	NULL	MAO0024	MAO0029	COA0004	HOS0007	NULL
9	9	3	NULL	MAO0003	NULL	NULL	HOS0002	NULL
10	10	3	NULL	MAO0017	MAO0032	NULL	HOS0003	NULL
11	11	3	NULL	MAO0018	NULL	NULL	HOS0010	NULL
12	12	3	NULL	MAO0022	MAO0028	COA005	HOS0009	NULL
13	13	4	NULL	MAO0038	NULL	NULL	HOS0014	NULL
14	14	4	NULL	MAO0035	MAO0030	NULL	HOS0012	NULL
15	15	4	NULL	MAO0041	NULL	NULL	HOS0015	NULL
16	16	4	NULL	MAO0039	MAO0042	NULL	HOS0012	NULL

Fig. 7 The list of combinations for the testing scenarios

Table 1 Result of the evaluation

Scenario	Parameter	Solution combinations
1	4	13, 14, 15, 16
2	1	1, 2, 3, 4
3	3	9, (10), 11, 12
4	2	5, 6, (7)

- Scenario 4: It is a day with event category casual. And 2 of the clothing combinations are not available because HOS0007 and MAO0029 are missing. After putting HOS0007 back in the closet the system shows the combination with the id 7 again.

9.1 Result

For the prototype the recommendation and cycle functionality were tested. The system passed all the test scenarios and the test were repeatable. The result showed that the control of the clothing cycle and the recommendation work. The test has to be extended with the amount of functionality of the system. The evaluation was done by using the command line interaction for every scenario the necessary commands were used. The Table 1 shows the result of the evaluation, the columns are: the Scenario ID, the combinations the system should give back as result, the parameters of recommendation, the status and output of the recommendation command.

The results in Table 1 were generated using the recommendation command of the system.

10 Conclusion and Future Work

The idea behind eKlarA was to support people in their daily tasks related to the clothing cycle. That means that the cycle has to be controlled by a computer system. The prototype of eKLarA showed that the basic system works. It is possible to observe and control the single stations. The style gives the possibility to make clothing recommendations. The implementation showed that the definition of style is a crucial element for the recommendations, in order to have a good user acceptance of the eKLarA system. The user interface will also be improved in the next prototype and studies with external user groups will be performed.

References

1. visionAware (2016) Organizing and labeling clothing when you are blind or have low vision—visionaware. http://www.visionaware.org/info/everyday-living/essential-skills/personal-self-care/organizing-and-labeling-clothing/1235
2. Roschinski H, Roschinski S (2016) Kleidung. http://www.asperger-wahrnehmung.de/alltag-kleidung.php
3. New Vision Interactive (2016) My fashion assistant—fashion for your lifestyle—closet organizer. Style manager. Shopping companion. http://www.myfashionassistant.com/mfa/
4. Finkenzeller K (2006) RFID-Handbuch: Grundlagen und praktische Anwendungen induktiver Funkanlagen, Transponder und kontaktloser Chipkarten, 4th edn. Hanser, München (u.a.)
5. Wong WK, Guo ZX (2014) Fashion supply chain management using radio frequency identification (RFID) technologies, woodhead publishing series in textiles, 152, Woodhead Publishing Ltd, Cambridge
6. Hug K (2001) Module, Klassen, Verträge: Ein Lehrbuch zur komponentenorientierten Softwarekonstruktion mit Component Pascal. Ausbildung und Studium, Vieweg + Teubner Verlag, http://books.google.de/books?id=1qXgefKrpo0C
7. Knoernschild K (2012) Java application architecture: modularity patterns with examples using OSGi. Prentice Hall PTR, Upper Saddle River NJ, Munich (u.a.)
8. Nierbeck (2014) Apache Karaf Cookbook. Packt Publishing
9. Calderin J (2010) Die Modebibel—alles, was Modedesigner wissen müssen, 1st edn. Stiebner, München

Part II
Embedded Systems for Biometric Data Acquisition

State of Charge Monitor for Wireless Sensor Networks

Mirko Carloni, Rocco d'Aparo, Pierpaolo Scorrano, Berardo Naticchia and Massimo Conti

Abstract Wireless sensor networks promise to become one of the most pervasive technologies in the next years. From Smart Cities to industrial safety and human health, from energy to environmental control, the potential of smart networks seems unlimited less until now. Some important features that a pervasive technology must face is its impact on the environment where the installation is made, how easy the installation is and the compatibility with existing technologies. These factors lead to realize, when possible, a network in which all devices from nodes (or routers) to end devices (or peripheral devices) are battery supplied and are as small as possible. In such a network, the batteries, the power management and the energy consumption monitoring, play a very important role, as relevant as the network functionality itself. In this work, we present a micropower battery monitor using a Coulomb to pulse frequency converter based on microcontroller. As a result, we will show that this solution can be well suited in low power applications, like wireless devices where low power consumption and wide dynamic range are important characters.

M. Carloni (✉) · R. d'Aparo · P. Scorrano
Gruppo Filippetti, Ancona, Italy
e-mail: mirko.carloni@gruppofilippetti.it

R. d'Aparo
e-mail: r.daparo@gruppofilippetti.it

P. Scorrano
e-mail: p.scorrano@gruppofilippetti.it

B. Naticchia · M. Conti
Dipartimento di Ingegneria dell'Informazione, Università Politecnica delle Marche, via Brecce Bianche, 12, 60131 Ancona, Italy
e-mail: b.naticchia@univpm.it

M. Conti
e-mail: m.conti@univpm.it

M. Conti et al. (eds.), *Mobile Networks for Biometric Data Analysis*,
Lecture Notes in Electrical Engineering 392, DOI 10.1007/978-3-319-39700-9_6

1 Introduction

Batteries can be classified in two main families: primary cells or not rechargeable battery and secondary cells, also known as rechargeable batteries. Some parameters must be taken into account when a battery is chosen to supply an electronic device:

- the nominal voltage, i.e. the voltage imposed by the battery, slightly different from the Open Circuit Voltage (OCV), that is considerably higher in a fully charged battery;
- the capacity, representing the amount of electric charge contained inside the battery;
- self-discharge current, representing the quantity of energy that the battery lose during its storage or when it is not used.

The battery capacity is a critical variable in the design of battery-powered systems and especially when non-rechargeable batteries are used. In fact, the choice of the batteries with a specific capacitance value determines the effective life of the device intended as the time of autonomous operation without need to intervention by an operator for the replacement of batteries. Examples are devices used in Wireless Sensor Networks (WSN) for Smart City applications. These WSN are often constituted by devices (sensors, routers, etc.) allocated in adverse environments and difficult to reach. The high installation and intervention costs require a guarantee of autonomous operation for long periods, up to several years. When a device must run for years without any human maintenance, the battery must have a sufficient energy and a limited self-discharge like in Lithium-Thionyl chloride Li-SOCl2 batteries. A monitoring system aimed to detect and to correct error conditions and malfunctions (corresponding into high power consumption states, that would alter the batteries life), is very useful in these applications [1].

The estimation of the remaining charge of the battery is fundamental for the correct energy management in every battery powered system from small dimension portable sensors, mobile phone or electric vehicles [1–3].

This work presents a novel Battery Monitor System (BMS) with the aim of monitoring constantly the energy consumption of a battery powered device. This monitoring system predicts the charge consumed by the device and estimates the remaining charge in the battery. Furthermore, the system is able to detect and correct anomalous behaviors related to high consumption profiles.

Moreover, information on the residual battery charge can be used to modify the routing table in a mesh network to maximize the life of a single node. The proposed BMS is able to dissipate few microampere, and it maintains a very wide dynamic range that allow to measure current consumption from IDLE to TX state of the device.

Section 2 presents the state of the art of the state of charge estimation. Section 3 presents the architecture of the proposed Battery Monitor System. Section 4 reports the experimental results.

2 State of Charge Estimation

The State of Charge (SoC) estimation is fundamental for the management of a battery. The State of Charge (SoC) of a battery, is defined as reported in Eq. 1, as the ratio between the residual charge available Q(t) and the nominal capacity Q_{nom}, that is given by the manufacturer. The nominal capacity is different from the maximum actual capacitance storable in the battery, because the latter has a complex dependence on the temperature, battery history and cycle life,

$$\mathrm{SoC}(t) = \frac{Q(t)}{Q_{\text{nom}}} \tag{1}$$

Knowing the SoC it is possible to obtain an indication of the time for which the battery will still be able to ensure sufficient energy before recharge or replacement. When estimating the state of charge of rechargeable batteries, it would be preferable to refer to the nominal capacity of a new cell rather than the actual capacity, because the latter gradually decreases with the aging of the battery itself.

Several methods are used for estimating the state of charge, each one is based on the evaluation of different parameters [4, 5]. The most common ones are the following:

1. Impedance values method [6–8]. The measurement of the impedance allows the knowledge of several parameters which depends on the SoC.
2. Open circuit voltage (OCV) method [9, 10]. The OCV method is based on the estimation of the relation between OCV and SoC. This is the easiest method but inaccurate and sometimes inapplicable, too. The charge-voltage curve is not linear and varies considerably according to the specific battery, to the specific manufacturer and to the operating conditions, such as life of the battery, temperature, loads, time to re-balance the charges, etc. Moreover, the curve of discharge voltage is rather flat for batteries constituted by cells of lithium-ion. This is a good characteristic for all those applications in which a constant voltage is required, but at the same time it represents a problem in the estimation of the charging state. Therefore, this method can be only indicative of the level of charge, but cannot be used for a precise estimation.
3. Coulomb Counting method [11–13]. This method has been used in the proposed system and will be described in detail in next section.

2.1 Coulomb Counting Method

The coulomb counting method measures the discharging current and integrates it over time. This amount must be subtracted to the previous capacity of the battery so obtaining the residual charge, as expressed in Eq. 2.

$$\mathrm{SoC}(t) = \mathrm{SoC}(t_0) - \int_{t_0}^{t} \frac{I(t)}{Q_{\mathrm{nom}}} dt \qquad (2)$$

This evaluation method presents some critical issues. A first issue is that the nominal capacity varies, in real situations, depending on the amount of current supplied. Furthermore, the measuring system itself increases the total consumption of the device.

Several commercial devices exist for measuring the state of charge and most of them use the method of Coulomb Counting. However, the measurement of the current can be made using different methods.

A simple method for current measurement uses an analog to digital converter, for example the device DS2780 of Dallas Semiconductor [14], that exploits the proportional relation between the current flowing through an external resistor with known value and the voltage drop across its terminals. This voltage value is then converted by an ADC (16 bit) to obtain the digital value of the output current from the battery. This value is stored in a specific register and updated every 3.515 s. The value of the accumulated charge is performed algebraically summing the contributions of positive and negative currents appropriately weighed in the period in which these are considered constant (i.e. for 3.515 s). This method is excellent in terms of precision and range of the measurement provided, but it is not efficient when the consumption profile varies quickly. This limit is stringent in applications such as WSN where high consumption states have very short duration time and some orders of magnitude higher then low power states (or stand-by). Moreover, the current consumption of DS2780 is declared between 65 and 95 μA in normal operation and it could be comparable to the current absorbed by the system to be monitored.

Similar principles of operation and similar problems occur in other devices, such as the Texas Instruments BQ27510-G2 and BQ3055 [15].

The solution adopted by the Linear Technology LTC4150 [16] uses a Pulse Frequency Modulation to measure the current absorbed. This device creates a current meter by converting the voltage drop across the sense resistor, proportional to the current flowing in it, in a sequence of pulses with variable frequency, which is associated at a specific amount of charge delivered by the battery. This method, differently from the previous one, provides measurements less accurate but it allows an assessment almost in real-time. Unfortunately, this device has rather high power consumption (the current is about 80–140 μA).

In conclusion, the integrated circuits available in the market are able to measure more or less precisely and quickly the current supplied by a battery, but they require a not negligible energy, making their use suitable in environments in which a power supply network circuit for recharging batteries is present. Finally, these devices do not include energy management control systems, such as hardware and/or firmware implementations of control policies, but only the possibility to send the measured values with a serial protocol.

3 Battery Monitor

The proposed system uses the coulomb counting method with a Pulse Frequency Modulation in an innovative way that allows a wide dynamic range, absorbing only 15 μA. First results are reported in [17]. The BMS has been applied as test application to the monitoring of a router in a wireless sensor network, as shown in Fig. 1.

Figure 2 reports the photos of the prototype of the battery monitor system. The dimension of the prototype are evident in the third photo in which the monitored battery is connected.

The schematic, reported in Fig. 3, highlights different sections of the system developed in different colors:

- Near Field Communication (NFC) and wakeup circuit
- Sense Circuit and Pulse Frequency Modulation (PFM) coulomb counter
- Microcontroller
- Voltage Regulator (BMS and Load Supply)
- Serial Level Adapter.

3.1 NFC and Wake-Up Circuit

The NFC circuit is a section that can receive a radiofrequency code (Manchester 125 kHz) coming from an external device (Smart Installer). This circuit is

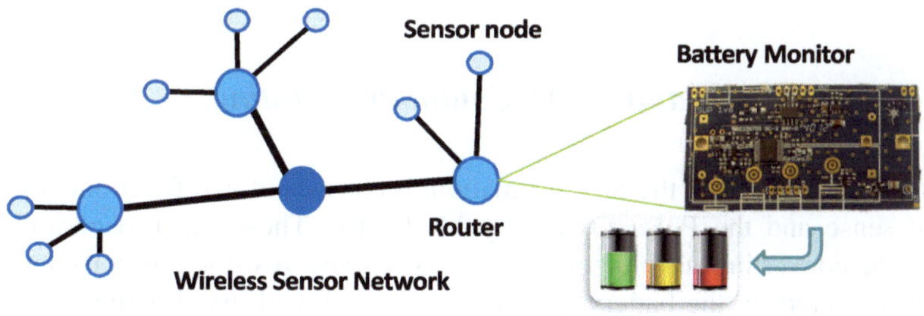

Fig. 1 Application example of the BMS in the monitor of a router in a wireless sensor network

Fig. 2 Photograph of the BMS

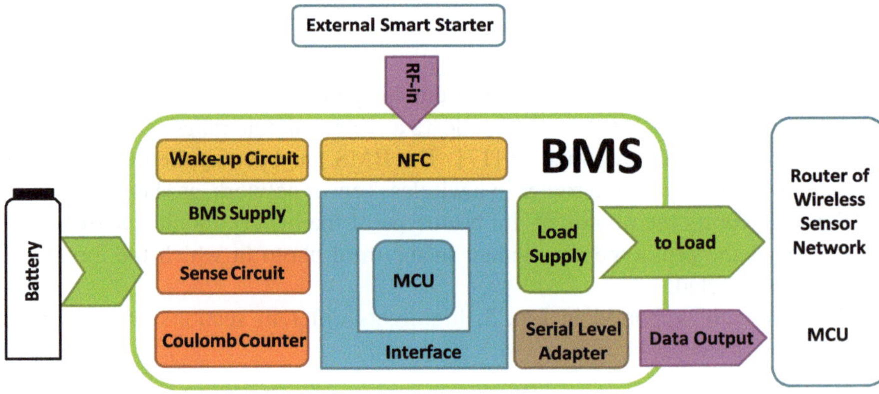

Fig. 3 BMS block diagram

composed by an inductive wake-up circuit (the Austriamicrosystem device AS3933) and a 125 kHz resonator. The Manchester modulated signal is captured by an on board SMD inductive antenna and decoded by the wake up integrated circuit. If the preamble and the pattern received is correct, an interrupt is generated, and the command coming from the external device is received by the battery monitor MCU, that goes from idle or deep sleep to operating state for the time necessary to perform the command requested, then it returns in a "deep sleep state".

If the command is "power on", the microcontroller supervisor goes in active state, and therefore it is ready to play its role of "supervisor" and "battery monitor". The NFC is used also to receive commands for installation purposes.

3.2 *Sense Circuit and PFM Coulomb Counting*

This section implements the SoC estimation. It consists of two functional units: the current sense and the Pulse Frequency Modulator. They realize a current meter capable of converting a voltage drop across a sense resistor (proportional to the current delivered by the battery) in a train of pulses with fixed width but variable repetition frequency.

The current sense implemented with a resistance in parallel to a super-capacitor, both with good temperature stability. The super-capacitor is used to obtain a wide dynamic range. Its effect is to smooth out high peaks of the current drained, spreading them on a wider time interval. In this way, it is measured not exactly the instantaneous current but its value averaged over the integration interval (depending on capacitance and resistance values).

A Current-Sense Amplifier is inserted after the R_1C_1 integrator: the TS1100-200 integrated circuit of Touchstone Semiconductor is used, it amplifies by a factor 200 the input voltage level with a very low current consumption. Therefore, the

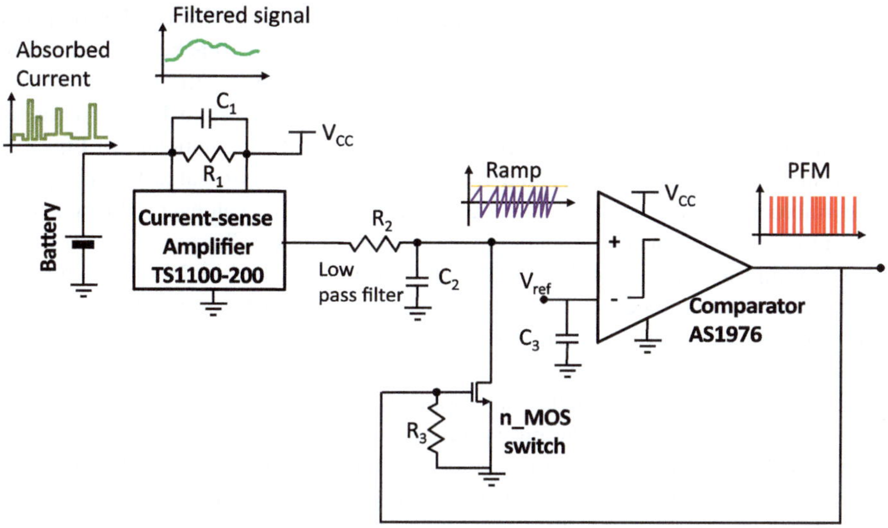

Fig. 4 Principle of operation of the Current Sense and PFM current counter

relationship between the voltage across R_1 and output voltage level of the amplifier is 200 V/V. The measurement of high instantaneous currents may saturate the measurement due to this high output-input ratio. For this reason, the capacitance C_1 has been inserted in parallel to the resistor R_1, as shown in Fig. 4.

Considering that the circuit is placed immediately after the supply batteries, the total current drained from the battery consists of two separate contributions: the effective current absorbed by the slave device and the current required by the BMS itself, which however is very low.

The output voltage of the current sense section is used to generate a voltage ramp connected to the not-inverting input of a voltage comparator (the AS1976 of AustriaMicroSystems with push/pull output stages).

The slope of the voltage ramp across the capacitance C_2 depends on the output voltage of the current sense. An appropriate threshold is connected to the inverting input of the voltage comparator. This threshold will be the maximum value the ramp can assume. The minimum obtainable threshold is obtained leaving the inverting input floating, thus using the internal offset of the comparator as comparing threshold. The datasheet of the voltage comparator AS1976 indicates that the internal offset is between 1 and 5 mV.

Finally, an electronic switch (an N-channel MOSFET) is connected to the output of the comparator, the switch connects the not-inverting input to ground when it exceeds the threshold value. This operation resets the ramp at the initial value, below the threshold.

In this way, a pulse train with fixed duration can be measured at the output stage of comparator. The pulse width is very short and it depends on the speed with which the electronic switch and the comparator are able to reset the non-inverting input, but variable pulse distance. On the contrary, the time distance between two pulses depends on the speed with which the input ramp reaches the threshold voltage. This speed is related to the output voltage from the current sense, in turn proportional to the current delivered by the battery. Therefore, a Pulse Frequency Modulation is obtained. These operations are described schematically in Fig. 4.

A specific energy value drained from battery can be associated to each pulse. The microcontroller, counting the pulses, estimates the values of the absorbed current and of the energy drained and finally the State of Charge.

The minimum value that the system can measure is related to the threshold of the comparator: for voltages at the not-inverting input of the comparator below the threshold the PFM will not generate pulses.

Conversely, the saturation of the system occurs when the output voltage of the Current-Sense Amplifier reaches the supply voltage (the Current-Sense Amplifier TS1100-200 has rail-to-rail output voltage).

The voltage at the not-inverting input of the comparator is amplified by the conversion factor of 200. These reasons justify the choice of a mediated rather than instantaneous measurement of the current drained from the battery. Thanks to RC's integrator, high instantaneous values of current are evaluable provided that their duration is limited.

3.3 Microcontroller

The PIC24F16KA101 from Microchip manufacturer has been chosen. It combines good performance (up to 16MIPS @ 32 MHz) at low current consumptions (from 4 to 40 μA in Low-Power state) not forgetting the presence of some key peripherals such as UART (Universal Asynchronous Receiver-Transmitter) for serial communication. Its role is to coordinate the BMS's operations, such as: forwarding the commands coming from the external Smart Installer through NFC to the controlled device through the serial interface (in the test example used in Sect. 4 the controlled device is a router of a wireless sensor network), counting of the pulses from the PFM generator, calculating the absorbed charge, estimating the State of Charge of the battery.

Furthermore, the microcontroller, monitoring of consumption of the router, can be used to diagnose malfunction states and to manage them by applying appropriate control policies, such as power reset.

These features can be customized modifying the firmware of microcontroller.

3.4 Voltage Regulator

The BMS presents on board circuitry capable to deliver power to the load, specifically, the Texas IC named TPS78225. It is an "Ultra-Low Quiescent Current, Low-Dropout Regulator", that is a voltage regulator with stable output fixed at 2.5 V.

The TPS78225 current consumption is typically of the order of 500 nA, that however rises to some 2 μA in presence of loads absorbing high currents. The TPS78225 is also equipped with an enable-pin directly managed by the micro-controller. In particular, when the microcontroller activates the enable, the voltage regulator TPS78225 cares to provide the supply voltage to the load. Otherwise when the entire BMS is turned off or when the management policies deem appropriate (for example when abnormal consumptions are detected) the TPS78225 is disabled, switching off the load.

3.5 Serial Level Adapter

The serial level adapter is necessary to integrate the functionality of serial communication between BMS and the controlled device. Specifically we used the integrated circuit NLSX4373 of ON Semiconductor, i.e. a level shifter for digital I/O lines, which allows to perform a matching between the I/O voltage levels from the microcontroller with those of the load. It must be considered, in fact, that the internal circuitry of the BMS works approximately at 3.4–3.6 V, while the slave device is precisely regulated at 2.5 V.

4 Experimental Results

4.1 Device Characterization

Analyzing the single components that realize functional blocks of the BMS it results that, from the consumption reported in manufacturer's datasheet [10–14], the entire device could absorb between 6.62 and 56.35 μA.

Ten prototypes of the BMS has been realized and the current absorbed has been measured. The values obtained from these measurements show clearly that *current consumption* of the BMS are between 14 and 15 μA except in a single case in which measured consumption is higher.

The *measured offset* values are almost always close to the maximum value stated (5 mV), except in a two cases where this value is greatly exceeded, probably due to defects on the IC, with a consequent inability to detect low values of delivered currents.

Another relevant parameter of the BMS is the *minimum value of detectable current*. As already said, this is directly linked to the offset voltage of the input stage of comparator that is declared by the manufacturer between 1 and 5 mV in standard environmental conditions. As a consequence, using ratio of 200 between the output voltage of the comparator and the current absorbed from the battery, the minimum current detectable should be approximately 25 μA. The measured offset values are between 4.6 and 5.2 mV, except in one case where this value is 15.2 mV.

The *maximum measurable value of the current* is related to the supply voltage of the comparator. The maximum value of the input voltage of the comparator is the supply voltage, 3.4–3.6 V in this project. Considering gain of the current sense amplifier, the maximum value of measurable current is about 17–18 mA. This limit is actually valid only for constant consumptions or with very high time duration. For peak of current higher than this limit, but of short duration, the presence of the R_1C_1 filter in the input stage allows not to saturate the measurement.

Other parameters that characterize the BMS, are the ratio between the amount of charge drained from the battery (Ah) and the frequency of output pulses of the PFM generator and the amount of charge (μAh) linked to each single pulse. The energy drained is therefore obtained counting the number of pulses and multiply it by the energy associated to each pulse. The characterization of each device has been performed in the following way maintaining disabled the output voltage regulator and the serial level adapter: (1) a constant resistive load has been inserted as slave device; (2) the current absorbed (load and BMS contributions) has been measured by an external instrument and compared to the output waveform of the PFM generator, the distance between the generated pulses has been measured; (3) the best fitting linear relation between the inverse of distance between two pulses expressed in seconds and the measured current expressed in Ampere has been derived. So, using this relationship and counting the number of pulses detected in one second, the amount of charge drained from the battery expressed in Asec has been obtained. Figure 5 shows the measured pulse frequency as a function of the constant current absorbed and the linear interpolation.

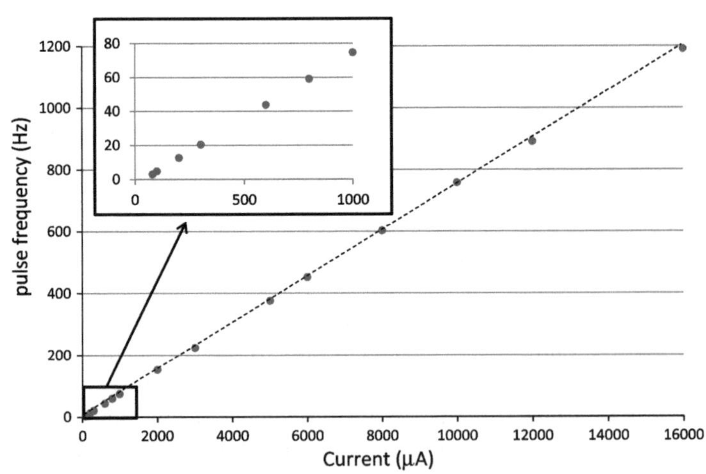

Fig. 5 Measured pulse frequency as a function of the current absorbed

4.2 Test in a Wireless Sensor Network Application

In a first real test condition, the BMS has been connected to the monitored device, i.e. a SmartSpace router used for WSN applications. This is an example of "real loads", when current profiles are not constant but rather strongly variable in amplitude and in time duration. The SmartSpace router allows a great variety of consumption's profiles linked to RF traffic conditions and the presence on board of UART sections for serial communication. The quiescent consumption of these routers is slightly less than 100 μA when the RX and TX are disabled, and only a nanopower wake-up on radio is enabled. Conversely, current consumptions are several orders of magnitude higher when RX or TX are enabled and data are sent and/or received. For example, the UHF section absorbs about 27 and 70 mA in RX and TX, respectively.

The test bench, shown in Fig. 6, is composed of a BMS, a SmartSpace router used as a slave load, a Lithium-Thionyl chloride Li-$SOCl_2$ battery (Tadiran) with nominal voltage of 3.6 V, a 1 Ω resistor in series to the battery to measure the current absorbed from the battery and finally a digital oscilloscope to acquire traces of the current profile and the pulses train (PFM) generated. Furthermore, some ad hoc Matlab scripts were implemented to convert traces acquired in numerical values respectively of the average current on 1 Ω resistor and the number of pulses.

As an example, Fig. 7 shows two screenshots of the digital oscilloscope in which are highlighted: in blue, the voltage drop on the sense resistor (1 Ω) caused by the current delivered by the battery; in green, the PFM pulses produced by the "Coulomb Counting" section of the BMS. Figure 8 shows, in another example, the screenshot of the digital oscilloscope with a higher time resolution.

From a first qualitative analysis, a "condensation" of pulses in correspondence of high current absorptions is evident (during the router start-up or when there is RF activity).

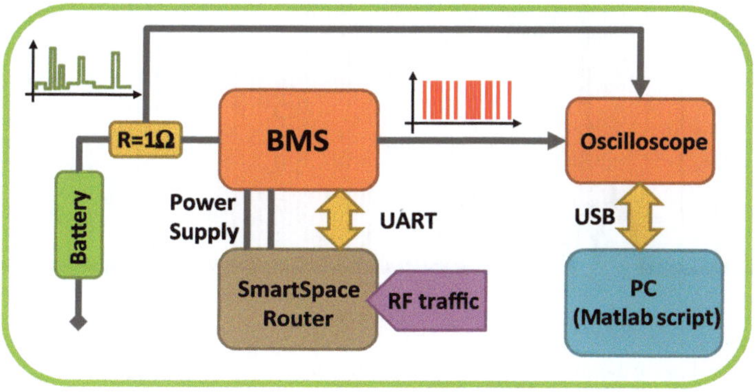

Fig. 6 Test bench for BMS in the wireless sensor network application

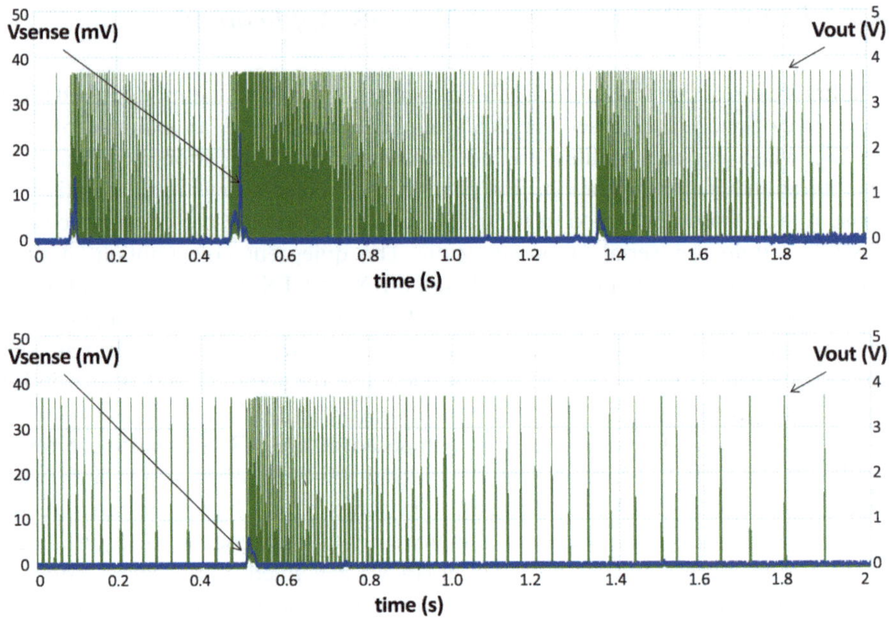

Fig. 7 Typical current consumption profile of SmartSpace's router and relative BMS's PFM

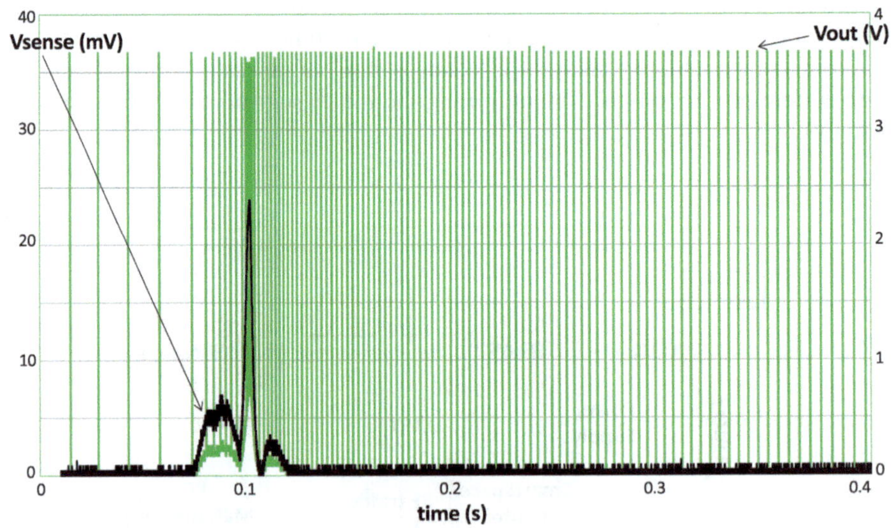

Fig. 8 Screenshot of the digital oscilloscope with a higher resolution with respect to Fig. 7

Table 1 Result of the monitoring tests on the SmartSpace router

	Pulse n.	Average current absorbed (μA)	Measured charge (μAs)	Estimated charge with BMS (μAs)	Error (%)
Low traffic condition					
1st min	1764	332.1	19.9255	19.0512	4.39
2nd min	459	89.9	5.3928	4.9572	8.08
Total	2223	211.0	25.3183	24.0084	5.17
High traffic condition					
1st min	3784	689.2	41.3542	40.8672	1.18
2nd min	2749	500.3	30.0176	29.6892	1.09
Total	6533	594.8	71.3718	70.5564	1.14

The time interval between pulses increases, when router returns in a quiet state and when its consumption is at quiescent value.

It can be noticed that high consumption's values (significantly above the threshold of 17 mA) but of short duration, do not saturate the PFM produced thanks to the inclusion of super-capacitor in parallel to the sense resistance, that spreads the impulses in a wide time interval.

Specific tests on the same router in two different operating conditions have been performed for a specific quantitative analysis. The first operating condition was a situation of heavy RF traffic, while the second was a situation of almost total absence of traffic. The total duration of the acquisition for each test was of 2 min, divided into blocks of 10 s.

The results are shown in Table 1. It can be seen, that in conditions of very low consumption, the estimation by counting the pulses differ by several percentage points from instrumental measurement.

The results in the test cases where heavy consumption is measured are more satisfactory. In this case estimated and measured values are similar and deviations are only by one percentage point. Moreover, the heavy traffic condition is closest to a real situation.

5 Conclusions

This work shows a novel Battery Management System using Pulse Frequency Modulation for the measurement of the current consumption in micropower applications. The BMS has been designed and a prototype has been developed and tested in a real application for the monitoring of the consumption of a router of a wireless sensor network. BMS is able to provide reliable results without interfering on measured values and on life duration of the supply batteries. Furthermore, the possibility to enable/disable using NFC of the entire device, serial peripheral and other ICs on board allows implementation of a wide range of control policies, that can be adapted to many different devices and operating conditions.

References

1. Zarrabi H, Al-Khalili A, Savaria Y (2011) Activity management in battery-powered embedded systems: a case study of ZigBee; WSN. In: 2011 18th IEEE international conference on electronics, circuits and systems (ICECS), pp 727–731
2. Scavongelli C, Franco F, Orcioni S, Conti M (2015) Battery management system simulation using system C. In: Proceedings of the IEEE 12th international workshop on intelligent solutions in embedded systems WISES2015, Ancona, Italy, pp 151–156, 29–30 Oct 2015
3. Conti M, Fedeli D, Virgulti M (2011) B4V2G: Bluetooth for electric vehicle to smart grid connection. In: Proceedings of the 9th international workshop on intelligent solutions in embedded systems WISES 2011, Regensburg, Germany, pp 13–18, 7–8 June 2011
4. Chang W-Y (2013) The state of charge estimating methods for battery: a review. ISRN applied mathematics, Hindawi Publishing Corporation, 2013, Article ID 953792, 7 p
5. Pang S, Farrell J, Du J, Barth M (2001) Battery state-of-charge estimation. In: Proceedings of the 2001 American control conference, 2001, vol 2, pp 1644–1649
6. Rodrigues S, Munichandraiah N, Shukla AK (2000) A review of state-of-charge indication of batteries by means of A.C. impedance measurements. J Power Sources 87(1–2):12–20
7. Huet F (1998) A review of impedance measurements for determination of the state-of-charge or state-of-health of secondary batteries. J Power Sources 70(1):59–69
8. Tairov S, Stevanatto L (2011) The novel method for estimating vrla battery state of charge. In: 2011 IEEE electronics, robotics and automotive mechanics conference (CERMA), pp 211–215
9. Ng K-S, Moo C-S, Chen Y-P, Hsieh Y-C (2008) State-of-charge estimation for lead-acid batteries based on dynamic open-circuit voltage. In: IEEE 2nd international power and energy conference, PECon 2008, pp 972–976
10. Coleman M, Lee CK, Zhu C, Hurley WG (2007) State-of-charge determination from EMF voltage estimation: using impedance, terminal voltage, and current for lead-acid and lithium-ion batteries. IEEE Trans Industr Electron 54(5):2550–2557
11. Ng K-S, Huang Y-F, Moo C-S, Hsieh Y-C (2009) An enhanced coulomb counting method for estimating state-of-charge and state-of-health of lead-acid batteries. In: 31st international telecommunications energy conference, 2009. INTELEC 2009. pp 1–5
12. Ng KS, Moo CS, Chen YP, Hsieh YC (2009) Enhanced Coulomb counting method for estimating state-of-charge and state-of-health of lithium-ion batteries. Appl Energy 86 (9):1506–1511
13. Lezhang L, Wang LY, Chen Z, Wang C, Lin F, Wang H (2013) Integrated system identification and state-of-charge estimation of battery systems. IEEE Trans Energy Convers 28(1):13–23
14. Maxim Integrated (2009) http://www.maximintegrated.com/datasheet/index.mvp/id/4560, Stand-Alone Fuel-Gauge IC (2009). Datasheet
15. Texas Instruments (2013) http://www.ti.com/lsds/ti/power-management/battery-management-productsproducts.page, Products for Battery Management (Consulted in 2013). Datasheets
16. Linear Technology (2003) http://www.linear.com/product/LTC4150, Coulomb Counter/Battery Gas Gauge (2003). Datasheet
17. Carloni M, d'Aparo R, Scorrano P, Naticchia B, Conti M (2013) A micropower supervisor for wireless nodes with a digital pulse frequency modulator battery monitor. In: Proceedings of SPIE 2013 microtechnologies, international conference VLSI circuits and systems, vol 8764. Grenoble, France, Paper 26, pp 0P.1–0P.12, 24–26 Apr 2013

Development of an Algorithm and a Sensor to Monitor the Heart Rate by Volumetric Measurement Techniques

Jens Gansloser and Ralf Seepold

Abstract The person's heart rate is an important indicator of their health status. A heart rate that is too high or too low could be a sign of several different diseases, such as a heart disorder, obesity, asthma, or many others. Many devices require users to wear the device on their chest or place a finger on the device. The approach presented in this paper describes the principle and implementation of a heart rate monitoring device, which is able to detect the heart rate with high precision with the sensor integrated in a wristband. One method to measure the heart rate is the photoplethysmogram technique. This method measures the change of blood volume through the absorption or reflection of light. A light emitting diode (LED) shines through a thin amount of tissue. A photo-diode registers the intensity of light that traverses the tissue or is reflected by the tissue. Since blood changes its volume with each heartbeat, the photo-diode detects more or less light from the LED. The device is able to measure the heart rate with a high precision, it has low performance and hardware requirements, and it allows an implementation with small micro-controllers.

1 Introduction

The influence of stress has been investigated in many domains. Each domain has specific conditions and stress may cause different behaviors over time. However, the overall result is that stress is destructive for a person's health: it causes consequences in the form of short-term behaviors and/or long-term diseases [1]. Describes the influence of stress on short-term behavior. Even for groups of people used to managing stress in their daily work, stress influences decisions that must be made

J. Gansloser (✉) · R. Seepold
University of Applied Sciences Konstanz, Brauneggerstr. 55, 78462 Konstanz, Germany
e-mail: jens.gansloser@htwg-konstanz.de
URL: http://uc-lab.in.htwg-konstanz.de

R. Seepold
e-mail: ralf.seepold@htwg-konstanz.de
URL: http://uc-lab.in.htwg-konstanz.de

M. Conti et al. (eds.), *Mobile Networks for Biometric Data Analysis*,
Lecture Notes in Electrical Engineering 392, DOI 10.1007/978-3-319-39700-9_7

under heavy time pressure. Furthermore, being stressed (not being busy) can hardly be detected by the person themselves [2]. In the long-term, stress may cause severe health problems, such as those documented in [3] and [4]. Stress may provoke negative effects in the workplace [5, 6], and it may lead to mental problems [7]. Due to the influence of stress on a person's nervous system, non-invasive stress detection (like analyzing the Heart Rate Variability, HRV) can contribute to detect stress and thus lower costs for health care in the long-term [8]. The approach presented in this paper describes the principle and implementation of a heart rate monitoring device (HRM), which is able to detect the heart rate with high precision with the sensor integrated in a wristband. It provides a convenient way to carry the sensor during daily activities and it supports long-term measurements. However, there are different ways to measure the heart rate or the HRV. The overall goal of this approach is to develop a sensor-based solution that is wearable, inexpensive, and highly accurate. Furthermore, it should use mobile architectures (like smartphones) as a support to analyze raw sensor data and to communicate to external servers, if it is required.

The technology used should be open source and well known. Additionally, it is important to only use a minimalistic hardware/software setup. A Raspberry Pi/Linux system[1] for example would be unnecessarily complex and oversized. All the requirements for the project are summarized in the following list:

- Capable of detecting the heart rate with high precision
- Low performance requirements
- Few and inexpensive hardware requirements (no special hardware)
- Only open source/free software should be used
- Hardware setup should be flexible to use (wristband, finger,…)
- Simplistic and understandable design
- Easy to use in further projects
- Easily expandable

In the following Section, a brief resume of the state of the art in this domain is presented. The principle and implementation of the heart rate monitoring device is discussed in Sect. 3 and two different approaches are shown. This is followed by a deeper explanation of the signal processing, the prototype implementation is presented, and the measurements are evaluated against the precision of a commercial device. Finally, a summary and a description of the future work is given.

2 Related Work

For the measurement of the heart rate, several techniques can be used. Common possibilities are pulse meters or an electrocardiogram (ECG). The ECG has a traditional placement of three electrodes on the body following the placement by

[1] http://www.raspberrypi.org

Einthoven [9]. This approach has the advantage of receiving data capturing the QRS complex and then analysing it via software [10]. Thus it offers much more than just detecting the heartbeat. The main disadvantage of this approach is that electrodes have to be placed on the body and typically wired to hardware. In comparison, a wristband is easier to wear, especially for long-term measurements.

Our approach should design a solution mainly for capturing the heart rate. While the ECG approach offers much more, it is beyond the scope of this work. Some commercial or some semi-professional hardware offers the possibility to capture the signal but in most cases the hardware or software is not fully accessible for deeper investigation; examples of devices with limitations on communication access are [11–13]. Other approaches (e.g. [14]), independently from being wearable or not, are excluded because the price should be within the range of consumer market products.

3 Model Approach

There are several ways to measure the heart rate. For this project, a technique is required which uses small and inexpensive hardware that can be mobile. The performance requirements should be low to allow an implementation with small energy-saving micro-controllers.

3.1 Principle of Operation

One method to measure the heart rate is the photoplethysmogram (PPG) technique. This approach measures the change of blood volume through the absorption or reflection of light. A light emitting diode (LED) shines through a thin amount of tissue. The wavelength of the light should be near infrared. A photo-diode registers the intensity of light that traversed the tissue or was reflected by the tissue. Since blood changes its volume with each heartbeat, more or less light from the LED gets detected by the photo-diode. If the heart pushes blood through the vessels, more blood flows between the LED and photo-diode (it has a higher volume) and more light gets absorbed by the blood. That means less light is registered by the photo-diode. If blood flows from the vessels to the heart (the volume declines), less light gets absorbed by the blood and the photo-diode detects more light. As a result, the registered intensity of light changes continuously with the pulse. By measuring the time between two intensity peaks the current pulse can be derived. Two possible settings for the heart rate monitor are displayed in Fig. 1. Besides detecting the light that traversed the tissue, the light reflected by the tissue can also be measured.

Based on these two techniques, the application offers high flexibility with the kind of measurement. The heart rate can be measured through the fingertip, at the

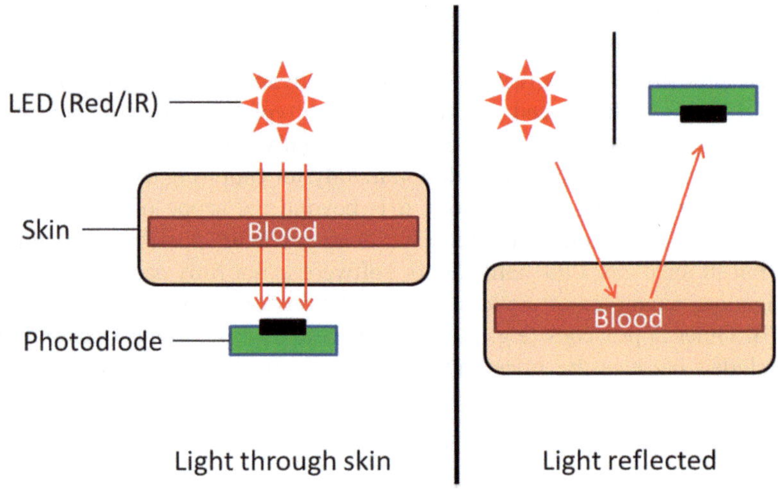

Fig. 1 Possible settings for HRM

ear, or at the wrist. Additionally, the hardware requirements are minimal. To implement this measurement technique, only a LED and a light sensor are required.

3.2 Measurement

Figure 2 shows the output signal of the light sensor. The diagram shows the different luminosity values over time, which represent the heart rate of a patient. It is measured by placing the finger between the photo-diode and the LED.

To get the exact heart rate, the time difference between two minima (or maxima) of the discrete sensor values needs to be determined. However, it is not trivial to determine the minima, because of the noise in the signal data. The noise comes from different environment settings and how the finger is placed on the sensor. Additionally, each person's blood volume changes different. These parameters result in changes in the data along the y-axis.

3.2.1 First Approach

A first naive approach is to try to determine the grade of each pair of sample points in the data and identify if the curve is declining or rising. With this knowledge, the minima could be identified. However, this approach is not optimal because it is impossible to separate different minima in the data and find the correct ones. Due to the noise, unwanted local minima could be detected. Introducing a static threshold is unsuitable too, because of the differences in the y-axis data in each sample. To conclude, this approach is not suitable for a reliable solution of the problem.

3.2.2 Second Approach

Taking a different approach, instead in the time domain the luminosity values (= signal) can also be viewed in the frequency domain. The time domain signal can be converted into a sum of sinusoids (with amplitude and phase) of different frequencies. The resulting functions are in the frequency domain and called the frequency domain representation: The phase spectrum shows the phase of each sinusoid, the amplitude spectrum shows the amplitude of each sinusoid. The mathematical operation to convert the input signal in the time domain to an output signal in the frequency domain is the Fourier Transform [16–18]. The resulting frequency domain representation shows, with which amplitude and phase each sinusoid (of certain frequency) is present in the time domain input signal—or in other words, of which frequencies the time domain signal consists.

As one can see in Fig. 2, the heart rate is represented by a periodic shift up and down of the luminosity values. These up and down values are distorted by some noise. The time difference between two minima (the period) can also be expressed as the reciprocal of the frequency of this periodic signal. From a signal processing view, these periodic up and downs contribute with the highest amplitude (in the desired frequency band) to the frequencies of which the time domain signal consists of. Therefore, to extract the heart rate the frequency which contributes most to the input signal (= heart rate) has to be determined.

The basic idea is to transform the input signal via the Fourier Transform and then determine the frequency that contributes most to the input signal. This means determining the peak of the output frequency signal. It is important, to only look at the frequency band of possible heart rates. Figure 3 shows the basic operation principle.

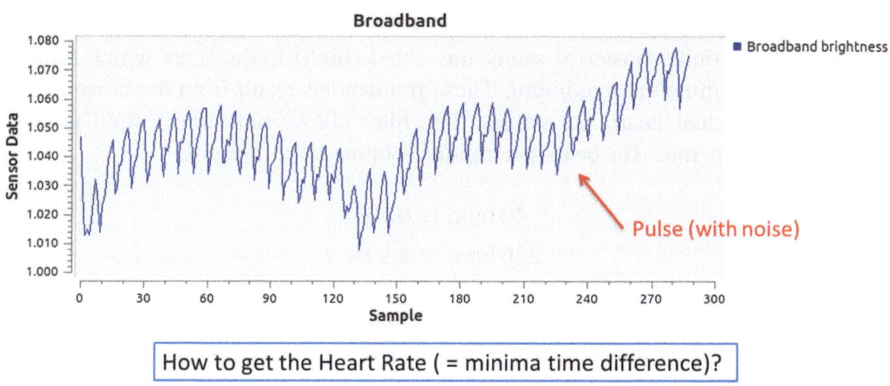

Fig. 2 Raw light sensor data

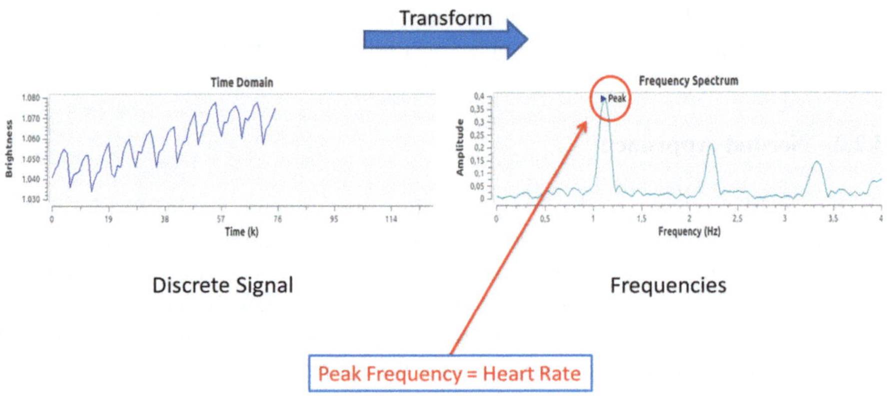

Fig. 3 Basic operation principle

The discrete input luminosity values are transformed to the frequency domain. The highest value of the resulting amplitude spectrum represents the heart rate frequency.

3.3 Signal Processing

To determine the heart rate from the input light signal, several processing steps are required. First, a window function has to be applied to reduce the leakage-effect which appears due to the usage of a finite (time limited) input signal. Therefore, a Hamming-Window is multiplied with the input signal. After that, zero padding is used to interpolate the resulting frequency spectrum. With this interpolating technique, the frequency resolution can be increased. Without zero-padding, it would be more difficult to determine the correct maxima due to the poor frequency resolution. Now the Fast Fourier Transform (FFT) is applied to transform the data into the frequency domain. The FFT is a fast implementation of the Fourier Transform. To remove unwanted frequencies, an ideal band-pass filter is used. Without a filter, the output signal would consist of many unwanted (high) frequencies which make it difficult to determine the maximum. These frequencies result from the noise and do not represent actual heart rate values. The filter allows only the desired possible frequencies/heart rates (in beats per minute—bpm) to be present:

$$\begin{aligned} 40\,\text{bpm} &\approx 0.7\,\text{Hz} \\ 230\,\text{bpm} &\approx 3.9\,\text{Hz} \end{aligned} \tag{1}$$

As the last step, the resulting (complex) values need to be converted (absolute value to get the amplitude spectrum) and scaled (to recover the attenuation due to

Fig. 4 Signal processing

windowing). After the previously mentioned steps, the correct (amplitude) frequency spectrum is available for detecting the heart rate. This is done by identifying the peak value (highest amplitude) in the frequency spectrum. Figure 4 summarizes the processing steps.

In addition to applying the signal processing steps, the correct parameters like sample rate or number of samples need to be chosen. There are several requirements to the signal processing with respect to the resulting precision and calculation speed.

- The heart rate should be measured frequently: The time until the user gets a new heart rate measurement has to be low. Therefore the number of samples used for the FFT should be small. Otherwise it would take too much time to acquire enough samples for the signal processing.
- The resulting frequency spectrum should have a high frequency/beats per minute (bpm) resolution (0.1 bpm).
- The data processing should be fast to allow an implementation on micro-controllers.
- The resulting frequency spectrum should be prepared to easily detect the peak frequency.

The most important parameters which heavily affect the precision and calculation speed are discussed in the following section.

Sample rate. To detect frequencies up to 4 Hz, a minimal sample rate of 8 Hz (according to the Shannon-Nyquist theorem) is required. This means the light sensor needs to support at least a sample interval of 125 ms. The prototype allows the sample rate to be dynamically changed. Good results have been achieved with a sample interval of 125 ms (8 Hz sample rate).

Frequency resolution. The optimal number of samples for the FFT is 128. It represents a good trade-off between frequency resolution and required time for the initial data acquisition. To further improve the frequency resolution, zero-padding has to be used. Additionally to that, another possibility to improve the frequency resolution is to use more samples, which would require more time to gather the data.

Sliding window. To measure the heart rate values continuously a sliding window is used which means that the FFT is calculated in periodic intervals with the last 128 samples. Like with the number of samples, the duration of the interval between two windows represents a trade-off between fast results and required

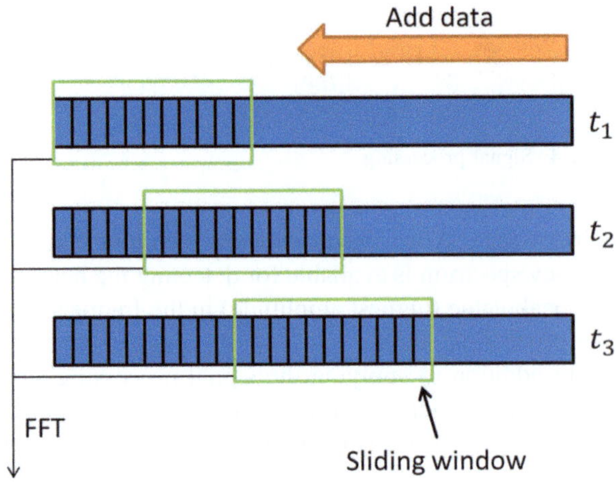

Fig. 5 Three sliding windows

computational time. Without this, the user would have to wait a long time until a measurement of the heart rate is available. A sliding window of five samples achieved good results. That means, after five samples have been acquired, the heart rate is calculated with the last 128 values. Figure 5 shows the incoming data and three successive sliding windows (green boxes) at three points in time labeled t_1, t_2 and t_3.

3.4 Data Flow

For measuring light, a Luminosity Sensor breakout board [15] is used which is connected to an Arduino.[2] The luminosity sensor driver was extended to support custom sample rates which are required by the signal processing. Because the implementation of the signal processing is not trivial and the debugging capabilities on the Arduino are limited, the calculation of the heart rate is done on a PC. The luminosity values are read by the Arduino via I^2C. After further processing, the light data is sent through UART to a PC.

The data flow between PC and light sensor can be separated into a control part and an application part. The application part contains the light sensors' luminosity values. The control part consists of setting or getting several Arduino and sensor configurations (for example sample rate). Figure 6 shows the hardware used and how it is connected.

[2]http://www.arduino.cc/

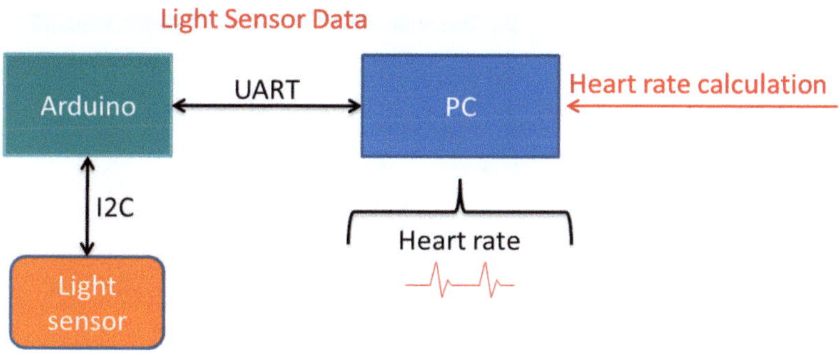

Fig. 6 Data flow

4 Prototype

The prototype is able to detect the heart rate with a very high precision (0.1 to 0.5 bpm). Most commercial devices only have a resolution of 1 bpm. Tests with commercial devices verified its functionality. This heart rate monitoring device requires only a few pieces of inexpensive hardware. The Fourier Transform is an elegant way to do the signal processing. It is quite robust against noise. Additionally, the HRM is very flexible. The implementation is based on well-known and easy to use technology. Also the hardware installation can be adapted to several uses. The only required hardware except the micro-controller/PC used for computations are the light sensor and the LED which are displayed in Fig. 7.

A graphical user Interface (GUI) on the PC part is used for displaying the various parameters and data. Additionally, it calls the signal processing module. Using a GUI simplifies the debugging and optimization process. Also it allows changes to the signal processing parameters on-the-fly which is useful for determining the optimal parameter settings. The GUI is shown in Fig. 8.

Fig. 7 Light sensor and LED

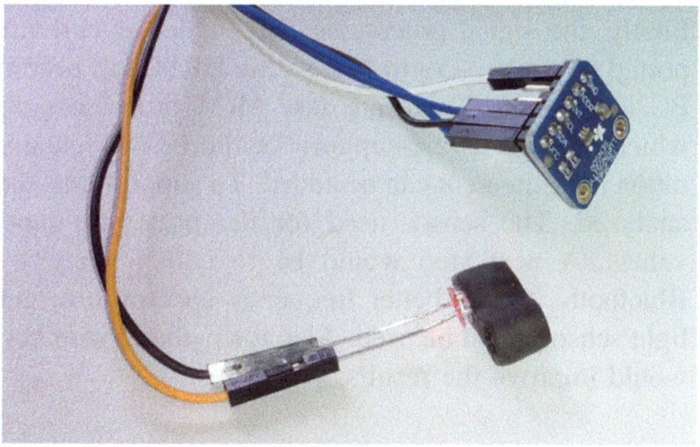

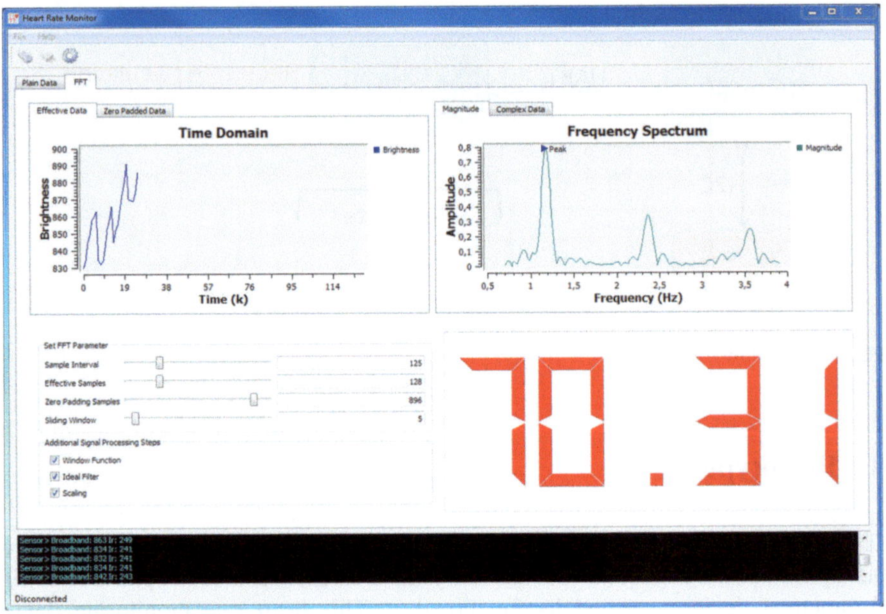

Fig. 8 HRM graphical user interface

5 Summary and Future Work

In summary, the approach demonstrates that a reliable HRM can be implemented with the support of a very inexpensive hardware implementation. The precision of 0.1 to 0.5 bpm offers several possibilities to analyze the signal in depth and to provide raw data for other approaches which need a wearable HRM. The use of digital signal processing methods simplifies the heart rate detection [19]. They allow measuring the heart rate with high precision. Compared to threshold methods, errors like noise from environment light can be easily removed.

Although the HRM works very well, some improvements can still be made: Firstly, the signal processing part, which is currently performed on a PC, can be ported to a micro-controller, so the HRM will become a stand-alone device and no PC is required. There are fast ARM processors with DSP capabilities available which offer hardware-support for the FFT calculation. Additionally, oxygen saturation measurement can be added. To support this, the infrared spectrum needs to be analyzed. The sensor used for the prototype supports measuring infrared light values. A next step would be to add a wireless communication protocol like Bluetooth. To get better frequency spectra, a brighter LED and a more sensitive light sensor could be used. Also a wristband with better environment light shielding would improve the results.

References

1. Martínez Fernández J, Augusto JC, Trombino G, Seepold R, Martínez Madrid N (2013) Self-aware trader: a new approach to safer trading, J Universal Comput Sc, 19(15): 2292–2319, ISSN: 0948-695X
2. Martínez J, Augusto JC, Seepold R,Madrid NM (2012) A sensor technology survey for a stress aware trading process, IEEE Trans Systems, Man and Cybernetics—Part C: Appl and Rev, ISSN: 1094-6977
3. Crowley OV, McKinley PS, Burg MM, Schwartz JE, Ryff CD, Weinstein M et al (2011) The interactive effect of change in perceived stress and trait anxiety on vagal recovery from cognitive challenge. Int J Psychophysiol 82:225–232
4. Pan RL, Li JK (2007) A noninvasive parametric evaluation of stress effects on global cardiovascular function. Cardiovasc Eng 7:74–80
5. Melillo P, Bracale M, Pecchia L (2011) Nonlinear Heart Rate Variability features for real-life stress detection. Case study: students under stress due to university examination. BioMedical Eng OnLine 10:1–13
6. Hjortskov N, Rissén D, Blangsted A, Fallentin N, Lundberg U, Søgaard K (2004) The effect of mental stress on heart rate variability and blood pressure during computer work. Eur J Appl Physiol 92:84–89
7. Chandola T, Heraclides A, Kumari M (2010) Psychophysiological biomarkers of workplace stressors. Neurosci Biobehav Rev 35:51–57
8. Scherz WD, Seepold R (2015) Physiologic parameter calculation supporting local and telemetric analysis. In: 6th Euro-pean Conference of the International Federation for Medical and Biological Engineering, International Federation for Medical and Biological Engineering (IFMBE), 45, ISSN 1680-0737
9. Einthoven W (1908) Weiteres über das Elektrokardiogram. Pflüger Arch. ges. Physiol, 517–548
10. Kohler B-U, Hennig C, Orglmeister R (2002) The principles of software QRS detection. IEEE Eng Med Biol Mag 21(1):42–57
11. Polar RS800cx, Available: http://www.polar.com/en/products/maximize_performance/running_multisport/RS800CX, Cited Sept 2014
12. Firstbeat Stress Monitoring, Available: http://www.firstbeattechnologies.com/index.php?page=3&sub_page=19&sub_page_2=93, Cited Sept 2014
13. Stresseraser, Available: http://www.stresseraser-uk.com, Cited Sept 2014
14. Mind Media, NeXus-10 MKII, http://www.mindmedia.info, Cited Sept 2014
15. TSL2561, Available: https://learn.adafruit.com/tsl2561/overview, Cited May 2014
16. Steven W, Smith (2014) The scientist and engineer's guide to digital signal processing. online book 1997 Available: http://www.dspguide.com/, Cited Jul 2014
17. Amidror I (2013) mastering the discrete fourier transform in one, two or several dimensions—pitfalls and artifacts. Springer
18. Smith JO (2014) Mathematics of the discrete fourier transform (DFT) with audio applications, online book 2007 Available: https://ccrma.stanford.edu/~jos/mdft/, Cited May 2014
19. Rusch TL, Sankar R, Scharf JE (1996) Signal processing methods for pulse oximetry. Comput Biol Med 26:143–159

Piezoelectric Energy Harvesting on Running Shoes

Enrico Camilloni, Giuseppe DeMaso-Gentile, Cristiano Scavongelli, Simone Orcioni and Massimo Conti

Abstract One of the strong requirements of wireless sensor networks is the low power consumption in order to extend the life of the network itself. This constraint is even stronger for body area networks. Energy harvesting is an interesting solution that captures the energy from the environment. This work presents an electromechanical model of a piezoelectric transducer in a cantilever configuration. The model has been used to design and optimize an energy harvesting system, that has been placed on a running shoe, that can be used to feed energy for sensors of body area networks.

1 Introduction

The continuous demand of energy saving has brought the research on new solutions for low power portable electronic devices, energy storage and energy sources.

Energy Harvesting Wireless Sensor Networks [1–3] are now possible due to the availability of low power transceivers and micro-controllers and new energy harvesting devices, Energy harvesting or energy scavenging is the process of extracting small amount of energy from ambient environment through various sources of energy. Energy harvesting is one of the most interesting solutions to energy generation, since it captures energy from the environment, energy that would otherwise be wasted.

Different types of energy harvesting sources exist, each one with its main application field. Three methods are used, for example, to convert mechanical strains into electrical energy: electromagnetic, electrostatic, and piezoelectric transducers. In this work we used a piezoelectric transducer, which has the capability of generating high voltage levels allowing to operate without extra electrical

E. Camilloni · G. DeMaso-Gentile · C. Scavongelli · S. Orcioni · M. Conti (✉)
Dipartimento di Ingegneria dell'Informazione, Università Politecnica delle Marche, via Brecce Bianche, 12, 60131 Ancona, Italy
e-mail: m.conti@univpm.it

M. Conti et al. (eds.), *Mobile Networks for Biometric Data Analysis*,
Lecture Notes in Electrical Engineering 392, DOI 10.1007/978-3-319-39700-9_8

power sources. Piezoelectric harvesters have been presented in literature and used in many applications like biomedical, industrial, RFId, sensing.

Paulo et al. [4] presented a review of the advantages, disadvantages and future trend of different energy harvesting methods with the specific application of portable medical devices. Caliò et al. [5] presented a review of the state of the art in piezoelectric energy harvesting. In [6, 7] a power converter specific for energy harvesting applications has been presented.

Hande et al. [8] studied and developed a vibration energy harvesting system for active RFId and microwireless devices used in monitoring applications. They have exploited the vibration of piezoelectric components mounted into vehicles to supply energy to a microcontroller, a sensor board, and a transceiver.

Lee and Youn [9] developed a vibrating skin structure and an additional thin piezoelectric layer in one device. This skin recovers energy created by multiple harmonic vibrations. They applied this design concept in an aircraft and a power transformer panel, but the skin can be applied to any engineering system with multimodal harmonic-vibrating skins.

Almouahed et al. [10] used energy harvesting in biomedical environment. They have developed a self-powered custom-designed tibial implant instrumented with piezoceramics. This allows to detect tibial complications by measuring the relative changes in pressure. In this case the piezoceramics are used as sensors and energy harvesters.

Many mechanism may be applied from energy scavenging from human body, but walking is the most obvious source of energy. Many energy harvesting devices are applied on shoes.

Paradiso et al. [11] explored a method to harness the energy dissipated in bending during a walk. They designed a system to harness foot strike energy by flattening curved, prestressed spring metal strips laminated with a semi-flexible form of piezoelectric lead zirconate titanate under the heel. Rocha et al. [12] described how to generate and energy harvesting from people walking. This application demonstrates how to delivery power from human body and it combines a piezoelectric film with a PDVF electrostatic generator to increase the generated power. The authors made and tested a prototype of a piezoelectric polymer, polyvinylidene fluoride (PDVF) foil integrated within a shoe; the energy generated ranges from tens to hundreds of milliwatts, depending on the area, the placement, the geometry, and the numbers of foils.

Wu et al. [13] researched to develop an harvesting floor with piezoelectric material, which can convert extra energy of walking motion into electrical energy. They have adopted a piezoelectric cantilever stream structure to increase efficiency. The energy produced in plucked method in fact is approximately 10 times than the forced method.

In [14] another insole pedometer has been proposed. The energy harvester is a sheet of polyvinylidene difluoride (PVDF). Since the sheet has been cut into small pieces and rolled to increase the total area. An organic charge pump circuit provides 12 μW power with 65 % efficiency for the 14 bit step counter.

Recently, dielectric elastomers (DEs) have emerged as the premier class of material capable of energy densities ranging from 5 to 40 times the densities of piezoelectric materials. In [15] the authors presented a methodology for the assignment of dielectric elastomers energy harvesters on medical shoe. They predict that each shoe could produce up to 34 mW at an average ambulation frequency is 2 Hz.

In [16] Shen et al. proposed a linear generator with three-phase armature windings fixed on the stator and a permanent magnet mover. During a walk, the generator stator fixed in a shoe moves forward, the mover has a relative displacement against the stator due to its inertia. Electricity is then generated by the back electromagnetic force.

In [17] Gatto and Frontoni developed an energy harvesting system incorporating different types of piezoelectric devices inside sole.

Meier et al. [18] provide a system mounted on a shoe for medical sensing using piezoelectric transducers (Mide Volture and Physik Instrumente Durract) for generating power. The electronics are integrated inside a conventional consumer shoe, measuring the pressure of the wearer's foot exerted on the sole. The harvester generates a maximum of 20 μJ of energy per step.

Saha et al. [19] used a piezoelectric transducer and a flexible polyvinylidene difluoride (PVDF) sheet to harness the heel strike energy and sole bending energy respectively. The system has been modeled and compared with the experimental results.

Fan and Yamamoto [20] developed a energy harvesting systems based on a giant magnetostriction material (GMM) Terfenol-D, which possesses higher energy density and durability than piezoelectric materials. The system was designed for automobile applications, but it was installed inside running shoes, too. The system is capable of producing more than 100 μJ of electric energy.

In [21] a mobile charging system is designed. A piezo electric generator is placed in a shoe. The power that is generated by piezo electric generator when a person walks is transferred to the device by using a wireless power transfer technique.

Zhao and You [22] developed a piezoelectric energy harvester for the parasitic mechanical energy in shoes originated from human motion. The harvester provides an average output power of 1 μW during a walk at a frequency of roughly 1 Hz.

This work presents an electromechanical model, the experimental analysis and the design of an energy harvesting system placed on a shoe and based on a piezoelectric cantilever. Section 2 presents the experimental analysis and an electric-mechanical model of a piezoelectric transducer in a cantilever configuration. The model has been characterized measuring the acceleration and the open circuit voltage of a piezoelectric cantilever subjected to a sinusoidal force with different values of frequency and subject to an impulsive force. Section 3 reports the design of the energy harvesting system consisting in the piezoelectric device, the power supply with an hysteresis comparator a microntroller and an ANT transceiver. Section 4 reports a real application on a shoe, used in a 10 km running competition.

2 Model Characterization and Experiments on Shoes

For the application of the piezoelectric device on a shoe we tested two piezoelectric elements: the MIDE Volture V22BL element [23] and the DuraAct Patch Transducer P-876 [24]. Both the devices work in a cantilever configuration and show similar performances, but the MIDE piezo has been finally chosen since the tight and elongated geometric form of this device is well suited to be installed under the sole of a running shoe or fixed aside the shoe.

A typical piezoelectric power harvester in cantilever configuration has a piezoelectric composite beam with a proof mass attached to the tip of the beam, as shown in Fig. 1a. An electromechanical model has been developed and characterized measuring the acceleration and the open circuit voltage of a piezoelectric cantilever subjected to a sinusoidal force with different frequencies and subject to an impulsive force. The model allows the design, analysis and optimization of the harvesting system before its physical implementation. When a force F is applied to the piezoelectric device as shown in Fig. 1a, an acceleration $\ddot{y}$ is induced with the relationship $F = m\ddot{y}$, where m is the equivalent mass of the piezoelectric element, this acceleration causes a deflection z of the cantilever. The equivalent electronic circuit, presented in [25, 26], reported in Fig. 1b, has been used to model the behavior of a piezoelectric transducer.

The quantities L, R and C are related to the kinetic energy, mechanical losses and elastic energy, respectively. C_L is the electrical capacitance measured between the surface electrodes of the transducer; V_L is the voltage generated by the piezoelectric device. The coefficients α and β model the inverse and the direct piezoelectric effects, respectively. The value of C_L is derived from the datasheet. Many experiments have been carried out and the coefficients of the model (L, R, C, α and β) have been estimated as reported in [26].

The piezoelectric device has been placed in a shoe and the vibrations are caused by periodicals impacts of the shoe with the ground. Therefore the piezo in the final application does not work in resonance with the sinusoidal force. For this reason in the experimental set up, reported in Fig. 2, the piezoelectric device was forced with a shaker by a pulse (equivalent to the impact of the shoe with the ground) in addition to the experiments with a sinusoidal force.

The piezoelectric device is then placed in a running shoe ad experiments were carried out. The position where the piezoelectric device is placed in the shoe is

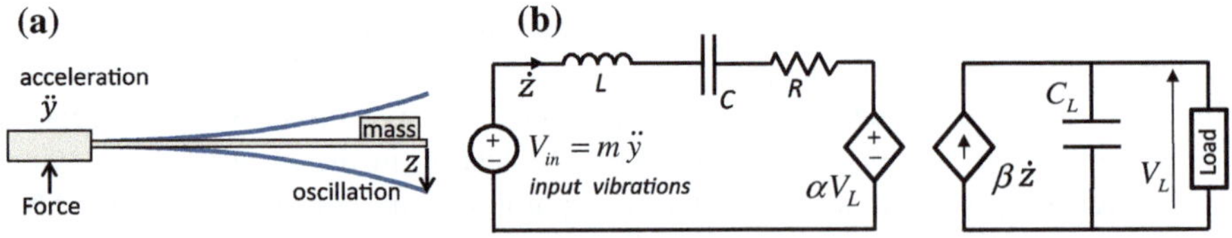

Fig. 1 Piezoelectric device in cantilever configuration (**a**), equivalent electric circuit (**b**)

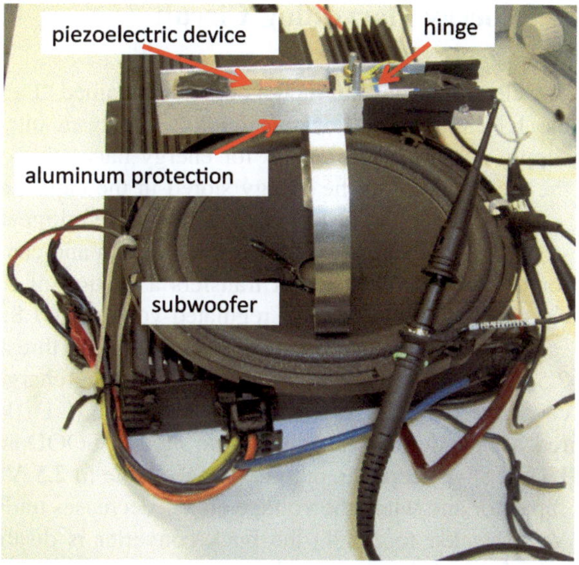

Fig. 2 Experimental set up for the piezoelectric characterization

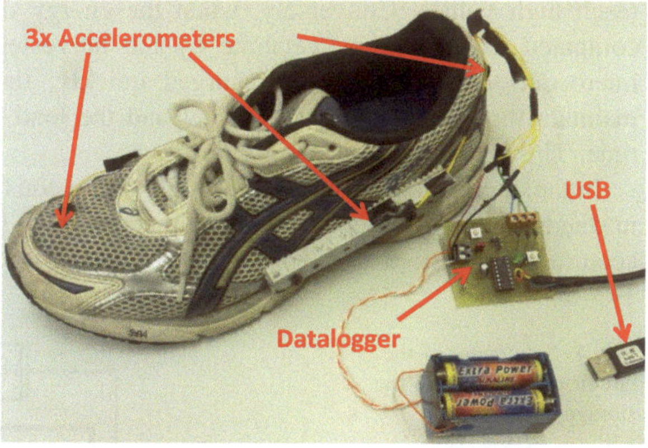

Fig. 3 Accelerometers placed in a running shoe in different positions: *tip*, *center* and *heel*

important for the comfort as well as to keep the maximum energy while running or walking.

Three accelerometers have been placed in a running shoe in different positions: tip, center (external side) and heel, as indicated in Fig. 3. Many experiments of outdoor walking and running were carried out with different speeds, uphill, downhill and plain [27]. The measurement of the acceleration waveform have been stored in a file to be subsequently used as input force in the electromechanical model of Fig. 1b.

3 Energy Harvesting Circuit

The electromechanical model of Fig. 1b, named "Piezo" in Fig. 4, was connected to the LTC3588-1 of linear technology [28], an ultralow quiescent current power supply designed specifically for energy harvesting, that converts the energy from the piezoelectric to the energy stored in the capacitor $C_{storage}$.

The LTC3588-1 has an internal full-wave bridge rectifier whose output current is stored on the capacitor $C_{storage}$ at the V_{in} pin and can be used as an energy reservoir for the buck converter that transfers a portion of the stored charge to the output (V_{out}) keeping a selectable regulated voltage of 1.8, 2.5, 3.3 or 3.6 V.

When the voltage on V_{in} rises above a rising threshold (4.4 V for V_{out} selected to 2.5 V) the buck converter is enabled and the charge is transferred from the input capacitor ($C_{storage}$) to the output capacitor (C1). When the output voltage V_{out} reaches the desired value, the logic signal PGOOD is set to 1 enabling the load. The LTC3588-1 keeps the output voltage stable to 2.5 V. The load discharges the input capacitor and when the voltage on V_{in} decreases under falling threshold (2.87 V for V_{out} selected to 2.5 V) the buck converter is disabled. Figure 5 shows a typical waveform of the circuit, when the energy dissipated by the load is low with respect to the energy generated by the piezoelectric device. The voltage across $C_{storage}$ can reach high values. Conversely, when the energy dissipated by the load is high compared to the energy generate by the harvester, as in a case the load is a microcontroller, $C_{storage}$ is discharged quickly, the LTC3588-1 is not able to maintain the output voltage on 2.5 V and the load can be active only for a short time. This second case is shown in Fig. 6.

In the application we considered, the system consists of a microcontroller that is in standby mode for great part of the time absorbing less than 10 μA and sometimes acquires data and sends them with a wireless connection absorbing about 5 mA.

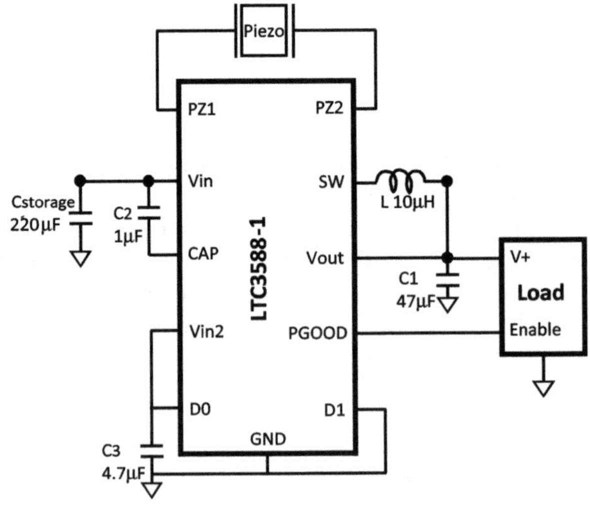

Fig. 4 Schematic of the circuit used to charge the storage capacitance

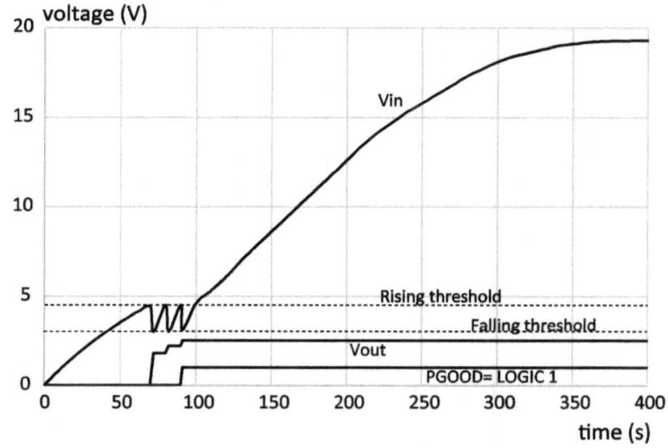

Fig. 5 Waveform of the circuit in Fig. 4 with low energy dissipation by the load

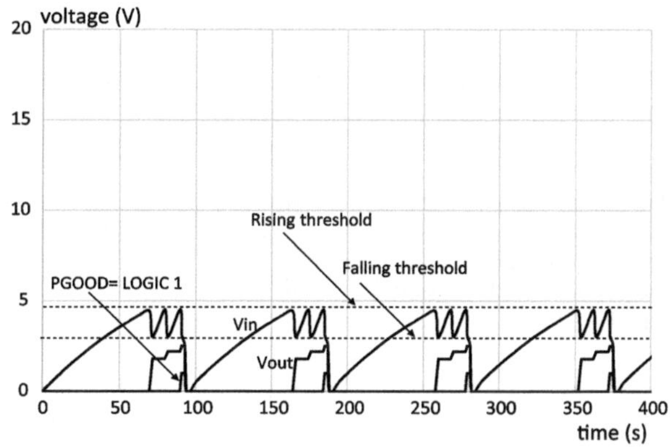

Fig. 6 Waveform of the circuit in Fig. 4 with high energy dissipation by the load

The power converter of Fig. 4 supplied by the piezoelectric device is not able to furnish a constant current of 5 mA, therefore we modified the architecture as reported in Fig. 4, adding a hysteresis circuit (LTC1540) that enables the load (the microcontroller for the data acquisition and the wireless transceiver) only when the voltage across $C_{storage}$ is high enough (about 18 V), so that the energy stored in the capacitance is enough to give current to the microcontroller for the time necessary for data acquisition and transmission.

The control circuit and hysteresis circuit is reported in Fig. 7. A typical waveform of the circuit is reported in Fig. 8. The load is supplied by the voltage Vhyst that reaches the desired value (2.5 V in our case), only when the energy accumulated in the capacitance $C_{storage}$ is high enough to activate the load for the desired time.

The falling threshold of the hysteresis has been fixed above the falling threshold of the LTC3588-1 so that V_{out} is stable to the desired value.

The design of the parameter values of the circuit in Fig. 7 are related to the energy generated by the piezoelectric device inserted in a shoe while the user is

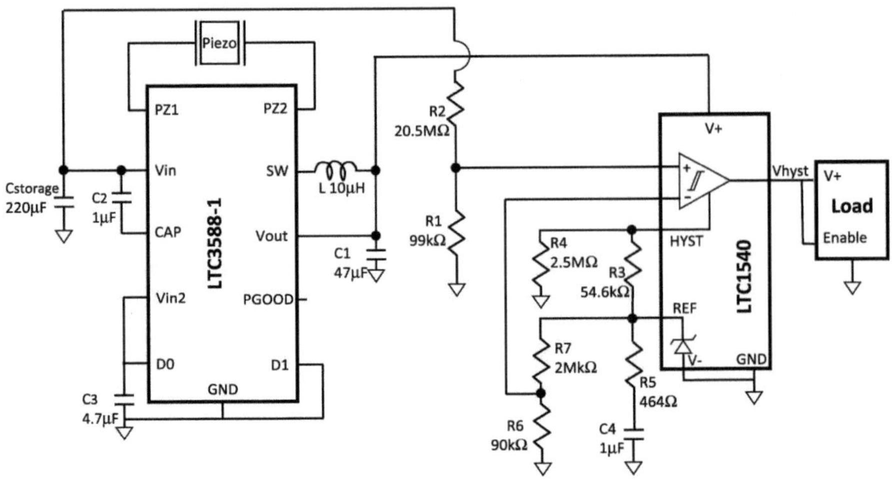

Fig. 7 Schematic of the circuit used to charge the storage capacitance with an additional hysteresis circuit

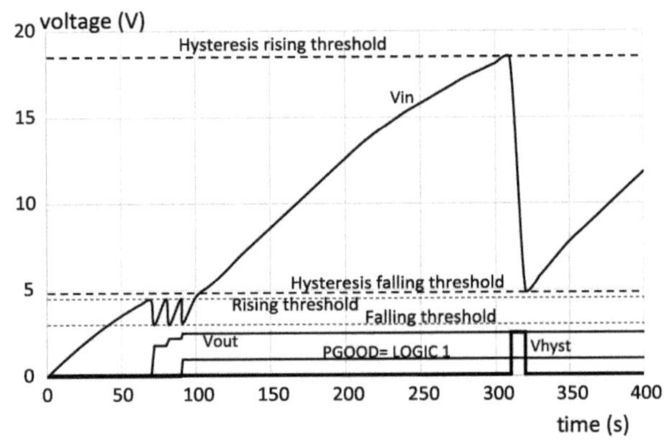

Fig. 8 Waveform of the circuit in Fig. 7 with high energy dissipation and with the addition of the hysteresis circuit

running and by the energy required by the load. We will investigate these aspects in the next subsections.

3.1 Charging Phase

The time required for the charge of the input capacitance have been estimated with SPICE simulations of the circuit of Fig. 7, using the electric model of the piezo-electric device in Fig. 1b and using, as input force, the acceleration data measured

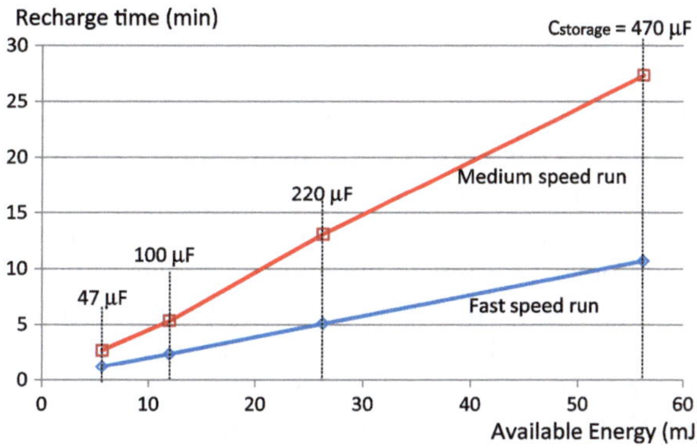

Fig. 9 Time required by the piezoelectric device to charge the storage capacitance

during the runs. The model and the experimental measurements of the acceleration are therefore essential to optimize the design of the circuit without modifying the hardware, connecting the circuit to the shoe and repeating the experiments. Four running conditions have been performed (fast speed run, medium speed run, slow speed run, and medium speed walk) for a duration of 10 min and with a sampling time of 1 ms. The results of the SPICE simulations are summarized in Fig. 9 that reports the time required by V_{in} to reach 18 V (hysteresis rising threshold) staring from 5 V (hysteresis falling threshold) as a function of the energy that will be available for the load (stored in the capacitance $C_{storage}$) for fast speed run and medium speed run. For slow speed run or walk the piezoelectric device is not able to charge the capacitance to 18 V in 30 min. The energy available for the load can be estimated by the following relationship,

$$E = \frac{1}{2}\eta C_{storage}\left(V_{in,max}^2 - V_{in,min}^2\right) \tag{1}$$

where η is the efficiency of LTC3588-1 that is approximately 80 % (it depends on temperature and the value of V_{out}, as can be seen in the datasheet), $V_{in,max}$ and $V_{in,min}$ are the rising threshold and falling threshold, respectively, of the hysteresis circuit as shown in Fig. 8. In this experiment $V_{in,max}$ and $V_{in,min}$ have been fixed to 18 and 5 V, respectively. The values of the available energy, estimated by (1) for different values of the storage capacitance are reported in Table 1.

Table 1 Energy available as function of the storage capacitance

C (µF)	V_{in} max(V)	V_{in} min (V)	η	E (mJ)
47	18	5	0.8	5.62
100	18	5	0.8	11.96
220	18	5	0.8	26.31
470	18	5	0.8	56.21

3.2 Discharging Phase

The energy harvesting system was designed to give energy to a device consisting of sensors (accelerometers), a microntroller that makes the data acquisition and a wireless transceiver, as in a typical application of body area network. In particular the microntroller MSP430 and the ANT transceiver PAN1323ETU have been used.

In order to estimate the power dissipated by the microcontroller and the transceiver we used the Texas Instruments experimental board MSP-EXP430F5438 and the transceiver PAN1323ETU. The board contains leds, one LCD display and other components not useful for our application. The transceiver Panasonic PAN1323ETU can be used for Bluetooth, Bluetooth Low Energy and ANT based on the IC CC256x.

The scheme of the complete system and the photo of the MSP-EXP430F5438 board are shown in Fig. 10.

Bluetooth Low Energy, ZigBee and ANT are protocols used a cyclic sleep scenario, in which a short-range and low-power wireless sensor node periodically sends a data packet to a remote 'hub' [29]. This is our typical application, since the node is in sleep mode for a great part of the time when the piezoelectric device is charging the capacitance, and for a short time the node has enough energy to transmit data.

We used the ANT protocol since this is a low power protocol used for body area network and it can be easily used in conjunction with smartphones and can be a good choice for a wide range of duty cycles [29], but similar results can be obtained using other low power protocols. The baud rate of the ANT protocol is 115,200 bps.

The packet sent via ANT is generally characterized by:

- 8 bytes of payload data (data that can be transmitted);
- 1 byte that indicates the frequency (2400–2485 MHz);
- 2 bytes for the network key;
- 4 bytes for the channel ID.

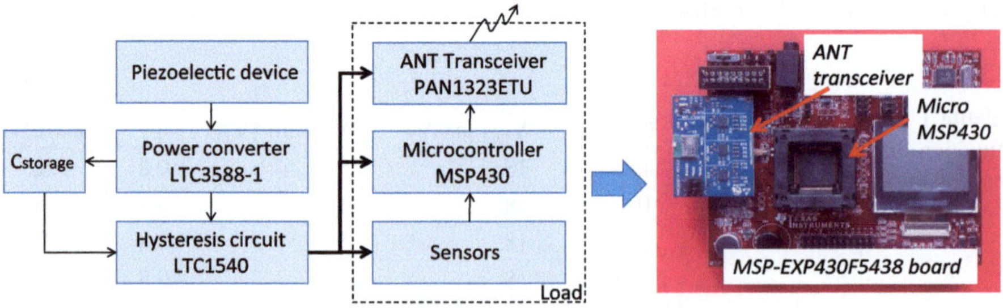

Fig. 10 Architecture of the complete system and the MSP-EXP430F5438 Experimenter Board, with the Micrcontroller MSP430 and the transceiver PAN1323ETU

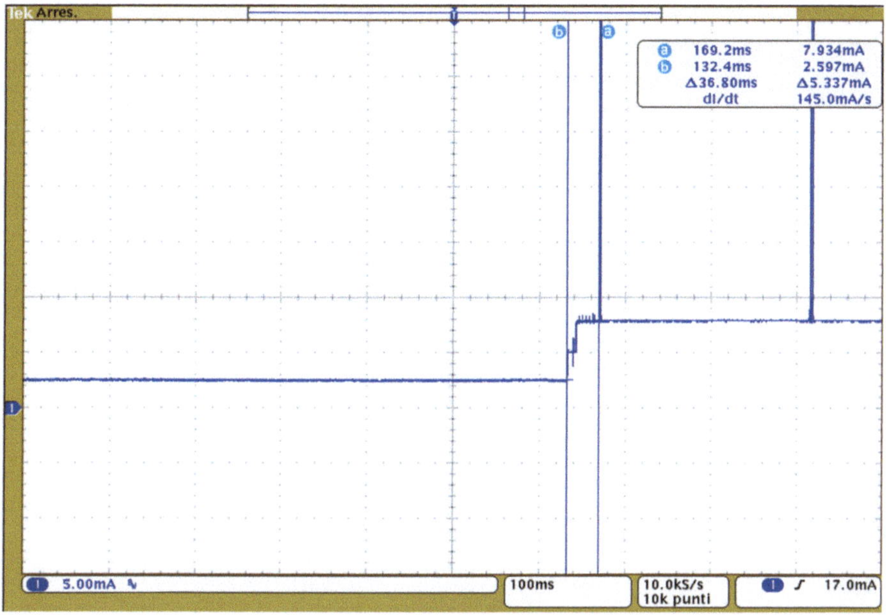

Fig. 11 Experimental measurement of the current absorbed by the ANT transceiver

The experimental set-up for the power dissipation consists in the following phases: the microcontroller is switched on, data from an external accelerometer are acquired using the ADC of the microcontroller, the data are sent to the ANT transceiver that sends successive packets to another board connected to a PC. Subsequently, the microcontroller and the transceiver are sent in sleep mode. A digital oscilloscope measured the current absorbed by the microntroller MSP430 and the ANT transceiver PAN1323ETU. As an example Fig. 11 reports the experimental measurement of the current absorbed by the ANT transceiver.

A model is derived from the measurements of the current absorbed by the microcontroller and by the transceiver in the different phases. Figure 12 reports the model in a case of the successive transmission of two packets of 8 bytes of data each.

The microcontroller has a start-up time of about 785 ms with a current consumption of 2.4, 0.35 and 5.5 mA. During this phase all the peripherals of the microcontroller are initialized. Now the system is in active mode and the code is executed: data are acquired with the internal ADC and they are sent to the transceiver using the UART. This phase has a duration of 36 ms with a current consumption of 3.4 mA. The microcontroller remains in active mode during each data transmission (1 ms) and for the time between each transmissions (250 ms). Then the shutdown is performed, it requires 200 ms with a consumption of 2.2 mA. The microcontroller absorbs 7 μA in sleep mode.

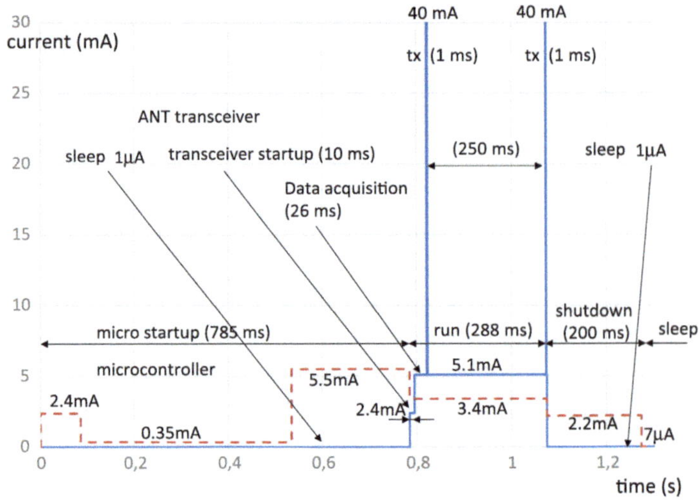

Fig. 12 Model of the current absorbed by the microcontroller and by the transceiver in the different phases

The ANT transceiver has a current consumption of 1 μA in sleep mode. The microcontroller wakes up the transceiver after 785 ms. The startup phase has a duration of 10 ms with a current consumption of 2.4 mA. At this moment the transceiver receives data via UART before transmitting (duration of 26 ms with 5.1 mA). The transmission is characterized by the peak of amplitude of 40 mA and a duration of 1 ms. The transmission rate of the ANT protocol is 115.2 kbps: an ANT package is composed of 15 bytes (8 bytes of data), therefore 1 ms is required to transmit 15 bytes. The transmission frequency is 4 Hz, consequently the transceiver sends a packet every 250 ms. In the experiment two successive transmissions have been performed. Therefore the transceiver absorbs 5.1 mA for 250 ms, and then it requires 40 mA for 1 ms to send the second packet. Then the transceiver goes in sleep mode dissipating 1 μA.

Considering the model represented in Fig. 12 and that the supply voltage of the system is 2.5 V, the energy required by the microcontroller and by the transceiver for the transmission of n packets can be estimated by the following relationships.

$$\begin{aligned} E_{micro} &= \int V_{dd} I_{micro} dt \\ &= 0.0025\{(2.4 * 85 + 0.35 * 450 + 5.5 * 250 + 3.4 * 37) \\ &\quad + 3.4 * 251 * (\mathrm{n}-1) + 2.2 * 200\} \\ &= 5.75575 + 2.1335 * (\mathrm{n}-1)[\mathrm{mJ}] \end{aligned}$$

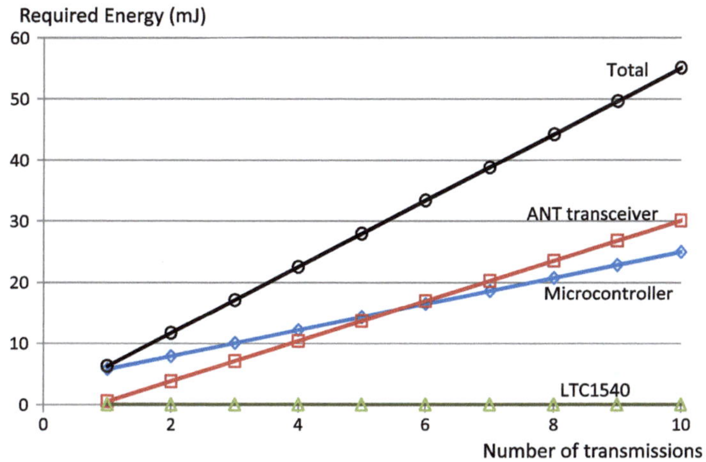

Fig. 13 Energy dissipated by the microcontroller, by the transceiver, by the LTC1540 and the total energy required as a function of the number of packets sent

$$\begin{aligned} E_{ANT} &= \int V_{dd} I_{ANT} dt \\ &= 0.0025\{(0.001 * 785 + 2.4 * 10 + 5.1 * 26 + 40 * 1) \\ &\quad + (5.1 * 250 + 40 * 1) * (\mathrm{n} - 1) + 0.007 * 200\} \\ &= 0.496963 + 3.287500 * (\mathrm{n} - 1)[\mathrm{mJ}] \end{aligned}$$

$$\begin{aligned} E_{LTC1540} &= \int V_{dd} I_{LTC1540} dt \\ &= 0.0025\{(0.0015 * 1022) + (0.0015) * (\mathrm{n} - 1)\} \\ &= 0.003833 + 0.000941 * (\mathrm{n} - 1)[\mathrm{mJ}] \end{aligned}$$

Figure 13 reports the energy dissipated by the microcontroller, by the transceiver, by the LTC1540 and the total energy required as a function of the number of packets sent.

Combining the results of Fig. 13 and the results of Fig. 9, that reports the time required by the piezoelectric device to recharge capacitance $C_{storage}$, we obtain the results of Fig. 14, that reports the number of ANT packets that the system has the energy to send and the corresponding time between two transmission cycles required to harvest the energy for the successive cycle.

For example if we use a $C_{storage}$ of 220 μF, the transceiver has the energy to transmit up to 4 packets of 8 data bytes each every 5 min running at fast speed, or every 13 min running at medium speed. Using a $C_{storage}$ of 100 μF, the transceiver has the energy to transmit 1 packet of 8 data bytes every 2 min running at fast speed, or every 5 min running at medium speed.

As an example the circuit of Fig. 7 has been simulated using SPICE with a load consisting of two current generators in parallel with a the profile of the model in

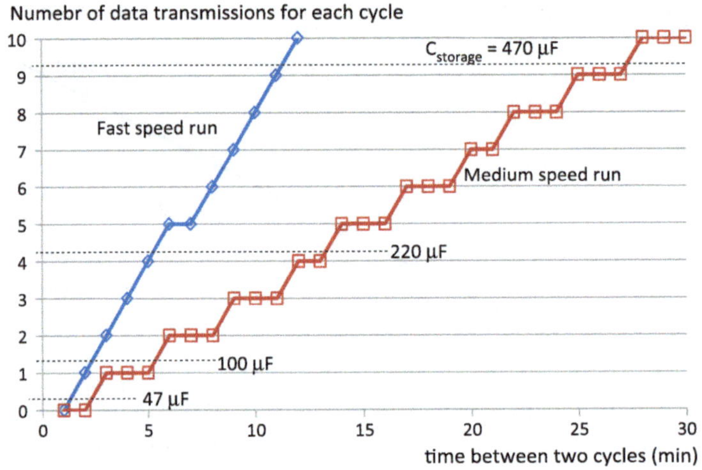

Fig. 14 Number of ANT packets that can be sent and the corresponding time between two transmission cycles required to harvest the energy for the successive cycle

Fig. 12, emulating the current consumption of the microcontroller and of the transceiver. Figure 15 reports the output voltage across the capacitance $C_{storage}$. It can be noticed that V_{in} remains higher than 5 V, the falling threshold of the hysteresis circuit, allowing the completion of the transmission of the two packets.

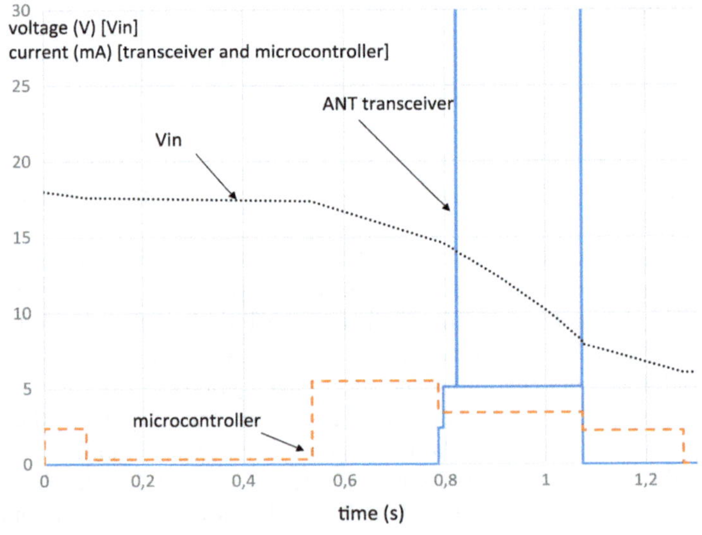

Fig. 15 Spice simulations: voltage V_{in} across the capacitance $C_{storage}$ and the current absorbed by transceiver and microcontroller

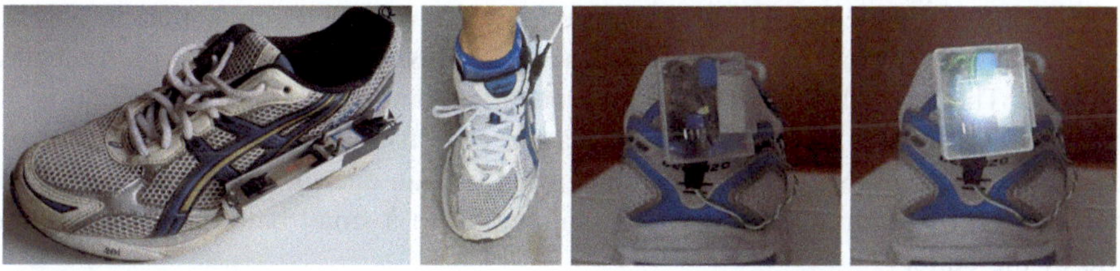

Fig. 16 Piezoelectric device placed in center of the shoe in an aluminum structure, the system with a led has been placed in the back of the shoe

Fig. 17 Denis Curzi (*on the right*), the winner of the Maratona del Piceno, Porto San Giorgio, Italy, April 29, 2012 and the author Massimo Conti who run the 10 km competition with the device proposed in this work reported in Fig. 16

4 Application in a Real System

A first prototype of the system has been implemented, with a led instead of a microcontroller and the ANT transceiver. It has been installed on a shoe, as shown in Fig. 16, and used in a 10 km running competition (Maratona del Piceno, Porto San Giorgio, Italy, April 29, 2012). Figure 17 shows the photo of Denis Curzi (on the right), the winner of the 42.195 km Maratona del Piceno (Porto San Giorgio, Italy, April 29, 2012) and the author Massimo Conti who ran the 10 km competition with the device proposed in this work reported in Fig. 16.

5 Conclusions

This work presents an electromechanical model, experimental analysis and the design of an energy harvesting system placed on a shoe based on a piezoelectric cantilever.

The work enabled us to design an energy harvesting circuits that is able to provide an energy of 5.6 mJ after a run of 75 s using a very small piezoelectric element. That is the harvester provides an average output power of 75 μW during a run. The harvesting system can supply a current of 4.5 mA for 1.3 s to a micro-controller and an ANT transceiver, thus providing an energy harvesting body area network. The system can be used in conjunction to a small battery improving the life of the battery.

Improvement can be made using new microcontroller and transceiver absorbing lower currents that will be available in the future. Additional work should be done on the piezoelectric material to improve the energy available.

References

1. Ramesh MV (2014) Design, development, and deployment of a wireless sensor network for detection of landslides. Ad Hoc Netw 13(Part A):2–18. doi:10.1016/j.adhoc.2012.09.002
2. Shen X, Bo C, Zhang J, Tang S, Mao X, Dai G (2013) EFCon: energy flow control for sustainable wireless sensor networks. Ad Hoc Netw 11(4):1421–1431. doi:10.1016/j.adhoc.2011.07.003
3. Severini M, Squartini S, Piazza F, Conti M (2015) Energy-aware task scheduler for self-powered sensor nodes: from model to firmware. Ad Hoc Netw 24(Part A):73–91. doi:10.1016/j.adhoc.2014.06.009
4. Paulo J, Gaspar PD (2010) Review and future trend of energy harvesting methods for portable medical devices. In: Proceedings of the world congress on engineering 2010, vol II, WCE 2010. London, June 30–July 2, 2010
5. Caliò R, Rongala UB, Camboni D, Milazzo M, Stefanini C, de Petris G, Oddo CM (2014) Piezoelectric energy harvesting solutions. Sensors 14:4755–4790. doi:10.3390/s140304755
6. d'Aparo R, Orcioni S, Conti M (2009) A digital controlled energy scavenger power converter. In: 2009 seventh workshop on intelligent solutions in embedded systems, pp 165–170
7. d'Aparo R, Orcioni S, Conti M (2011) A sigma-delta controlled power converter for energy harvesting applications. In: Conti M, Orcioni S, Martinez Madrid N, Seepold ER (eds) Solutions on embedded systems. Springer, Dordrecht, Netherlands, pp 257–270
8. Hande A, Bridgelall R, Zoghi e B (2010) Vibration energy harvesting for disaster asset monitoring using active RFID tags. In: Proceedings of the IEEE 98.9 (2010), pp 1620–1628
9. Lee S, Youn B (2011) A new piezoelectric energy harvesting design concept: multimodal energy harvesting skin. IEEE Trans Ultrason Ferroelectr Freq Control 58(3):629–645
10. Almouahed S et al (2010) Self-powered instrumented knee implant for early detection of postoperative complications. In: 2010 annual international conference of the IEEE engineering in medicine and biology society (EMBC), pp 5121–5124
11. Paradiso J (2006) Systems for human-powered mobile computing. In: 2006 43rd ACM/IEEE design automation conference, pp 645–650
12. Rocha J, Goncalves L, Rocha P, Silva M, Lanceros-Mendez S (2010) Energy harvesting from piezoelectric materials fully integrated in footwear. IEEE Trans Ind Electron 57(3):813–819
13. Wu TT, Yao WS, Wang SH, Tsai MC (2010) Analysis of high efficiency piezoelectric floor on intelligent buildings. In: Proceedings of SICE annual conference 2010, Taipei, pp 1777–1780
14. Ishida K, Huang TC, Honda K, Shinozuka Y, Fuketa H, Yokota T, Zschieschang U, Klauk H, Tortissier G, Sekitani T, Toshiyoshi H, Takamiya M, Someya T, Sakurai T (2013) Insole pedometer with piezoelectric energy harvester and 2 V organic circuits. IEEE J Solid-State Circuits 48(1):255–264

15. Wendt JB, Goudar V, Noshadi H, Potkonjak M (2012) Spatiotemporal assignment of energy harvesters on a self-sustaining medical shoe. IEEE Int J Sens 1–4
16. Shen JX, Wang CF, Luk PCK, Miao DM, Shi D, Xu C (2013) A shoe-equipped linear generator for energy harvesting. IEEE Trans Ind Appl 49(2):990–996
17. Gatto A, Frontoni E (2014) Energy harvesting system for smart shoes. In: 2014 IEEE/ASME 10th international conference on mechatronic and embedded systems and applications (MESA), pp 1–6
18. Meier R, Kelly N, Almog O, Chiang P (2014) A piezoelectric energy-harvesting shoe system for podiatric sensing. In: 2014 36th annual international conference of the IEEE engineering in medicine and biology society (EMBC), pp 622–625
19. Saha P, Goswami S, Chakrabarty S, Sarkar S (2014) Simulation and model verification of shoe embedded piezoelectric energy harvester. In: 2014 6th IEEE power india international conference (PIICON), pp 1–6
20. Fan T, Yamamoto Y (2015) Vibration-induced energy harvesting system using Terfenol-D. In: 2015 IEEE international conference on mechatronics and automation (ICMA), pp 2319–2324
21. Paul PJ, Tutu RSD, Richards WK, Jerome VM (2015) Project power shoe: piezoelectric wireless power transfer. A mobile charging technique. In: 2015 IEEE global humanitarian technology conference (GHTC), pp 334–339
22. Zhao Jingjing, You Zheng (2014) A Shoe-embedded piezoelectric energy harvester for wearable sensors. Int J Sens 14:12497–12510. doi:10.3390/s140712497
23. Mide Volture Datasheet, web site: http://www.mide.com/products/volture/v22bl.php
24. Physik Instrumente DuraAct Patch Transducer P-876 Datasheet, web site: http://www.physikinstrumente.com/en/pdf/P876_Datasheet.pdf
25. Romani A, Sangiorgi E, Tartagni M, Paganelli RP (2013) Joint modeling of piezoelectric transducers and power conversion circuits for energy harvesting applications. IEEE Sensors J 13(3):916–925
26. Camilloni E, Carloni M, Giammarini M, Conti M (2013) Energy harvesting with piezoelectric applied on shoes. In: Proceedings of SPIE 2013 microtechnologies, international conference VLSI circuits and systems, vol 8764. Grenoble, France, paper 4, pp 05.1–05.12, 24–26 Apr 2013
27. Camilloni E, Demaso-Gentile G, Scavongelli C, Conti M (2014) Energy harvesting for body area networks. In: Proceedings of the international workshop "mobile networks for biometric data analysis" mBiDA, Ancona, Italy, pp 113–126, 30–31 Oct 2014
28. LTC3588 Datasheet, web site: http://www.linear.com
29. Power Consumption Analysis of Bluetooth Low Energy, ZigBee and ANT Sensor Nodes in a Cyclic Sleep Scenario, A Dementyev, S Hodges, S Taylor, J Smith - Wireless Symposium (IWS), 2013 IEEE International, 2013

Implementing Steganocryptography in Scanner and Angio-Scanner Medical Images

Indrit Enesi and Blerina Zanaj

Abstract Telemedicine is a field of research in modern telecommunication and information technologies. It provides remote clinical care to help many people. Communication through internet and its services are becoming less expensive every day. This aspect of a positive growth of internet communication brings more possibilities to intercept and have access into the information as well. Data privacy and integrity are the main aims in data communication, nowadays. Many methods exist to keep secure the data that are transmitted through internet but none of them is perfect. To increase the efficiency of data protection, different methods and combinations are used. This paper analyzes the amount of hidden information by combining the data encryption and steganography techniques in medical images. The images taken into account are obtained from scanner and angio-scanner. The patient data are encrypted previously, then included in the scanner and angio-scanner images substituting the image LSBs. The results of the experiment are assessed by doctors, they admit that there is no visual difference between the original and the cover images. This paper analyses the sum in percentage of intensities, SSD and PSNR of the images used. The results show that this method achieves a higher performance in the scanner images than in the angio-scanner.

Keywords Steganography · Cryptography · Scanner · Angio-scanner · Medical images · LSB

I. Enesi (✉)
Telecommunication and Electronic Department, Faculty of Information Technology, Polytechnic University of Tirana, Sheshi “Nënë Tereza”, Tirana, Albania
e-mail: ienesi@fti.edu.al

B. Zanaj
Department of Mathematics and Informatics, Faculty of Economics and Agribusiness, Agricultural University of Tirana, Kodër-Kamëz, Tirana, Albania
e-mail: bzanaj@ubt.edu.al

M. Conti et al. (eds.), *Mobile Networks for Biometric Data Analysis*,
Lecture Notes in Electrical Engineering 392, DOI 10.1007/978-3-319-39700-9_9

1 Introduction

Telemedicine [1, 2] is widely used with the development of the communication technologies. Telemedicine is making a great progress in the communication between patients and hospital or between two hospitals. The problem which receives special attention is that of protecting the privacy of the patient's data. The insurance company will avoid insuring a patient, or the big companies will substitute an unhealthy director.

Both Steganography and Cryptography have something in common. They do not allow an unauthorized person to access the private information. Both methods have good performance in preventing intruders to access communication. But none of these technologies are perfect and no one of them is unbreakable. Using them in combination with each other gives a higher performance. In Table 1 is shown the comparison between different techniques for private communications [3]. Encryption allows the secure communication by using only one key and it is needed only to read the information at the receiver side. An intruder cannot make the decryption, but he can modify the file. By doing so the intruder has made it unreadable from the receiver part. Digital signature allows inserting the copyright in a document [4]. The signature can be easily erased but every change in the document would make the signature unavailable, so the document integrity is preserved.

Steganography provides a secret communication which can not be removed without changing substantially the data where it is hidden. The hidden data will remain secure as long as the penetrators have not found a way to detect it.

Previous work has been done on assessing authenticity of the medical images [5–8]. In this paper, is explored the steganographic method for embedding encrypted textual information into medical images by using the least significant bit insertion [6].

2 Advanced Encryption Algorithm

AES is a specification for the encryption of electronic data. The algorithm used by AES is a symmetric algorithm, meaning that the same key is used for both encryption and decryption. AES is based on a principle known as substitution—permutation. It is fast in both software and hardware. It does not use a fiestel

Table 1 Comparison of the different private security communications

	Confidentially	Integrity	Unchangeable
Encryption	Yes	No	Yes
Digital signature	No	Yes	No
Steganography	Yes/No	Yes/No	Yes

network. AES has a fixed block size of 128 bits and a key size of 128,192 or 256 bits. The block size has a maximum of 256 bits and the key size has no theoretical maximum [8].

3 Steganography

To hide secret data in a cover image, the image should have a great quantity of data redundancy or enough noise. This is needed because the hiding process used by steganography substitutes these redundant data with our secret message. This limits the quantity of data that we can use in steganography.

In Fig. 1 it is shown a simple example of the encoding and decoding process. In this example the image is trying to be hidden into another image that will serve as a cover to produce the stego-image. As mentioned above in the coding process we need a key. It can be a public or a private one, needed to encode the secret message at the sender side. The receiver will decode the stego-image by using the public key. Combining the information in such a way reduces the possibility of a third part to get the stego-object, to decode it and to detect the secret information from it. After passing through the decoder, the stego object is taken as an output. The stego object is the original cover object that keeps the secret information within. This object should look almost identical with the cover object otherwise a third party can understand and receive the hidden information. In the decoding process the stego object serves as an input to the system. The public or the private key is also needed.

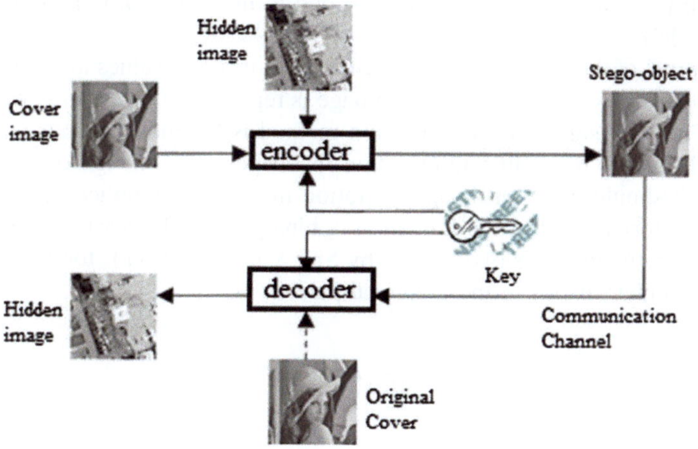

Fig. 1 The overall process of coding and decoding

Based on the used encryption technique, during the decoding process, sometimes, it is necessary the original cover object [9].

After the decoding process is finished, the secret information can be extracted from stego object and is readable.

4 Image Steganography

The process of hiding a message inside a digital image is one of the most used techniques. This comes from the limited visual perception ability of humans (HVS —human visual system). 8-bit images are appropriate to use them for their relatively small size. The only disadvantage is that only 256 colours available can be used, and it can be a great problem during the encoding process. A set of colours is frequently used with the grey scale when we have 8-bit images like gif, because the gradual variation in colour can be noticed more hardly as the secret message in it. A 24-bites image gives more flexibility when it is used in Steganography. The wider range of colours (over 16 millions) that may be used, goes beyond what our visual system can perceive and it is impossible to distinguish the changes as the secret message is encoded.

4.1 The Least Significant Bit Insertion

This method assumes that information in a 24-bit digital image is represented as an array of triplets, and these triplets correspond to intensities of red, green and blue (RGB model).

Each pixel of the image can be described by a triple of values associated to each color. In the case of 8-bit images, the image is represented by an array of grayscale values. The least significant bit insertion method is the most obvious and also the most well-known for hiding information in images. The change of the least significant bit should cause an image alteration that is barely noticeable. By considering the information S to be hidden as a binary one. The least significant bit of each pixel in the image is overwritten by $Si \in X = \{0,.., 2^{nc}\text{-}1\}$, for $1 \leq i \leq |S|$ and n_c is the number of bits in the graphical palette. So

$$SS_i = \sum_{k=1}^{n_c} b[i,k] \cdot 2^{n_c} - 1 \tag{1}$$

where (b[i, 1] … $b[i,n_c]$) is the binary representation of Si, and $b[i,n_c]$ is the least significant bit. For 24-bit images, the changes are minimal and virtually imperceptible to the human eye when |S| is reasonable. The 8-bit images, however, undergo significant changes and are easily detected by histogram attack, where pixel intensity frequencies have changed their distribution [6].This method is quite vulnerable to geometric transformations and filtering as well as compression schemes (such as JPEG) because it involves changes in the least significant bits and the steganographic information is destroyed [10, 11].

4.2 Encoding Process

The encoding process passes through five phases:

In the first phase, the function STEGANOCODER "hides" a message inside an image that is chosen from the user. At the output we have an image file that is protected from coding and encryption.

Second phase: Encryption using XOR key.

Third phase: Preparing the coverage.

Fourth phase: hiding data, this is accomplished by using in turn RGBBGRRG. The data will be hidden in the columns by passing from left to right in the source image.

Fifth phase: The final output from encoding process.

4.3 Decoding Process

In the decoding process, the function STEGANODECODE extracts the hidden message by reversing the encoding process. The five phases are as below.

First phase: Recover the Header set, analyzing the header, decrypt and define the size of message.

Second phase: Isolates the potential message, use the MODULO mathematic.

Third phase: the decrypting process.

Fourth phase: preparing the message.

Fifth phase: the final output.

5 Experimental Results

Several scanner and angio-scanner images are taken as well some text files containing the history of the patient data. While the history of the patient data grows up, the size of the text file grows up as well. These data are encrypted first and then they are included within the angio-scanner and scanner images. Some measuring techniques are used to analyze the ability of medical images to carry text files inside of them.

5.1 Scanner Images

Scanner images are taken from different patients. These patient's data are saved in a text file and are encrypted with a symmetric key and the text file is implemented in LSBs of scanner image, which serves as a cover image. Three approaches are used to assess steganographic methods, the difference of the images, the sum of squares of differences (SSD) [7] and the PSNR. The formula for calculating SSD is:

$$\mathrm{SSD} = \sum_{M,N} \left[I_1(m,n) - I_2(m,n) \right]^2 \tag{2}$$

The formula for calculating PSNR is:

$$\mathrm{PSNR} = 10 \log_{10}\left(R^2/\mathrm{MSE}\right) \tag{3}$$

where MSE = SSD/(M * N)

Figure 4 represents the difference image between the original (Fig. 2) and cover image (Fig. 3) for the historic text file of 2 kB. Figure 5 represents the difference images between original and cover images for the historic text file of 19 kB. Image differences in appearance do not represent visual values, the difference image looks black. We convert the difference image to grayscale through the function rgb2gray (). The amount of pixel intensity values of the difference image is estimated. The results of differences for the cover images with text varying from 1 to 19 kB, are presented in Table 2. In this table we also present the ratio of the sum of intensity values with the total pixel values of the image in percentage.

Fig. 2 Original image

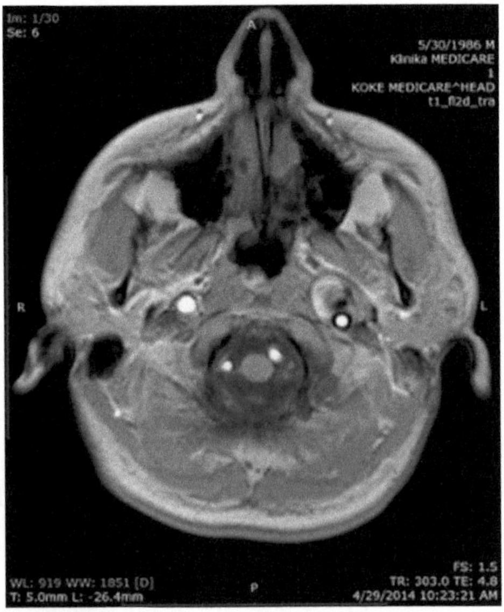

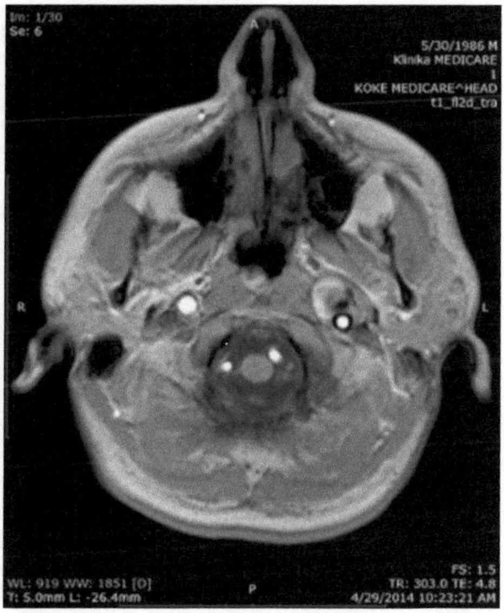

Fig. 3 Cover image for text 19kB

Fig. 4 Difference image between original and cover for text 2kB

The total value of the original image (after converted to grayscale) is 34348615, so we can calculate the percentage of the pixel differences.

In the Table 3 are shown the values of sum of difference squares (SSD) for the text files with different size.

In the Table 4, are presented PSNR values for original image compared with cover images.

Fig. 5 Difference image between original and cover for text 19kB

Table 2 Sum of image pixel values for images of differences

Size of text file	1kB	2kB	7kB	13kB	19kB
Sum of intensity values	7821	8043	10612	14001	17363
Value in percentage	0,02276	0,02341	0,03089	0,04076	0,05054

Table 3 Dependencyof SSD from text file size

Size of text file	1kB	2kB	7kB	13kB	19kB
SSD	75997	77742	88817	102555	116654

Table 4 Values of PSNR

Size of text file	1kB	2kB	7kB	13kB	19kB
PSNR	53.9840	53.8924	53.3510	52.7532	52.2224

5.2 Angio-Scanner (CT) Images

Angio-scanner images are taken from different patients. As above the patients data are saved in a text file, encrypted with a key and then implemented in LSB of CT image, serving as a cover image.

Based on file size from 1 kB text to 19 kB, we code the information and insert it into the LSBs of the original image forming the cover images, then the differences

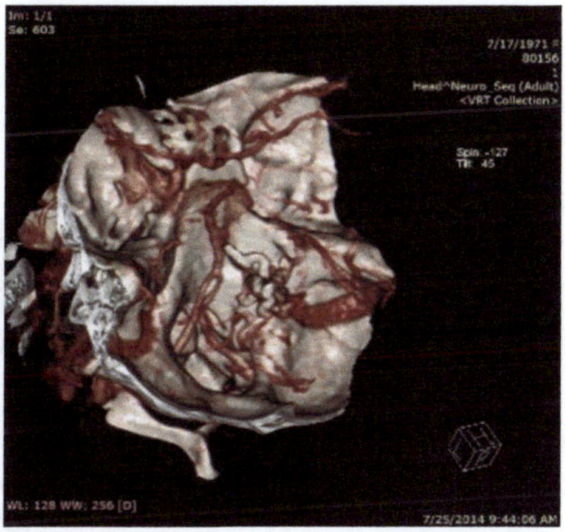

Fig. 6 Original Image

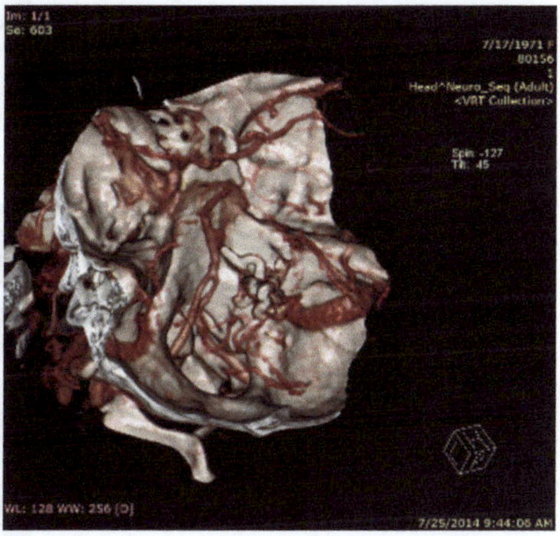

Fig. 7 Cover image

Table 5 Sum of imagepixelvalues for difference images

Size of text file	1kB	2kB	7kB	13kB	19kB
Intensity sum (for the three plans r,g,b)	599573	599723	600923	603033	604898
Values in percentage	0.52376	0.52389	0.52494	0.52678	0.52841

between the original image and cover images are calculated. Figure 6 represents the original angio-scanner image and Fig. 7 presents the cover image for 19 kB encrypted information inside it. As in scanner images, the image differences in appearance do not represent visual values, the differences image looks black. Three approaches are taken in consideration to assess the results of steganographic methods, the difference of the images, sum of squares of differences (SSD) [6], [12] and PSNR values. The results are presented in Tables 5 and 6.

Table 6 Dependency of SSD from text file size

Size of text file	1kB	2kB	7kB	13kB	19kB
SSD	2313774	2313968	2316152	2321083	2325568

Fig. 8 Difference image between original and cover for text 2kB

Fig. 9 Difference image between original and cover for text 19kB

Below are presented the PSNR values of the original image compared with cover images (Figs. 8, 9).

From the experimental results we see that the impact of the changed LSB bits due to the insertion of encoded text information is smaller in scanner images than in color images of angio-scanner. Also, the PSNR values are smaller in the angio-scanner images than in scanner images (Table 7).

Table 7 Values of PSNR

Size of text file	1kB	2kB	7kB	13kB	19kB
PSNR	44.0754	44.0737	44.0626	44.0487	44.0347

6 Conclusions

Privacy of the patient data is a crucial element in telemedicine. The combination of encryption techniques with steganographic techniques offers very good ways to protect information in telemedicine. This paper shows that LSB Method provides an efficient technique to hide the information within the image pixels. The text file that keeps the patient history grows in size, it also increases the sums of intensity values in difference images as well as SSD and PSNR between the original image and the cover one. The percentage of sums of intensity values remains relatively very small, PSNR values are relatively very large, and by increasing the size of coded text files, it does not bring any visual damage to the medical image. The constraints of legal requirements of safety and confidentiality are fully satisfied. By analyzing the results of measurements for both categories of images, we conclude that the steganocryptography technique performs better in scanner images than in angio-scanner images.

References

1. Maheu MM, Whitten P, Allen A (2001) E-Health, Telehealth, and Telemedicine: a guide to startup and success, Jossey-Bass Inc. A Wiley Company
2. Rao HN, Pandya M, Prasad KMVV (2013) An evaluation of data security for telemedicine application development. Int J Comput Appl 79(1): 0975–8887
3. Frye L et al Network management of a wireless sensor networkwww.lehigh.edu/images/userImages/jgs2/Page_7287/LU-CSE-07-.pdf
4. Katz J (2010) Digital Signatures, Springer
5. Mortazavian P, Jahangiri M, Fatemizadeh E (2001) A low-degradation steganography model for data hiding in medical images. In: Proceedings of the fourth IASTED international conference Visalization, Imaging, And Image Processing September 6–8, Marbella, Spain
6. Sampaio RA, JackowskiMP Assessment of steganographic methods in medical imaging. (Online). Available http://www.ucsp.edu.pe/sibgrapi2013/eproceedings/wuw/115732_1.pdf
7. Keerthana L., Venkataramanaiah B (2014) Ecg steganography based privacy protection of medical datas for telemedicine application. IOSR J VLSI and Signal Process (IOSR-JVSP) 4 (2), 46–51 e-ISSN: 2319–4200, p-ISSN No: 2319–4197
8. Thomas SE, Philip ST, Nazar S, Mathew A, Joseph N (2012) Advanced cryptographic steganography using multimedia files. In: International Conference on Electrical Engineering and Computer Science (ICEECS-2012), Trivendum, ISBN Number: 978-93-81693-58-2
9. Pandey V, Singh A, Shrivastava M (2012) Medical image protection by using cryptography data-hiding and steganography. Int J Emerg Technol Adv Eng 2(1), Website: www.ijetae.com ISSN 2250-2459
10. ORTI J Advanced JPEG steganography & detection. (Online) Available https://www.blackhat.com/docs/asia-14/materials/Ortiz/Asia-14-Ortiz-Advanced-JPEG-Steganography-And-Detection.pdf

11. Jokay M, Moravcik T (2010) Image-based JPEG steganography. (Online) Available https://www.sav.sk/journals/uploads/0317153109jo-mo.pdf
12. Chugh G, Yadav R, Saini R (2014) A new image steganographic approach based on mod factor for RGB images. Int J Sig Process, Image Process and Pattern Recogn. 7(3)

Part III
Biometric Data in Automotive and Home Applications

Stress-Aware Generation of Recommendations in a Driving System to Increase User Acceptance

Emre Yay and Natividad Martínez Madrid

Abstract Besides the optimisation of the car, energy-efficiency and safety can also be increased by optimising the driving behaviour. Based on this fact, a driving system is in development whose goal is to educate the driver in energy-efficient and safe driving. It monitors the driver, the car and the environment and gives energy-efficiency and safety relevant recommendations. However, the driving system tries not to distract or bother the driver by giving recommendations for example during stressful driving situations or when the driver is not interested in that recommendation. Therefore, the driving system monitors the stress level of the driver as well as the reaction of the driver to a given recommendation and decides whether to give a recommendation or not. This allows to suppress recommendations when needed and, thus, to increase the road safety and the user acceptance of the driving system.

1 Introduction

As a consequence of several oil crises and the climate change, society's awareness of saving energy and protecting the environment increased. Furthermore, statistics showed [5] that the growing number of cars and drivers resulted in an increase of accidents and fatalities on the road. There is the potential to increase the energy-efficiency and road safety by adapting the driving behaviour. Fan [15] showed that the driving behaviour has a great factor on safety. This is also verified by the German Statistical Office that showed that 86 % of the accidents happened in Germany in 2013 because of driver mistakes [5]. Furthermore, according to [8, 6] there is the possibility to increase the energy-efficiency up to 30 % by adapting the driving behaviour to the driving situation.

E. Yay (✉) · N.M. Madrid
Reutlingen University, Alteburgstr. 150, 72764 Reutlingen, Germany
e-mail: emre.yay@reutlingen-university.de

N.M. Madrid
e-mail: natividad.martinez@reutlingen-university.de

M. Conti et al. (eds.), *Mobile Networks for Biometric Data Analysis*,
Lecture Notes in Electrical Engineering 392, DOI 10.1007/978-3-319-39700-9_10

There are already several driving systems [3, 10] with the goal to optimise the driving behaviour in energy-efficiency or safety. However, these driving systems show recommendations to the driver without considering the condition of the driver. In contrast, the driving system, which is in development, tries not to distract or bother the driver by giving recommendations in stressful driving situations. Furthermore, this can prevent the increase of driver's mental load in a stressful driving situation. Thus, this allows to increase the road safety as a high mental load can lead to accidents [2].

2 Related Work

There are already existing driving systems whose goal is to educate the driver in energy-efficient or safe driving. The driving system of Kia [3] monitors the car and activates a green lamp when an energy-efficient driving behaviour is detected. A red lamp indicates that the driver has broken a driving rule and is therefore driving energy-inefficient. A white coloured lamp represents the stand-by of the driving system or a normal fuel consumption of the car. Kia evaluated the driving system during an eco-driving event on a 7 km route with 100 test drivers. The result showed an increase of the energy-efficiency between 10 and 20 %.

Another energy-efficient driving system that tries to improve the driving behaviour comes from Fiat [4]. In contrast, to the driving system of Kia, Fiat collects telemetric information during the journey and saves it on a flash pen. By plugging in the flash pen in a computer allows the servers of Fiat to analyse the driving behaviour in terms of energy-efficiency. Fiat defines an energy-efficient driving behaviour as steady acceleration, steady deceleration, early gear changes and constant speed. The result of the analysis is a so called eco:Index that indicates the energy-efficiency of the journey. Additionally, driving recommendations are shown to the driver in order to improve the driving behaviour in terms of energy-efficiency. Fiat analysed the eco:Index of the drivers over a 30-day period. The results showed an average improvement of 2.25 % in terms of energy-efficiency.

The safety relevant driving system of Lotan and Toledo introduced in [10] tries to educate the driver in safe driving by monitoring and analysing the car movement during the journey by collecting information about the acceleration, speed of the vehicle and the position of the car. The driving system warns to the driver in real-time to prevent an aggressive driving behaviour or when the current driving behaviour does not match the typical driving behaviour of the driver. The warnings are presented as a text message or using the in-vehicle display unit. Besides the warning of the driver, the driving system generates reports over the trip safety based on the collected information. Within the reports the driving system indicates the trip safety by using green, yellow and red colours that stand for cautious, moderate and aggressive driving behaviour. The evaluation of the driving system was done with 33 test drivers during 5 months without using the real-time feedback. The results showed that the driving system has an effect on the driving behaviour as the risk indices dropped in the first month in which the reports were used as a feedback. This effect continued for three months. However, in the 5th month the driving risk indices were back to the initial values. According to Lotan and Toledo, this suggests that the initial impact of the system is significant, however it decreases over time without routine follow-up maintenance efforts.

The current available driving systems cover either the area of energy-efficiency or safety. Furthermore, they only show warnings to the driver or indicate an energy-efficient driving behaviour using a coloured lamp. Thus, the feedback of the driving systems are not sufficient, as they do not show the wrong-doings of the driver or give driving recommendations to the driver. This would allow to eliminate precisely the bad driving habits of the driver, as bad driving habits are the main cause of energy-inefficient or unsafe driving. Furthermore, they do not consider the driver condition or the individual driving behaviour. For example, they do not suppress warnings or recommendations when the driver is not able to drive energy-efficient or safe, as the driver may be in a stressful driving situation or as some driving recommendations may be irrelevant or not acceptable for the driver in specific driving situations.

On the basis of the findings, a driving system is in development that tries to improve the driving behaviour in energy-efficiency and safety while considering the driver condition and the individual driving behaviour. Therefore, the driving system monitors the driver, the car and the environment and shows recommendations to the driver, when it detects that the driver is not driving energy-efficient or safe. However, before the driving system shows a recommendation, it decides on the basis of the driver condition and the individual driving behaviour if the recommendation should be shown. This allows the elimination of bad driving habits, while considering the driver condition and the individual driving behaviour of the driver.

3 The Driving Rules

Energy-efficient and safe driving is described by a set of driving rules that are the basis for the energy-efficiency and safety relevant driving system. The driving rules are used to check if the driver is driving energy-efficient and safe. According to Barkenbus [1] an energy-efficient driving behaviour involves for example a smooth acceleration between 2000 and 2500 revolutions per minute or the anticipation to the traffic flow. The energy-efficient driving rules, which are also part of the dutch eco-driving programme, were evaluated by van Mierlo et al. [14]: (1) to shift as soon as possible, (2) press the throttle quickly and vigorously to keep up with the traffic and (3) shift down as late as possible to keep the car rolling in the highest gear without engaging the clutch. The result of the evaluation showed that the correct interpretation of the driving rules increased the energy-efficiency. However, some drivers had problems by applying the driving rules (1) and (2) as these rules were too contradictory to the drivers. Thus, van Mierlo et al. recommend to use only the driving rules (1) and (3). Furthermore, van Mierlo et al. observed that the adherence of the driving rules leaded to a reduced driving speed and, thus, to a higher road safety [6].

The driving rules for a safe driving behaviour are trying to prevent mainly an aggressive driving behaviour. An aggressive driving behaviour includes such things as speeding or driving to close to the car in front [13]. Furthermore, statistics [5] showed that speeding and insufficient distance to the car in front caused also accidents with personal injury. The United Nations Economic Commission for Europe [13] and the New Zealand Transport Agency [11] defined driving rules for a safe driving behaviour that prevent speeding, distraction, fatigue and so on. On the basis of the facts from [5, 11, 13], driving rules were derived and used in the driving system for the detection of an inefficient or unsafe driving behaviour. In the following an excerpt of the energy-efficiency and safety relevant driving rules that are used in the driving system are listed:

- Shift as soon as possible at a maximum of 2500 r/min (2000 r/min for a diesel)
- Drive at a steady speed using the highest possible gear to keep the engine speed down
- Turn off the engine when the engine idles longer than 1 min
- Keep enough distance to the preceding car (minimum distance is equivalent to distance travelled by a vehicle in two seconds or the half of the speed in meters)
- Adapt your speed to the given situation and don't exceed the speed limit
- Avoid any distractions (i.e. don't use the mobile phone during the journey)

4 The Driving System

In order to show recommendations, the driving system is separated in two main steps: gathering the data and generating the recommendations. First the driving system acquires data from the car, the driver and the environment using the in-vehicle serial-bus systems and attached sensors. The information about the driver condition is gathered using an ear sensor. The ear sensor provides information about the heart rate, which can be used to calculate the heart rate variability. The heart rate variability allows to indicate stress [9]. On the basis of the gathered data from the car and the driver, the driving system generates a driver profile, which includes for example information about the average stress level of the driver. Furthermore, the driving system stores driving rules that describe the energy-efficient and safe driving behaviour. The gathered information, the generated driving profile and the driving rules are the basis for the generation of the recommendations.

In the second step, the driving system generates recommendations and decides if a recommendation should be shown to the driver. Therefore, it compares the driving rules against the gathered data to find an inefficient or unsafe driving behaviour. Furthermore, it analyses if the current driving behaviour differs significantly from his typical driving behaviour. For example, a recommendation to calm down or to have a rest is generated when the stress level of the driver is significantly higher than his typical stress level. When a broken driving rule is found, the driving system decides if a recommendation should be given to the driver. The decision is made on the basis of the current stress level of the driver and the individual driving behaviour, for example the reaction of the driver to already given recommendations. For example, if the driver is in stress or has ignored a recommendation repeatedly, the driving system will suppress the driving recommendation in order not to distract the driver in a stressful driving situation or not to bother the driver with recommendations which are not interesting in the opinion of the driver. If the driving system decides to show the recommendation to the driver, it will be displayed for example on the in-vehicle display unit.

5 Architecture

The architecture of the driving system is based on the multi-tier architecture and is separated in three main components: the Data Layer, the Processing Layer and the Graphical Layer. Figure 1 shows the architecture of the driving system including the three layers and the modules of the Data Layer and the Processing Layer.

First, the information that is needed for deciding whether to show a recommendation is gathered by the Data Layer from the driver, the car and the environment. Therefore, the Interface module of the Data Layer is connected to the in-vehicle serial-bus systems like CAN and to additionally attached sensors that provide the needed information. After the collection of the information, it is passed to the Data Aggregated module in order to aggregate the data. On the basis of the aggregated data and the gathered information from the Interface module the Profile Update module generates a driving profile that represents the typical driving behaviour of the driver. The driving profile includes for example information about

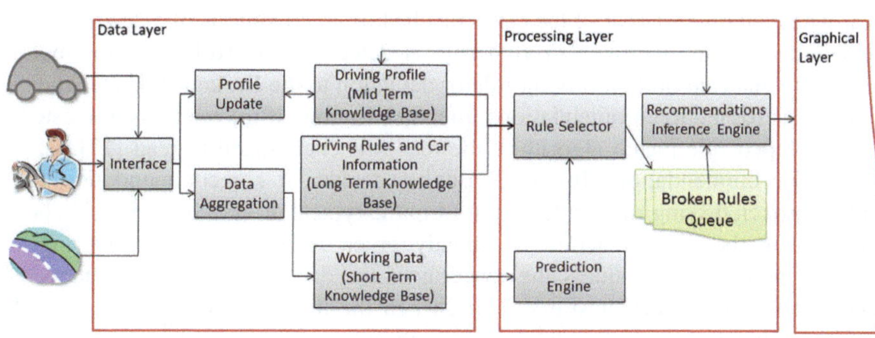

Fig. 1 The architecture of the driving system including the modules of the data and processing layer

the average driver stress level or the average manner of driving. However, after the first start of the driving system, the driving profile needs several update cycles until it has enough data for representing the typical driving behaviour of the driver. There are three different knowledge bases in the Data Layer to store the information needed for detecting a broken rule and generating corresponding a recommendation: Short-, Mid- and Long-Term. The Short-Term Knowledge Base consists of the working memory that stores all relevant data from the driver, the car and the environment. The generated driving profile is stored in the Mid-Term Knowledge Base. Finally, the Long-Term Knowledge Base consists of the driving rules, which are used to represent an energy-efficient and safe driving behaviour and the car facts, which are used for example for calculating the fuel consumption.

The Processing Layer is responsible for analysing the current driving behaviour and deciding whether to show a recommendation. Furthermore, it predicts the state of the car in the Prediction Engine module based on the data stored in the Working Memory. The Prediction Engine module considers the values speed, engine speed and distance to the car in front as the car state. The prediction of the car state allows to show recommendations before a breaking of a driving rule occurs and, thus, to avoid the breaking of the driving rule by the driver. If the reader is interested in the Prediction Engine module, the reader is encouraged to read [16].

The driving behaviour has to be analysed in order to find a broken driving rule. This is done in two steps, first the Rule Selector module gathers all relevant data from the Prediction Engine module, which also passes the data from the Working Memory to the Rule Selector module. Then it checks if the driver has broken an energy-efficiency or safety relevant driving rule and if the current driving behaviour deviates significantly from the typical driving behaviour, for example if the current stress level differs significantly from the typical driver stress level. In order to match the data against the driving rules a rule match algorithm was developed to find broken driving rules. The rule matching algorithm represents a driving rule in a tree structured network, in which the corresponding data is pushed for checking the conditions of the driving rules. For example, the conditions of the driving rule "shift as soon as possible at a maximum of 2500 r/min" would be current revolutions per minute >2500 r/min and current gear <highest gear, as is not able to shift to the next gear when the highest gear of the car is reached. If all conditions of a driving rule or deviation are true, a broken driving rule is detected. The Rule Selector module passes the detected broken driving rule or deviation to the Broken Rules Queue that is the basis for the Recommendations Inference Engine. If the reader is interested in the developed rule match algorithm, the reader is encouraged to read [17].

The Recommendations Inference Engine decides if a recommendation for a certain broken driving rule or deviation from the typical driving behaviour is shown and detects the drive reaction to already given recommendations. During the decision process, it considers the reaction of the driver to already given recommendations and the driver stress. Thus, when the driver is in stress or ignored a recommendation repeatedly the recommendation is suppressed in order not to distract the driver in stressful driving situations or not to bother the driver with recommendations that are not interesting in the sense of the driver. If the

Recommendations Inference Engine decides to show a recommendation, it is passed from the Recommendations Inference Engine module to the Graphical Layer that renders the recommendations.

The Graphical Layer is the interface between the driver and the driving system and is responsible for rendering the recommendations. Furthermore, the driver has the opportunity to choose his driving profile using the graphical user interface. The incoming recommendations from the Processing Layer are rendered for example on the in-vehicle display unit or by an audio signal in order to improve the driving behaviour of the driver.

6 Monitoring the Driver Stress

The driver stress is monitored using the physiologic signals of the driver, as the "physiologic signals provide a viable method of measuring a driver stress level" [7]. For the stress detection the autonomic nervous system of the driver is for interest, as it is concerned to regulate the heart rate, breathing rate or other visceral activities [9]. The autonomous nervous system consists of two activities that modulate the heart rate: the sympathetic—and parasympathetic activity. The sympathetic activity is responsible for preparing the body for stressful situations, whereby the parasympathetic activity is related to restore the body to a resting state. Under normal conditions the two activities are in balance. However, in stressful situations the sympathetic activity is increased that leads to an unbalance between the two activities. This unbalance can be detected using the heart rate. Therefore, the heart rate of the driver is monitored using an ear sensor. On the basis of the monitored heart rate, the heart rate variability (HRV) of the driver can be derived. The HRV is the variation of the interval between the two heart beats and represents the ratio of the sympathetic and parasympathetic activities [7]. Thus, the HRV allows to detect the stress level of the driver [7, 9, 12], as high stress leads to an increase of the sympathetic activity and, thus, to an high HRV. In contrast, a low HRV indicates a balance of the sympathetic and parasympathetic activity, why a low stress level can be assumed.

The driver stress level in the first prototype of the driving system is monitored using the software emWave[1] and an ear sensor. The software derives the HRV on the basis of the monitored heart beat, respectively the interval between two heart beats, also called inter-beat interval (IBI), and separates the stress level in the categories low or no stress, medium stress and high stress. As the first prototype of the driving system is developed on a driving simulator the CAN-bus of the driving simulator is used to provide the information about the driver stress level to the driving system. Figure 2 shows the hardware set-up for monitoring the driver stress in the environment of the driving simulator. The ear sensor that monitors the heart

[1]emWave is a software of the company HeartMath: http://www.heartmath.com/.

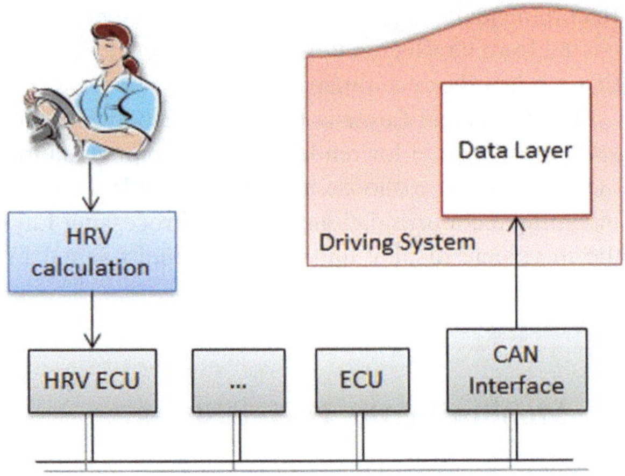

Fig. 2 The hardware set-up for detecting the driver stress

rate of the driver is connected to the software. The gathered heart rate from the driver is then processed in the software to get the driver stress level. To provide the calculated driver stress level to the driving system, the software is connected to the HRV electronic control unit (ECU) of the driving simulator CAN-bus. The HRV ECU collects the stress level from the software and forwards it into the CAN-bus. This allows the driving system to gather the stress level of the driver from the CAN-bus using the Interface module of the Data Layer. Finally, the Interface module of the Data Layer provides the gathered stress level for further processing.

7 Decision of Showing a Recommendation

The decision whether to show a recommendation is done by the Recommendations Inference Engine on the basis of the broken driving rules, the driver stress level and the driver reaction to already given recommendations. However, before the decision to show a recommendation can be made, the Rule Selector module has to find a broken driving rule or a deviation from the typical driving behaviour. In case a broken driving rule or a deviation is found, the Rule Selector module puts the broken driving rule or deviation into the Broken Rules Queue. Additionally, the corresponding information that caused the breaking or deviation and the driver stress level at that time is also added to the Broken rules Queue. When the Broken Rules Queue contains broken driving rules or deviations from the typical driving behaviour, the Recommendations Inference Engine starts to process the broken driving rules or deviations using the first come, first served principle.

Figure 3 illustrates the decision process of the Recommendations Inference Engine. First, it gets a broken rule or deviation from the Broken Rules Queue and checks if a similar recommendation was given to the driver shortly before. If a recommendation was given shortly before, the Recommendations Inference Engine

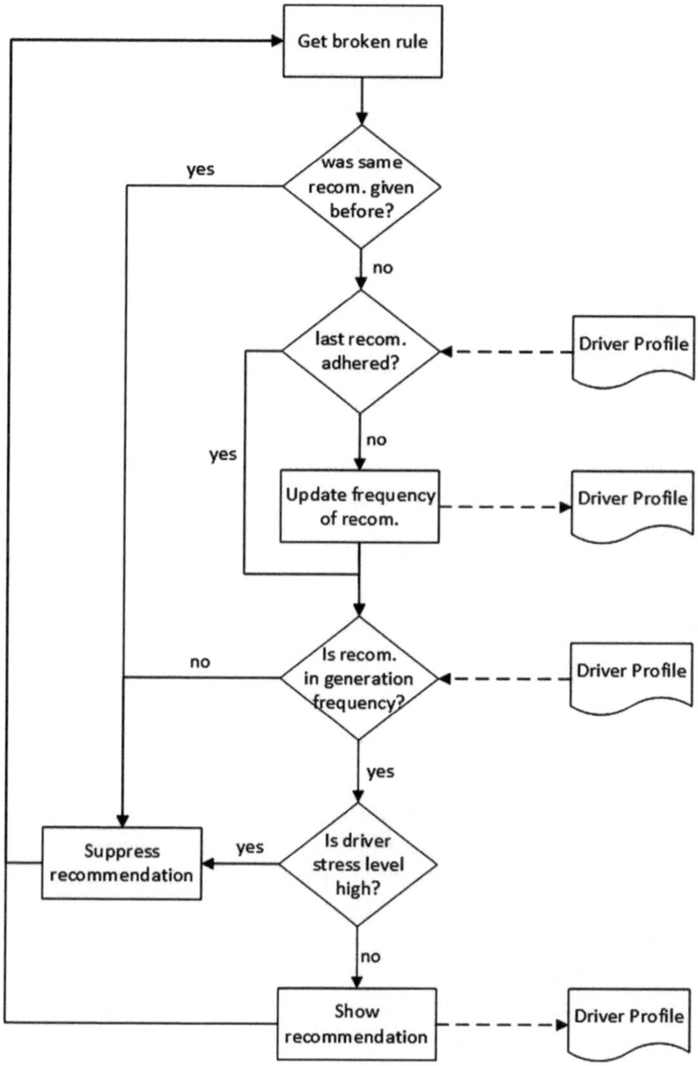

Fig. 3 The decision process to show a recommendation

discards the broken rule and waits a certain time until the same broken rule is processed again. Thus, the recommendation based on this broken rule will be suppressed. This allows to give the driver time to react appropriate to the already given recommendation and, thus, not to bother the driver by showing the same recommendation over and over within a short period. However, if there was no recommendation given shortly before, the Recommendation Inference Engine checks next, if the last given recommendation was adhered. Therefore, the Recommendations Inference Engine proofs if the last given recommendation is the same as the current generated recommendation. The last given recommendation is stored within the driving profile of the driver. In case the last recommendation was the same as the current recommendation, the Recommendations Inference Engine assumes that the driver did not adhere the last given recommendation. Thus, the Recommendations Inference Engine counts that the recommendation was ignored. If the recommendation was ignored repeatedly by the driver the Recommendations Inference Engine decreases the generation frequency of that recommendation in the driver profile. The generation frequency indicates how often a certain recommendation should be shown to the driver. On the basis of the generation frequency, the Recommendations Inference Engine decides whether to show the recommendation or not. This allows to increase the user acceptance, as the driver is not bothered with recommendations, which are not relevant for them and, thus, the driver would ignore. If the current recommendation is acceptable regarding the frequency, the Recommendations Inference Engine proofs if the stress level of the driver at the time of the breaking or deviation is acceptable to show the recommendation. The recommendation will be suppressed if the stress level of the driver is high, as the driving system does not want to distract the driver in a stressful driving situation by showing recommendations. In contrast, if the stress level is low or medium the recommendation is shown to the driver and the recommendation is stored in the driving profile as the last given recommendation.

8 Conclusion and Further Work

In this paper a driving system is introduced, which is in development. Its goal is to educate the driver in energy-efficient and safe driving. To accomplish this, it collects information about the car, the driver and the environment using the in-vehicle serial bus systems and a vital sensor. On the basis of the acquired information, it analyses the driving behaviour as well as the stress level of the driver and gives adequate recommendations to the driver. In contrast to other driving systems, the presented driving system considers the areas of energy-efficiency and safety as well as driver condition and the individual driving behaviour. This allows the driving system to create adequate recommendations, which are not distracting the driver in stressful driving situations or bothering the driver. Thus, this leads to an increase of the road safety, to a decrease of the fuel consumption and to a higher acceptance of the driving system. As the driving system is in development, the future work comprises the consideration of additional vital information of the driver in order to recognise, besides the stress level of the driver, for example emotions or the fitness to drive. This would allow the driving system to react appropriate to the individual driving behaviour for example by customising the recommendations more precisely or giving additional recommendations on the basis of the emotions or the fitness to drive.

References

1. Barkenbus JN (2009) Eco-driving: an overlooked climate change initiative. Energy Policy 38 (2):762769
2. Brookhuis KA, de Waard D (2010) Monitoring drivers mental workload in driving simulators using physiological measures. Accid Anal Prev 42(3):898–903
3. Cho HJ (2008) Eco driving system. URL http://kia-buzz.com/eco-driving-system/, Last visit: 07.10.2014
4. Fiat (2010) Eco-driving uncovered: the benefits and challenges of eco-driving, based on the first study using real journey data
5. German Statistical Office (2014) Verkehr- Verkehrsunfälle 2013
6. Haworth N, Symmons M (2001) Driving to reduce fuel consumption and improve road safety. In: Proceedings road safety research, policing and education conference, vol 1, pp 7
7. Healey J, Picard R (2000) Smartcar: detecting driver stress. In: Proceedings of the 15th international conference on pattern recognition, vol 4, pp 218–221
8. Helms H, Lambrecht U, Hanusch J (2010) Energieeffizienz im Verkehr. Energieeffizienz 1:309–329
9. Kumar M, Weippert M, Vilbrandt R, Kreuzfeld S, Stoll R (2007) Fuzzy evaluation of heart rate signals for mental stress assessment. IEEE Trans Fuzzy Syst 15(5):791–808
10. Lotan T, Toledo T (2006) An in-vehicle data recorder for evaluation of driving behavior and safety. Transportation Research Board of the National Academies, Paper No. 061607, pp 1–14
11. New Zealand Transport Agency (2007) Your safe driving policy: helping you to manage work-related road safety and keep your employees and vehicles safe on the roads
12. Salahuddin L, Kim D (2006) Detection of acute stress by heart rate variability using a prototype mobile ECG sensor. In: Proceedings of the 2006 international conference on hybrid information technology, vol 2, pp 453–459
13. UNECE—United Nations Economic Commission for Europe (2004) Aggressive driving behaviour (background paper). URL http://www.unece.org/trans/roadsafe/rs4aggr.html, Last visit: 07.02.2015
14. van Mierlo J, Maggetto G, van de Burgwal E, Gense R (2004) Driving style and traffic measures—influence on vehicle emissions and fuel consumption. Proc Inst Mech Eng Part D J Automobile Eng 218:43–50
15. Xiaoqiu F, Jinzhang J, Guoqiang Z (2011) Impact of driving behavior on the traffic safety of highway intersection. In: Third international conference on measuring technology and mechatronics, vol 2, pp 370–373
16. Yay E, Madrid NM (2013) Seedrive—an adaptive and rule based driving system. In: The 9th international conference on intelligent environments, Athens, Greece, pp 262–265
17. Yay E, Madrid NM, Ramírez JAO (2014) Using an improved rule match algorithm in an expert system to detect broken driving rules for an energy-efficiency and safety relevant driving system. Procedia Comput Sci KES 35(1):127–136

Stress Map Based Information System for Increasing Road Safety

Patrick Datko, Ralf Seepold and Natividad Martínez Madrid

Abstract Stress is becoming an important topic in modern life. The influence of stress results in a higher rate of health disorders such as burnout, heart problems, obesity, asthma, diabetes, depressions and many others. Furthermore individual's behavior and capabilities could be directly affected leading to altered cognition, inappropriate decision making and problem solving skills. In a dynamic and unpredictable environment, such as automotive, this can result in a higher risk for accidents. Different papers faced the estimation as well as prediction of drivers' stress level during driving. Another important question is not only the stress level of the driver himself, but also the influence on and of a group of other drivers in the near area. This paper proposes a system, which determines a group of drivers in a near area as clusters and it derives the individual stress level. This information will be analyzed to generate a stress map, which represents a graphical view about road section with a higher stress influence. Aggregated data can be used to generate navigation routes with a lower stress influence to decrease stress influenced driving as well as improve road safety.

P. Datko (✉) · R. Seepold
HTWG Konstanz, Brauneggerstr. 55, 78462 Konstanz, Germany
e-mail: patrick.datko@htwg-konstanz.de
URL: http://uc-lab.in.htwg-konstanz.de

R. Seepold
e-mail: ralf.seepold@htwg-konstanz.de
URL: http://uc-lab.in.htwg-konstanz.de

N.M. Madrid
Reutlingen University, Alteburgstr. 150, 72762 Reutlingen, Germany
e-mail: natividad.martinez@reutlingen-university.de
URL: http://iotlab.reutlingen-university.de/

M. Conti et al. (eds.), *Mobile Networks for Biometric Data Analysis*,
Lecture Notes in Electrical Engineering 392, DOI 10.1007/978-3-319-39700-9_11

1 Introduction

An essential part of daily routine of every human being is mobility. Whether in the private or business sector, the automobile is one of the most used means of transportation. Modern requirements on individual's mobility as well as an increasing traffic volume, based on a higher amount of used vehicles every year, cause new challenges for road safety. The automobile industry is focusing on this area trying to produce new vehicles with modern Advanced Driver Assistance Systems (ADAS) such as Lane change assistance, Intelligent speed adaption, Night Vision, etc. as well as the development of autonomous vehicle, to improve the driving process and increase the road safety. However, the number of annual traffic accidents remains practically constant, in spite of the usage of these modern technologies. The main reason for this is the misbehavior of drivers on the road [1], which can be caused by stress. Driver's capabilities to perform a vehicle are significantly influenced by stress. Processes such as problem solving and decision-making are affected as well as an altered cognition [2, 3]. This impact is a high risk in rapid and unpredictable environments like driving, not only for the affected driver but also for others in the immediate environment.

Stress can be seen as a biological reaction induced by a confrontation of psychological or spiritual stressor [4]. The reasons for stressed circumstances can be different especially related to a driving situation, they have to be separated. Factors can be classified in two groups non-driving related and driving related. Non-driving related factors can already occur, before the driver is entering the vehicle and therefore can be seen as a static initial state. Sources for this kind of factor could be time-pressure as well as private and/or job-related concerns. Whereas driving-related factors dynamically increase drivers' stress level while driving, based on sources such as information overload (e.g. check traffic notifications, usage of smartphone, observing navigation system, various traffic signs, etc.), the behavior of other drivers, environmental influences (e.g. building lots), but also time pressure based on external factors such as traffic jams. Driving related factors represent a major challenge, because it is hard to predict.

However, considering typical volume of traffic, periodic patterns can be determined. For example the drive to the workplace, when on weekdays in the morning and in the afternoon a higher number of vehicles trying to enter a city than in the nights. Furthermore, at the beginning of school holidays a higher number of vehicles will leave the city than commonly or at the weekends when the traffic volume increase because people go to shops which are usually in the same area. Situations like these can result in a higher stress level for a driver. This paper proposes a system, which uses the stress level of the drivers and their corresponding geo locations, to generate so called "heat maps" representing areas of higher stress and therefore a potential risk for road safety. Based on determined heat maps, "less stressful routes" can be calculated for navigation systems as well as dynamically notify drivers about a upcoming increase of stress on the road. Moreover, for long-term route planning, future heat maps can be used to estimate the best time of travel. The basic goal of the system is to improve road safety by decreasing the stress level of drivers and therefore a potential misleading behavior.

In the following section, related works and approaches in the area of determining and prediction of drivers' stress level as well as intelligent in-vehicle navigation systems will be shown. An abstract model of the system, the architecture and the concrete components for clustering, analyzing and prediction as well as the information processing model will be explained. Concluding an outlook will presented on the future work according to a short conclusion.

2 State of the Art

Stress measurement and detection become an interesting topic for researchers, industries as well as public organizations in the area of health care.

Many papers with different techniques addressing this area. In [5] a method was introduced to estimate drivers' stress level based on differential skin temperature measurement. Another approach was described in [6], using unobtrusive wearable sensors to detect mental stress. The authors used accessed information to estimate the state of the autonomic nervous system adapted from analysis of heart rate variability (HRV). Likewise [7] extend the basic idea of mental stress detection based on HRV, by relating given stress factors and the age of the person to find a corresponding influence on the status of the mental stress level. Information overload such as road signs and traffic lights motivates a potential raising of drivers' stress level, which is described in [8].

Numerous scientific papers attempt to combine stress detection or the prediction of drivers' stress level as a base for intelligent car assistance systems (ICAS).

In [9] an analysis of HRV parameters aiming to detect the most promising HRV parameters for workload measurement during real world driving was published. A reasoning-based framework was introduced in [10] to predict stress level of drivers due specific driving events. Based on salivary amylase as biomarkers the authors in [11] evaluated drivers' stress potential using a driving simulator. All described references used only an individual driver and an individual driver's stress level for stress detection or prediction. The influence of other drivers' stress level or the monitoring of a group of drivers were not considered.

In [12] the authors propose an intelligent in-vehicle Route Guidance System (RGS) usable by a speech interface to configure different kind of operation modes such as map searching, locating, shortest path representing, 3D navigation as well as route guidance. Despite the target of decrease drivers' stress level by minimizing the direct interaction between the driver and the RGS, this approach is not using other input parameters such as the current stress level or volume of traffic to dynamically adapt the system. For example, a driver with a higher stress level should not be confronted with the same number of information, provided by the RGS than a driver in a moderate state. In [13] a model to analyze the psychological stress factors is described. The authors focused on the relationship between facial expression intensity and Stress Response Scale. However, determined results are only representing a single driver not an aggregation for a complete or multiple groups, which can be used for further analysis.

As proposed in [14, 15] an approach was introduced for interchanging information between cars in close areas vehicular communication systems (VCS) as a base for intelligent transport systems (ITS) targeted the increase of road safety. Vehicular AdHoc NETworks (VANETs) provide information exchange among each vehicle, called Vehicle-to-Vehicle (V2V) or with a given infrastructure (V2I). This solution becomes less effective when the group of vehicles, which build a network, is sparse so the number of network nodes is low. A proposal to close this gap is given in [16]. The authors used a VANET as one cluster respectively as a group of vehicles, to interchange information with another cluster not near, e.g. a reason like low density of equipped vehicles, a satellite system is used as a notification dispatcher. With this approach, different clusters (of grouped vehicles) can share information among each other. However, distributed information examines only the cluster in its entirety but not with other groups as well as provide information for further processing such as a prediction of stress related road sections.

In the following section, a basic architecture will be proposed for clustering a group of vehicles into groups based on their location. Corresponding analyze additional information given by the vehicle itself and the driver such as the potential detected stress level. Aggregated information will be used to compare it with the own cluster and other clusters in the near area as well as distribute information and provide the possibility for navigation requests.

3 Architecture Model

The basic system is based on a client-server architecture. The server part is represented by a backbone of multiple cluster nodes accessible via Internet. Load-Balancers are used to dispose application resources to achieve a higher rate of concurrent connections as well as computation possibilities. The clients are represented by drivers' smartphones, which is gathering current location based (e.g. GPS) and external sensor data (e.g. heart rate, stress level) while driving. Collected information will be forwarded periodically to the server, either via Global System or Mobile Communications (GSM), Universal Mobile Telecommunications System (UMTS) or Long Term Evolution (LTE). In Fig. 1 the basic system architecture and the corresponding information exchange is demonstrated.

The architecture of the server software is shown in Fig. 2. The software architecture is layered-based with different service components aggregating and interchanging information among each other.

As explained the system is divided in several components for clustering, analyzing and notifications. The communication interface is the connection part between client (e.g. vehicle), the logical service components (software parts) and therefore the communication access point. It works as information dispatcher,

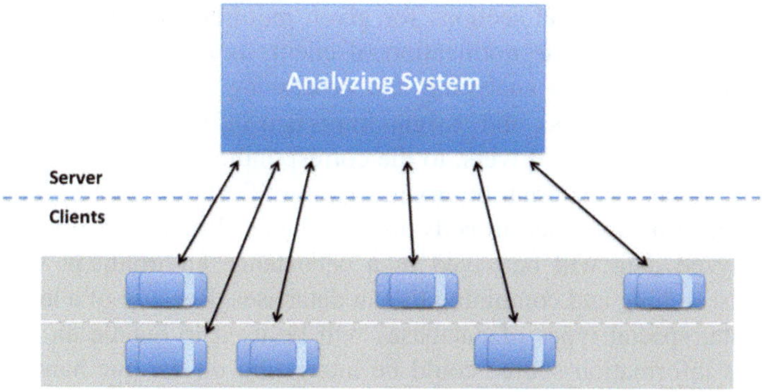

Fig. 1 Basic system architecture and information exchange

Fig. 2 Layered architecture of server system

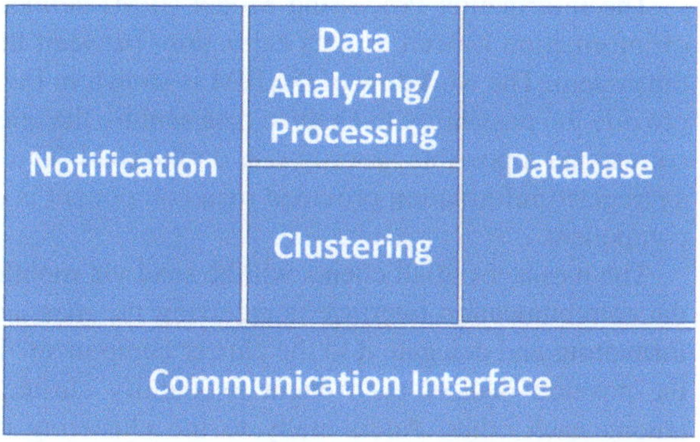

which has to assimilate all communication between client and server. Therefore, the communication interface uses basic techniques such as sockets to comply the given requirements. The clustering service uses received location-based information for clustering individual clients into concrete groups. For the estimation of clusters, known algorithms of the area of data mining are used. For generating heat maps, representing a dense region of potential stress, density-based clustering algorithms are used. The calculated clusters will be forwarded with all gathered individual's information to the analyzing and processing component.

The main component of the system is the data analyzing and processing component. This component uses accessible information of the group to compare one group with another and monitor individuals against the group. If the majority of group members are identified as stressed, the geo location area of the group will be marked as potential stressful for the current weekday and time. This aggregated data set represents the base for a navigation system trying to avoid potential stressful roads. If a change of a groups' stress level will be detected, following groups will be notified about a potential risk as well as external observers like traffic supervisors or intelligent driving assistance systems. Additionally processed data can be used to predict future stressful road sections for given requirements. As well as the analyzing of data, the requisite notification of clients is required. Based on the calculated results of the data analyzing and processing component, the notification service is adopting this task and assigns important information, like an increasing stress level of surrounding drivers, to the corresponding receivers, like in this case groups, individuals or external observers. The notification will be broadcasted via the communication interface directly to the clients. Long-term observations raw and/or analyzed data will be saved in a persistence environment with ordinary methods of searching and combining like in databases. Because of a high amount of received data, special types of databases will be used to handle an easy and fast access. This information source could be interesting on the one hand for external persons like traffic supervisors to analyze anomalies in a given area and on the other hand as base for further calculation resulting in predictions of the data analyzing and processing component.

The Information Processing Model (IPM) defines the delegation and workflow of information as well as data extraction, between the different components within the system. The workflow of the IPM is shown in Fig. 3. This model is necessary to provide the possibility to analyze data sent by the clients and to notify if a detection correspond a possible security risk on the road. The IPM can be seen as a pipeline aggregated information provided by a component is used as input for the following component.

The messages of all clients will be send via mobile internet to the system, where the communication interface is receiving the messages, validate the correctness of formatting and delegate it to the cluster component. The Clustering Component use the position data of each client to identify clusters of vehicles, which are near among each other. Additionally to the clustering of groups, all clusters will be sorted in a list; thereby it is possible to represent a sequence of the clusters. In fact, one vehicle or node, which cannot be assigned to any cluster, will not be referred as a cluster.

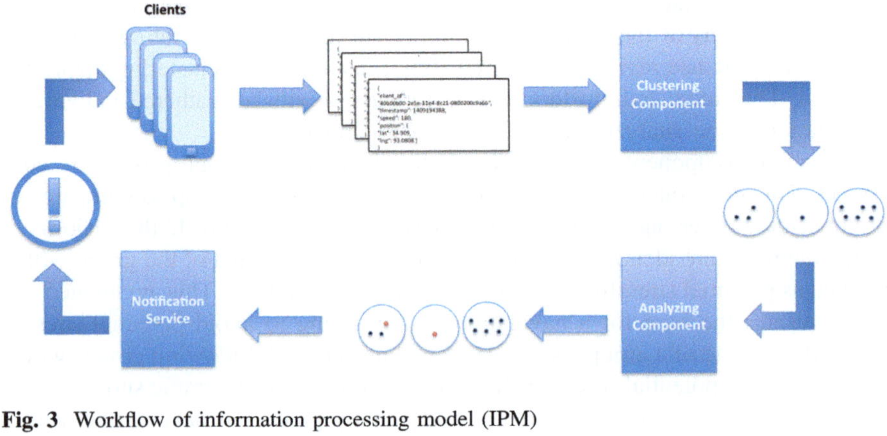

Fig. 3 Workflow of information processing model (IPM)

However based on its location data, it will be arranged in the cluster sequence, because this vehicle can be entering a clustering in the future, which is an important information for further analyzing of the sequence. All generated clusters, and single nodes, as well as their status information, are delegated as a sorted list to the Analyzing component. Based on the aggregated information received from the Clustering component, the analyzing component tries to detect anomalies in the stress-level of drivers within a cluster as well as process the stress-map generation, with the corresponding information. Both procedures are necessary to estimate a potential risk on the road. If the Analyzing component detects any risk, it will delegate a warning message with the additional information who has to be informed to the Notification service. This component sends a notification based on the input data of the analyzing component to all affected groups as well as to concrete single clients.

This abstract model was designed to facilitate the usage of it in different kind of fields or domains like for example crowd clustering in events. Due the component-based architecture and defined interfaces, it is possible to extend the system or replace components with others such as different algorithms for the clustering. In the following subsection, the basic processing of the stress map generation will be explained.

3.1 Stress Map Processing

The generation of the stress map is based on the methodology of so called "heat maps". This kind of maps is a graphical representation of data, which are represented by colors stored in a matrix as 2D displays. Based on different colors, regions with a higher amount of similar data can be graphically determined. The usage of heat maps is widely spread in several different domains for further analysis. Typical application areas are professional sports activities, web analytics, biology and weather forecasts. The advantages of heat maps will be used as a base for the processing and generating of stress map. In Fig. 4 the stress map processing over a period is presented. As mentioned previously, clients data such as the current geo location as well as meta-data like weekday and time will be gathered and send to the server periodically.

In the first step, groups of clients will be clustered in groups based on their geo location and density on the road. After that, the pre-processed data will be used for the generation of the stress map. All potential stress levels of each member of a group will be compared with the rest. If the majority of the group members are identified being stressed, than the geo location of the group will be marked as potentially stressful, for the given meta-data (t_2). If a previous record exists (t_0, t_1), within a given period, overlapping parts or the whole area of the actual group, this data will be compared. If the previous record can also be classified as, potential stressful than the area has a priority for potential stressful driving, which does a more abstruse colored area on a heat map represent. If it was marked as

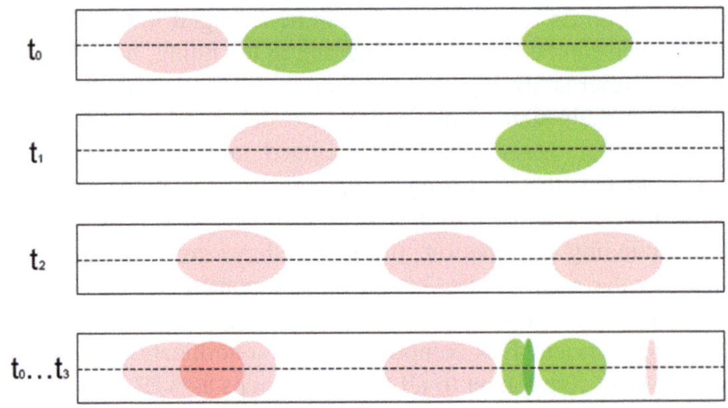

Fig. 4 Stress map processing based on gathered history information

non-stressful, for example a stressful behavior is only related with the group not the environment, than the value of the area will be equalized, which a brighter colored area represents. When more records, for a given time period, exists all data will be used for the calculation of the relevance of an area and the representation of a heat map. Additionally to the categorization of road sections, a changing in the stress levels of groups will be monitored, to notify following groups about a potential risk in the future.

Aggregated data can be represented as routes on typical road maps for users as well as professions such as highway supervisors or public authorities. Furthermore, the stress map can be used for a new mode of navigation systems, where the user is able to plan routes based on a lower stress level, which could additionally result in a relief of traffic volume and therefore an increasing of road safety. Also provides the possibility to predict future road section with a potential higher stress level. Also possible is the usage of a recommendation system for users. However, a change in paradigm of traveling is necessary. Typically, driver has a destination, which should be reached at a defined time. The driver will calculate the average time, needed from the start to the destination. Based on this the start time is defined by the defined arrival time less the duration of the driving. For example if an employee want to be at his workplace at 9 am and the duration from his home to the job is 1 h, than the user will start at 8 am. The proposed system changes this typical traveling pattern. The user has to define the start and the destination as well as the desired arrival time and an interval for departure. If there is no fixed time for arrival, also an interval can be used. The advantage of this paradigm is that the system can calculate the best time for departure to arrive the destination on time and with increased stress influences. For example, the employee want to be in work at 9 am and he is defining a departure interval from 7 am to 8 am. Based on the input parameters the system will calculate the best time maybe 7.25 am to reach the destination in time as well as with a lower stress influence as if he would start earlier or later.

Due to the fact that this is a basic idea, we should bear in mind that it is also possible that circumstances changes during a driver is already traveling therefore also the calculation of a navigation route has to be adapted dynamically while driving. Another aspect a particular time periods such as school holidays, that can have a complete different influence and volume of traffic than under typical circumstances. We deal with these conditions after a first prototype.

4 Conclusion and Future Work

Individual's behavior such as attention level or decision-making are strongly influenced by stress. This could result in a higher risk for accidents in a dynamic environment like automotive. To decrease the hazard, this approach proposes to relate individual stress levels to clustered dynamic groups of nearby, creating a group dependent stress level in their corresponding area. The main target is to provide a so called stress map which should be used to identified road sections which has a higher stress influence for a given time period. This concept has not been tackled by other studies. Automatic detection of stress influences or a common stress rising will be used for notifications to the affected drivers/groups as well as generate routes with less stress influence to improve driving safety with current technologies. A first prototype was developed based on the explained architecture and methods. For the next step, a simulation software will be developed to test the system and the implemented functionalities, which will be the best for future field tests. According to this, the prototype will be extended to handy dynamic changes on a calculated navigation route while driving as well as integrate a recommendation system to avoid stressful driving situations. Finally it has to be evaluated how intelligent traffic control system can be incorporated, as well as the proposed approach may have influence on typical driving behavior and increase the safety on the road as expected.

References

1. Asbridge M, Smart RG, Mann RE (2006) Can we prevent road rage? Trauma Violence Abuse 7:109–121. doi:10.1177/1524838006286689
2. Lisetti CL, Nasoz F (2004) Using noninvasive wearable computers to recognize human emotions from physiological signals. EURASIP J Appl Signal Process 2004:1672–1687
3. Fernandez JM, Ochoa EP, Madrid NM, Seepold R (2009) A distributed management platform to support trading decisions under panic behavior. In: 2009 seventh workshop on intelligent solutions in embedded systems, pp 141–147 (2009)
4. Yata T, Sannohe H, Nakasako M, Tao M (1993) How to cope with stress. Yuhikaku
5. Yamakoshi T, Yamakoshi K, Tanaka S, Nogawa M, Park SB, Shibata M, Sawada Y, Rolfe P, Hirose Y (2008) Feasibility study on driver's stress detection from differential skin temperature measurement. In: 30th annual international conference of the IEEE Engineering in Medicine and Biology Society. EMBS 2008, pp 1076–1079

6. Choi J, Gutierrez-Osuna R (2009) Using heart rate monitors to detect mental stress. In: Sixth international workshop on wearable and implantable body sensor networks. BSN 2009, pp 219–223
7. Salahuddin L, Jeong MG, Kim D, Lim SK, Won K, Woo JM (2007) Dependence of heart rate variability on stress factors of stress response inventory. In: 9th international conference on e-health networking, application and services, pp 236–239
8. Nakamura Y, Yamanaka K, Park MK, Kawakami M (2013) Basic study for promoting driving safety support systems among elderly drivers. In: 2013 international conference on Biometrics and Kansei Engineering (ICBAKE), pp 109–112
9. Eilebrecht B, Wolter S, Lem J, Lindner H, Vogt R, Walter M, Leonhardt S (2012) The relevance of HRV parameters for driver workload detection in real world driving. Comput Cardiol (CinC) 2012:409–412
10. Rigas G, Katsis CD, Bougia P, Fotiadis DI (2008) A reasoning-based framework for car driver's stress prediction. In: 16th mediterranean conference on control and automation, pp 627–632
11. Yamaguchi M, Wakasugi J, Sakakima J (2006) Evaluation of driver stress using biomarker in motor-vehicle driving simulator. In: 28th annual international conference of the IEEE Engineering in Medicine and Biology Society. EMBS'06, pp 1834–1837
12. Yang Z, Zhang L, Wang J, Wang Y, Guan Q, Feng J (2006) Design of intelligent in-vehicle navigation systems for dynamic route guidance with real-time information. In: IEEE international conference on vehicular electronics and safety. ICVES 2006, pp 184, 188, 13–15 Dec 2006. doi:10.1109/ICVES.2006.371579
13. Sato K, Otsu H, Madokoro H, Kadowaki S (2013) Analysis of psychological stress factors by using bayesian network. In: 2013 IEEE international conference on mechatronics and automation (ICMA), pp 811, 818, 4–7 Aug 2013. doi:10.1109/ICMA.2013.6618020
14. IEEE Std 802.11p (2010) IEEE standard for information technology—local and metropolitan area networks—specific requirements—part 11: wireless lan medium access control (mac) and physical layer (phy) specifications amendment 6: wireless access in vehicular environments. IEEE Std 802.11p-2010 (Amendment to IEEE Std 802.11-2007 as amended by IEEE Std 802.11k-2008, IEEE Std 802.11r-2008, IEEE Std 802.11y-2008, IEEE Std 802.11n-2009, and IEEE Std 802.11w-2009), pp 1–51
15. Lee GY, Park HM, Cho HG, Choi SH, Park SH (2011) The implementation of the intelligent transport system for the real-time roadside environment information transfer. In: 2011 13th international conference on advanced communication technology (ICACT), pp 76–81
16. Kloiber B, Strang T, Spijker H, Heijenk G (2012) Improving information dissemination in sparse vehicular networks by adding satellite communication. In: Intelligent vehicles symposium (IV), 2012 IEEE, pp 611–617

An OpenCV Based Android Application for Drowsiness Detection on Mobile Devices

Laura Montanini, Ennio Gambi and Susanna Spinsante

Abstract Modern mobile devices, such as smartphones, are typically equipped with many different sensors able to capture biometric data, that can be used in a number of different applications and services. Further, the availability of onboard navigation capabilities makes it possible to replace with smartphones other devices, like GPS navigators, inside the vehicle, and new use cases may be evaluated and tested. In this paper, a software application for mobile devices equipped with Android O.S. is presented, as a tool for automatic driver's drowsiness detection, based on computer vision techniques implemented through the OpenCV library. Experimental results show the effectiveness of the application, despite the limited computational resources required and the varying ambient conditions.

1 Introduction

The striking diffusion of smart mobile communication devices among users, and their pervasiveness in people's daily life, open the way to the effective implementation of a number of advanced applications aimed at entertainment, productivity, but also safety and health monitoring. Modern mobile devices are typically equipped with many different sensors able to capture biometric data, that can be used in a number of different applications and services [1, 2].

Due to bad drivers' habits, the use of mobile devices when driving is mentioned as one of the main causes for car accidents and crashes [3–5]. However, thanks to the great advances in the technology embedded in mobile devices, they can provide many useful functions for drivers, such as maps, traffic warnings (through proper applications), and assisted navigation. In this paper, we propose to use a mobile device (e.g. a smartphone) as a tool for automatic driver's drowsiness detection, aimed at monitoring the driver's fatigue to prevent possible accidents. As a matter

L. Montanini (✉) · E. Gambi · S. Spinsante
Dipartimento di Ingegneria dell'Informazione, Universita' Politecnica delle Marche, via Brecce Bianche 12, 60131 Ancona, Italy
e-mail: laura.montanini@univpm.it

M. Conti et al. (eds.), *Mobile Networks for Biometric Data Analysis*,
Lecture Notes in Electrical Engineering 392, DOI 10.1007/978-3-319-39700-9_12

of fact, embedded tools or sensors for automatic fatigue detection, and driver's assistance [6, 7], are usually available only in top-class vehicles. Moreover, most of the existing literature on automatic drowsiness detection relies on complex algorithms, such as power spectrum analysis of EEG signals [8], or Hidden Markov Model (HMM) based dynamic modeling [9], or require quite uncomfortable interfaces the driver has to interact with, such as brain-computer interfaces [10], or wearable devices [11]. The CarSafe application presented by You et al. [12] is an interesting, though quite complex application, that tries to exploit both the front and rear cameras of a smartphone, to monitor the driver behavior and to track the road conditions, at the same time. Hardware limitations make it not possible to have a simultaneous use of both the cameras, and the switching time requested to move control from a sensor to the other one has to be accurately estimated to limit so called blind spots, and ensure a timely response to possibly dangerous conditions. The review by Sahayadhas et al. [13], shows that researchers have attempted to determine the driver drowsiness using vehicle-based measures, behavioral, and physiological measures, to afford the complexity of the problem. By designing a hybrid drowsiness detection system that combines non-intrusive physiological measures with other measures, it would be possible to accurately determine the drowsiness level of a driver. The software application proposed in this paper may represent one of the tools used in a hybrid system, to detect alarming behaviors shown by the driver, and possibly to generate feedback.

This paper focuses on the possibility to implement a low-complexity algorithm that, by exploiting computer vision techniques and the video frames captured by a camera sensor on board a smartphone, is able to automatically detect potential drowsiness by a subject, and possibly to generate acoustic alarms. Detection of driver's fatigue involves processing a sequence of images of a face, to detect the eyes; once the position of the eyes is located, the system is designed to determine whether the eyes are opened or closed, and detect fatigue. The algorithm is implemented as a software application running on Android O.S. based mobile devices, and builds upon the OpenCV software library [14]. The paper is organized as follows: Sect. 2 briefly discusses the adoption of Computer Vision (CV) techniques in mobile devices; Sect. 3 presents some basic algorithms and operations designed for face and eyes detection, whereas Sect. 4 focuses on the algorithm for eyes state analysis. The software implementation of the application is discussed in Sect. 5, together with related experimental tests; finally, Sect. 6 draws the main conclusion of the work.

2 Computer Vision in Mobile Devices

The drowsiness detection algorithm presented in this paper exploits CV techniques for face and eyes detection, and state of the eyes recognition. The main purpose of CV is to replicate the human vision, by acquiring a photograph of an area and interpreting its contents. To this aim, the image acquisition and interpretation should be performed in real time in order to simulate as much as possible the human sight. Unfortunately, CV algorithms are usually very complex and require high computational capability: this is a strong challenge, especially when dealing with mobile devices, characterized by power constraints and sometimes limited computational resources. However, the significant development of mobile technology in recent years has led to a rapid growth of CV applications on mobile devices, such as image editing, augmented reality, object recognition and similar [15–17]. For example, one of the most common operating system for mobile devices, the Android O.S., starting from version 4.0 provides specific classes for face and features detection within the frame captured by the camera sensor. These functionalities allow the automatic localization of the eyes and the central point of the mouth within the image [18]. In the proposed project, it is necessary to recognize not only the position of the subject's eye, but also if it is open or closed. To this aim we do not exploit the native O.S. functionalities, but we use the OpenCV open source software library. OpenCV is released under a BSD license, and focused on computer vision. It has C++, C, Python and Java (Android) interfaces and supports Windows, Linux, Android, iOS and Mac operating systems. The purpose of the OpenCV library is to collect the more widely used functionalities and algorithms in the CV field. The library includes a wide range of functions, which cover the most different needs: image processing, optimized math functions, and a complete package of matrix algebra, developed to support the other CV primitives.

3 Algorithm for Face and Eyes Detection

In this section we discuss the theoretical basis of an algorithm developed for the face and eyes detection from an RGB frame, and selected for our software implementation. The algorithm, designed by Viola and Jones [19], and subsequently refined by Lienhart and Maydt [20], takes advantage of feature-based techniques, built upon the concept of *feature*, that are faster than techniques based on skin color model.

First, from the greyscale image, an intermediate image called *integral image* is calculated, through some simple steps. The integral image is obtained by applying the following formula:

$$ii(x, y) = \sum_{x' \leq x, y' \leq y} i(x', y') \qquad (1)$$

where x and y represent the coordinates of the block $ii(.)$ in the integral image, and $i(.)$ represents a pixel in the grayscale image. In other words, each block in the integral image is calculated from the grayscale image, by addition of all the intensity values whose indexes are lower than the indexes of the considered block.

Once the integral image is computed, the algorithm proceeds with feature extraction. The features, called Haar-like features, allow to determine if a face is present or not in a frame. Figure 1 pictorially shows the set of Haar-like features

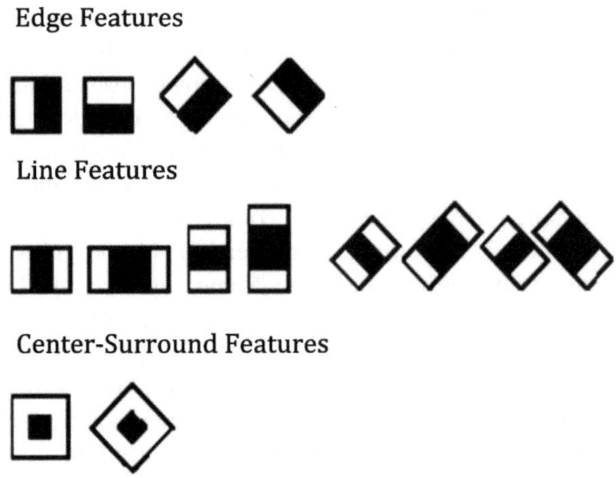

Fig. 1 Sample features used by the *Haar* boosted tree classifier

used. Although it is not immediately obvious, they represent the differences of intensity between two or more adjacent rectangular areas in the image. The value of the feature is the result of summing all pixel intensities both in the white and the black side of the rectangle, and then computing their difference. Considering all the possible dimensions and positions of each feature, the total number of features potentially needed grows very quickly. However, among all the possible characteristics, most of them are irrelevant. For example, consider Fig. 2. The upper row shows two good features: the former selected feature reveals that the region of the eyes is darker than the region of the nose and cheeks. The latter reveals that the eyes are darker than area of the nose. However, the same masks applied on the image area of the cheeks, or anywhere else, are irrelevant. Viola and Jones presented the so-called *adaboost* algorithm. It allows to select, through a learning process, among

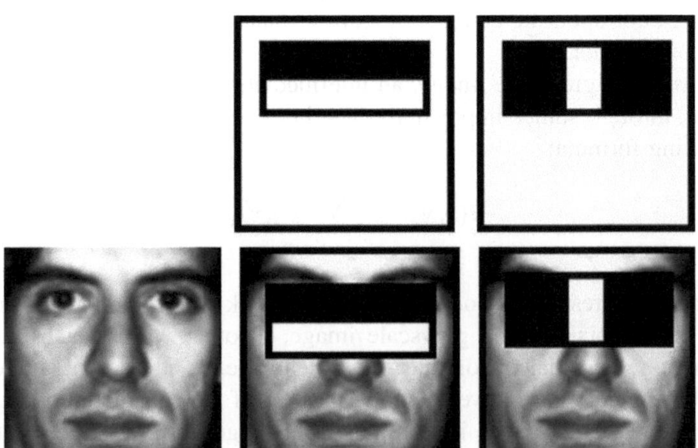

Fig. 2 Relevant features for face detection

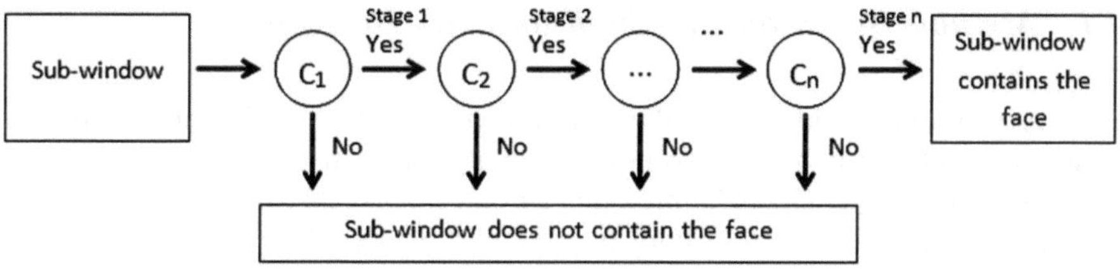

Fig. 3 Example of classifier cascade

all the features extracted, only the ones providing the strongest ability to discriminate between two classes. From a practical point of view, a search window of a given size is moved over the image, looking for regions of the image that have a similarity degree with a face object. The classifiers used are instructed by selecting specific features that are used to discriminate the target objects inside the video frames. The complete algorithm provides a cascade of weighed classifiers, in order to make the procedure the most selective as possible (see Fig. 3). In fact, a classifier is extremely fast in discarding unpromising windows, so the algorithm can quickly determine if a region does not contain the desired object. When the decision about the presence or not of a face in a given region of the image is not clear, the information is passed from a classifier to the next one. At the end of the cascade, if the final classifier has not rejected the selected region, the algorithm concludes that it contains a face.

4 Algorithm for Eyes State Analysis

Since the Viola-Jones algorithm allows to locate the face and eyes within a video frame, the next step is to identify if the eyes are open or closed. The recognition of the state of the eye is one of the recurring problems faced in CV. To solve this problem there are several ways. At the design stage, different procedures have been evaluated:

- Locate the pupil and/or iris within the region of interest: if the pupil is detected, the eye can be assumed open, otherwise it is closed. This method exploits the *Hough transform* [21] which allows to detect circles within an image.
- Observe the distribution of darkest points in the restricted sub-image around the eyes. When the eye is open, dark spots are distributed mainly around the iris, and concentrated in the central part of the eye. But when the eye is closed, the darker pixels are those represented by the cilia, assuming an elongated shape, easily identifiable. This is the method effectively used in our application.

Some preliminary tests have been performed in order to evaluate the reliability of the selected strategy in recognizing the status of the eyes.

4.1 Method Based on the Pupil and Iris Detection

The first solution considered has the aim of identifying the pupil. Since the pupil is perfectly circular, it is possible to develop an algorithm that searches for circles in the region of the eye. OpenCV provides an object class that implements methods for the geometric figures detection; in particular, it uses the *Hough transform* to detect circles. The detection algorithm is based on the idea that each point belonging to the image gives a contribution, or *vote*, in the search for a solution in a different space respect to the image, called the *parameter space*. Points with the highest number of votes located in the parameter space are shortlisted to belong to the center of the circle.

In order to evaluate the effectiveness of this approach, preliminary tests carried out using commercial mobile devices have been performed, showing a negative result, that fails to recognize the pupil or the iris. The problem is due to the poor resolution of the front camera the devices are equipped with: the image of the eye is grainy and the algorithm cannot find the circles.

4.2 Methods Based on the Distribution of Dark Pixels

An alternative solution consists in evaluating the different distribution of dark pixels in the frame, for both open and closed eye. Let us consider the grayscale eye region detected by the Viola-Jones algorithm. In Fig. 4 it is possible to see the eye image, in which 300 darker points were colored white, to provide visual feedback. As clearly visible, when the eyes are open, dark pixels are distributed approximately in the center of the eye, while when the eyes are closed, pixels assume a well-defined elongated form, due to the eyelashes. Therefore, a method to extract a comparable characteristic must be identified. The first idea was to calculate the diagonal; it starts from the first lower left white pixel, to the last white pixel in the top right position. The diagonal so defined should be longer when the eye is closed. However, the distribution of dark pixels sometimes is not uniform (see for example Fig. 5), and a few pixels away from the others may adversely affect the calculation of the diagonal.

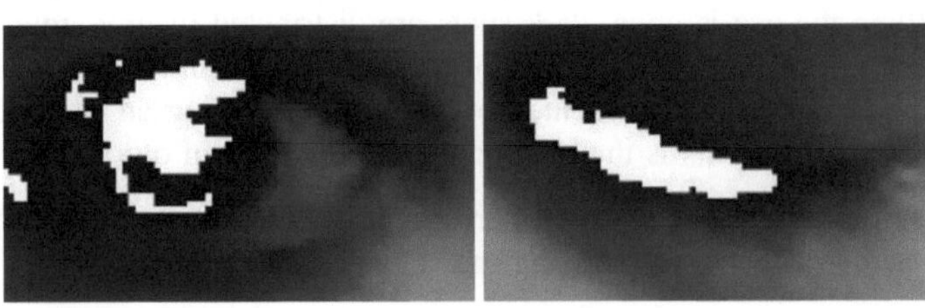

Fig. 4 Sub-images representing the eyes in which the 300 darker pixels have been colored *white*

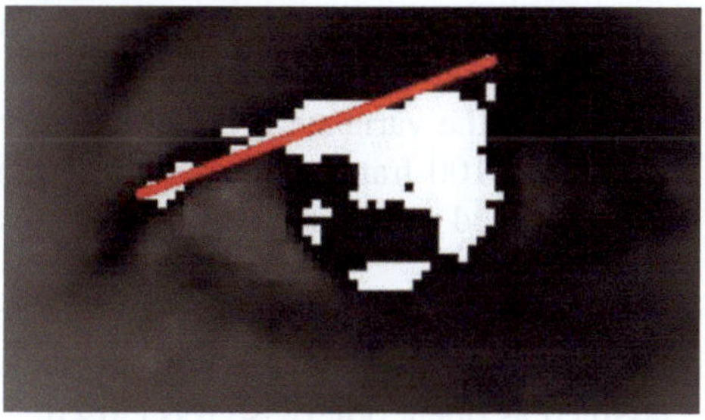

Fig. 5 Open eye with non-uniform distribution of dark pixels

To eliminate this problem, a median filter was applied. The main idea of the median filter is to replace each input pixel by the median of the neighboring entries. Taking a square window of predefined size that slides over the whole image, values within the window are sorted in ascending order, and the median value is assigned to the midpoint of this window. Due to the median filter, the algorithm based on the diagonals is more robust (see Figs. 6). The results, however, are not satisfactory, because when the eyelashes are particularly long or thick, the median filter is unable to remove all the white pixels away from the center.

An alternative way to discriminate the state of the eye consists in counting, for each column of the matrix representing the image, the number of adjacent white pixels (see Fig. 7). These values are defined as *lengths*. As discussed earlier, when the eye is closed the shape of the white spot is elongated and compact. On the

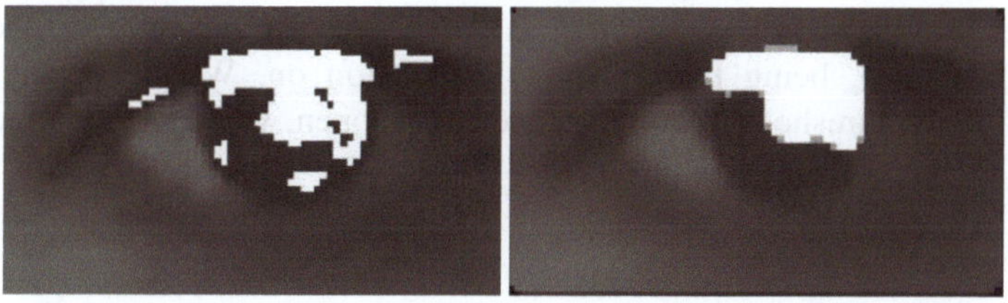

Fig. 6 Comparison of two eye images with and without the application of the median filter

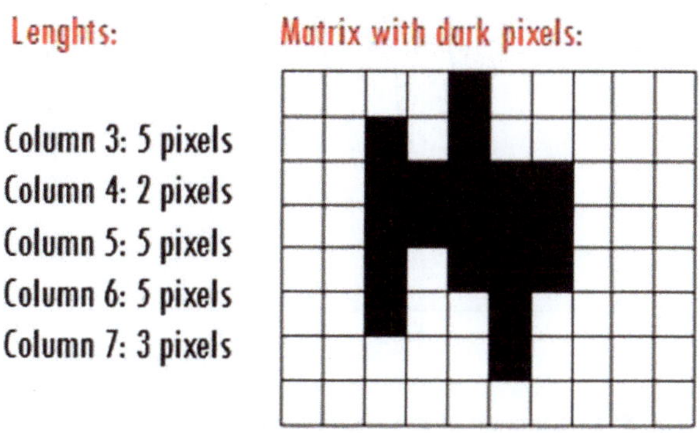

Fig. 7 Example of calculation of the *lengths* of a matrix

contrary, in case of open eye, the stain has an irregular and less compact shape. Therefore, when the eye is closed, *lengths* have a small variance, while in the case of open eye the variance is greater. Figure 8 shows the distribution of the variance values over 100 frames captured from a mobile device, representing the sequence open-to-closed eye. It is clearly shown how the variance is reduced when the eye is closed.

The next step is to define a threshold for the value of the variance which allows to discriminate the state of the eye: if the variance exceeds the threshold, the eye is classified as open, otherwise it is closed. The definition of this threshold is a critical stage in the algorithm development. The variance depends on many factors, such as the length of the eyelashes, the size of the matrix, and the environmental brightness. The most reasonable solution consists in developing a learning algorithm able to define an adaptive threshold. To this aim, a preliminary calibration phase is introduced. The adaptive threshold is obtained using the following formula:

$$threshold = \left(\frac{max(v_{left}) + min(v_{left})}{3} + \frac{max(v_{right}) + min(v_{right})}{3}\right) * \frac{1}{2} \qquad (2)$$

where v_{right} and v_{left} are arrays representing, respectively, the variance values of the right and left eye, calculated over the last 25 frames. The value of the divider within the parentheses has been chosen empirically: a threshold defined as a precise arithmetic mean between the minimum and maximum of the two arrays provides worse results. After the calibration step, the formula is applied frame by frame, so the threshold value varies depending on the environmental context.

Figure 9 shows how the *length* metric variance and the adaptive threshold values change over time, being the detecting application on. When the variance gets greater than the threshold, the eye is detected as open, whereas when the variance

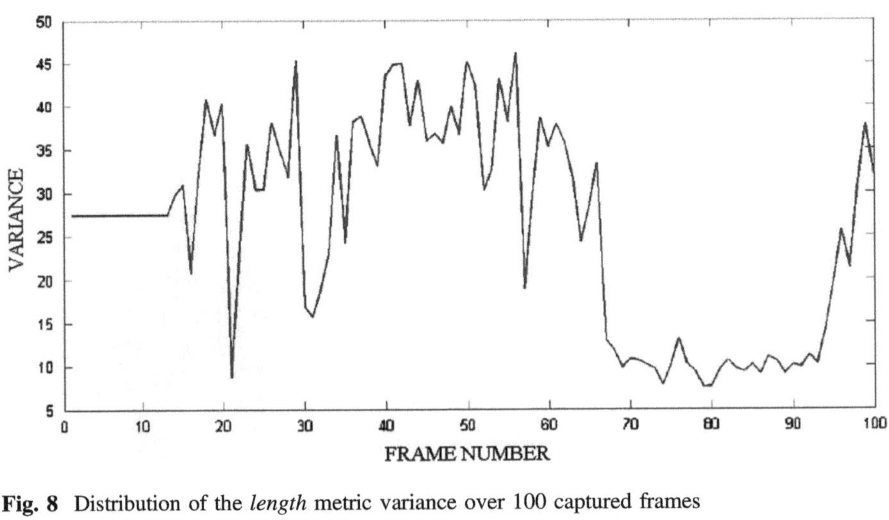

Fig. 8 Distribution of the *length* metric variance over 100 captured frames

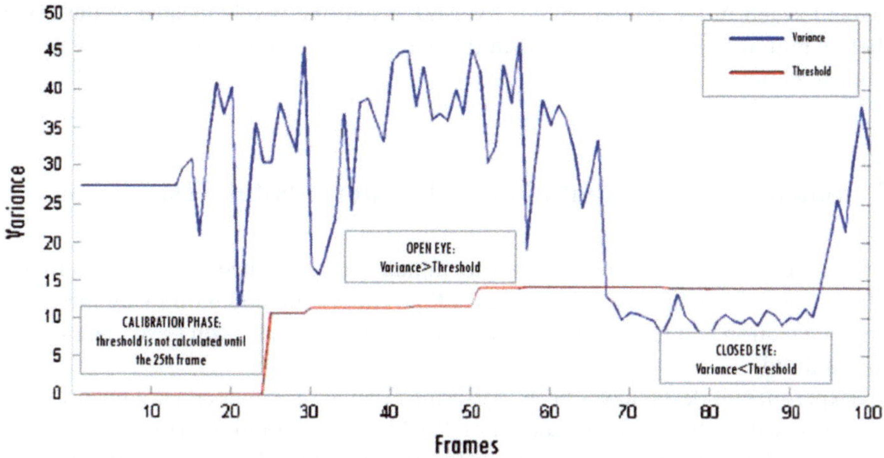

Fig. 9 Distribution of the *length* metric variance and adaptive threshold value, over 100 captured frames

becomes smaller than the threshold, the eye is classified as closed. A first problem with the method described above occurs when the gaze is not directed towards the camera sensor. In that case, the metric variance remains as low as it is when the eye is closed: the algorithm detects a false state of the eye (see Fig. 10).

To solve this problem a further decision-making process is introduced. The region of the eye is divided into 6 areas, as illustrated in Fig. 11. Once the variance is calculated, the algorithm checks in which area the majority of the 300 dark pixels are located. If the variance is below the established threshold, the percentage of

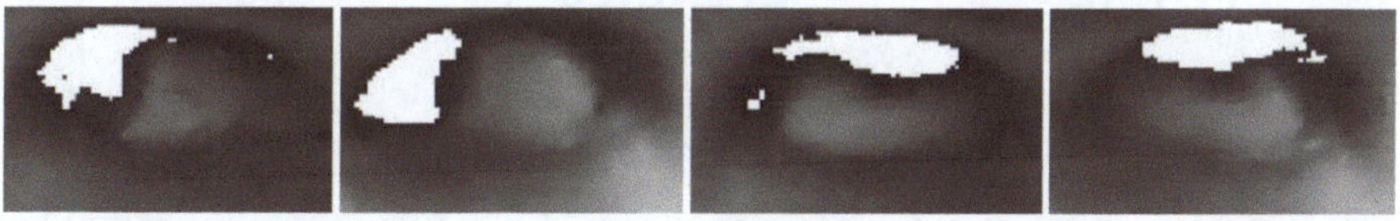

Fig. 10 Different positions of the eye producing a low variance and a false detection of closed eye

Fig. 11 Detail of the eye image subdivided into 6 areas

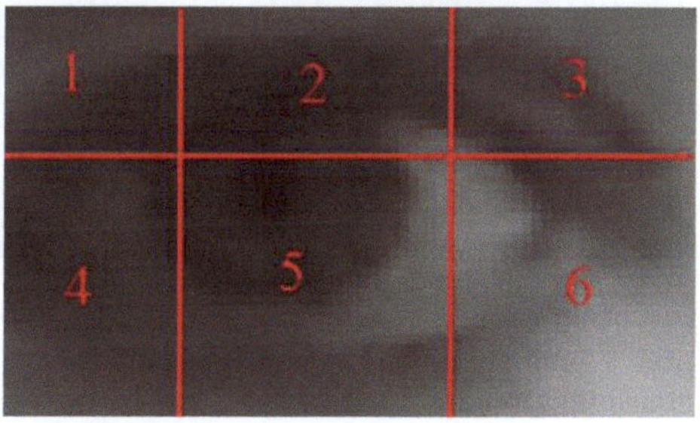

pixels located in the bottom area is over 60 % and the pixels located in the center (below region 5) exceeds 40 %, then the eye is classified as closed, otherwise it is labeled as open.

5 Software Implementation and Experimental Tests

The algorithm discussed in Sect. 4 was initially implemented as an offline tool in Matlab, in order to test the effectiveness of the chosen method. Following the offline implementation and test of the algorithm, a new version has been developed in Java for Android.

Experimental tests have been performed using a commercial tablet device, equipped with a front-facing camera of 1.3 Mpixels resolution. Tests were carried out in a laboratory environment in order to verify the effectiveness of the application, under different light conditions. Positioning the device 50 cm far from the face, the tests have been performed under different light conditions (direct light from a lamp, diffuse artificial light, light off), as shown in Fig. 12. In all the

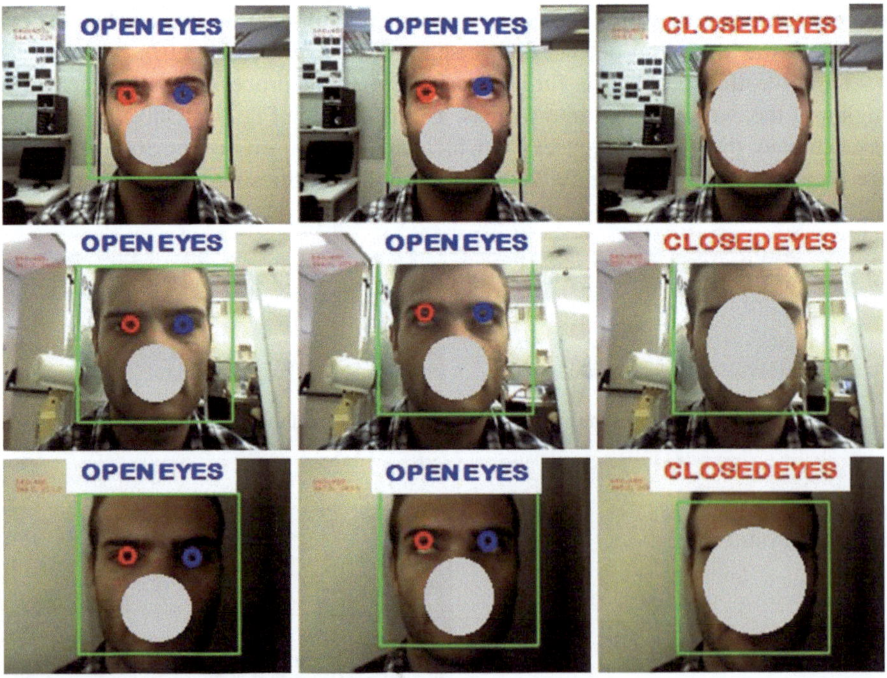

Fig. 12 Tests performed in different light conditions (respectively from *top* to *bottom*: direct light from a lamp, diffused light from the ceiling neon, light off) on (from *left* to *right*) open eyes, eyes turned upward, and closed eyes. Some parts of the face have been masked for privacy reason

different light conditions, the system has been able to successfully recognize the status of the eyes.

Since the application is designed to be used in an automotive context, characterized by a rapid variability of ambient light conditions, due, for example, to the headlights of cars proceedings in opposite direction, or to the presence of street lamps, and since the device may experience vibrations due to road bumps, additional tests have been performed to better simulate a realistic working condition. The system capability in correctly recognizing the state of the eye is closely related to the capability of the OpenCv library in detecting the eye position within the video frame. In darkness or semi-darkness conditions, the library, in fact, is not able to identify the face or it can identify only some characteristics, such as one eye only, as depicted in Fig. 13. The presence of a light source may solve the problem, although it is not very practical in a realistic context. However, in low or highly non-uniform brightness conditions, despite some rare misinterpretations, the system continues to function properly, as shown in Fig. 14.

A further test has evaluated the performance of the system when both the device and the user are moving. Such a situation is characterized by vibrations, that may cause a misalignment between the camera sensor and the user's face, and rapid variations in brightness. The first expected effect of such a condition is the acquisition of frames not in focus; in such frames it is not easy to detect and recognize the state of the eye. Within the video stream, however, the failure in recognizing some single frames does not interfere significantly with the overall functioning of the application. The vibrations and sudden movements seem therefore not significantly affect the system performance. Conversely, at every sudden change in brightness, the system can make mistakes, and starts over to work correctly only after a while. This can be explained by recalling that, as stated in the previous section, it was decided to use an adaptive threshold to discriminate the state of the eye: after 25 frames, the threshold will be updated by taking into account the new lighting conditions.

Moving away the device from the user's face, the algorithm stops working. This limitation is due to the low resolution of the front camera, which does not allow to

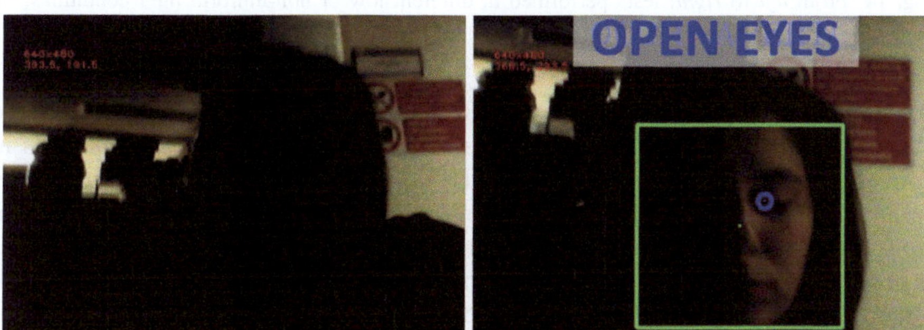

Fig. 13 From *left* to *right:* test performed in darkness and semi-darkness conditions

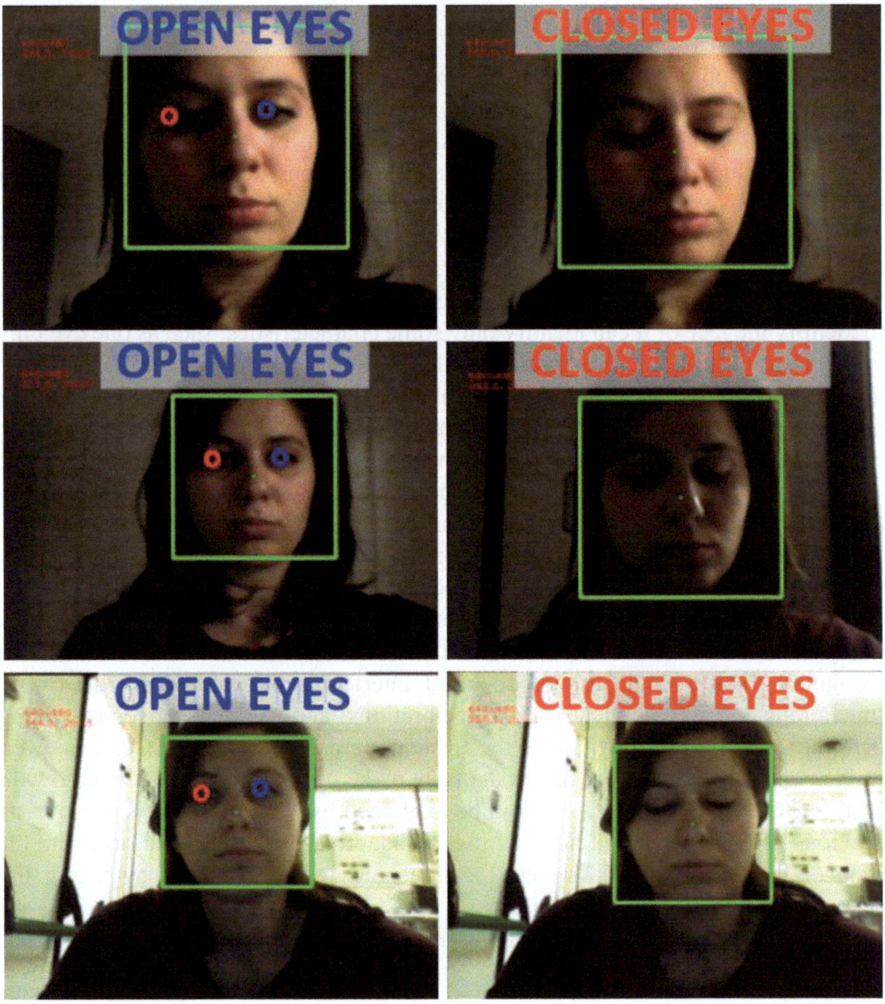

Fig. 14 From *left* to *right*: tests performed in different low or non-uniform light conditions, on open eyes and closed eyes

acquire sufficiently detailed frames, at greater distances. In fact, the region of the eyes extracted from these frames is considerably dark, thus the algorithm fails in correctly inferring the state of the eyes. However, the distance limit is acceptable, assuming that inside a car the device can be located at a distance of about 50 cm from the driver's face, as for a sat-nav device bracket located on the windshield. It is possible to improve the application performance by using a device equipped with a more powerful camera. During the execution of the algorithm, the frame rate is about 2.5 fps, indicating the computational complexity is not too high, and allowing the mobile device to run the software in real time.

Further test sessions are planned to be performed in order to assess the robustness of the application, when users with diverse facial features are captured by the camera, possibly wearing glasses or a beard.

6 Conclusion

A software application for mobile Android based devices has been presented that, exploiting computer vision techniques, is able to automatically monitor face and eyes conditions, to detect potential drowsiness events. Preliminary experimental tests provided promising results, that could be further improved by the application of sensor fusion techniques monitoring different parameters, such as the driver's heart rate, to increase the detection rate and reliability of the proposed solution.

References

1. Xing S, Hanghang T, Ping J (2014) Activity recognition with smartphone sensors. Tsinghua Science and Technology 19(3):235–249
2. Nickel C, Wirtl T, Busch C (2013) Authentication of smartphone users based on the way they walk using k-NN algorithm. In: 2012 International conference on intelligent information hiding and multimedia signal processing (IIH-MSP), 18–20 Jul 2012, pp 16–20
3. Center for Accident Research & Road Safety—Queensland (2015) Mobile phone use and distraction while driving, September 2015, http://www.carrsq.qut.edu.au/publications/corporate/mobile_phones_and_distraction_fs.pdf
4. Distracted Driving: Facts And Statistics, http://www.distraction.gov/stats-research-laws/facts-and-statistics.html
5. World Health Organization (2011) Mobile phone use: a growing problem of driver distraction. http://www.who.int/violence_injury_prevention/publications/road_traffic/distracted_driving_en.pdf?ua=1
6. Polidori L, Gambi E, Spinsante S (2008) Proposal of a driver assistance system based on video and radar data fusion. In: Proceedings of 16th international conference on software, telecommunications and computer networks (SoftCOM 2008), Split, pp 300–304
7. You CW et al (2013) Carsafe app: alerting drowsy and distracted drivers using dual cameras on smartphones. In: Proceeding of the 11th annual international conference on mobile systems, applications, and services. ACM
8. Kim D, Han H, Cho S, Chong U (2012) Detection of drowsiness with eyes open using EEG-based power spectrum analysis. In 2012 7th International forum on strategic technology (IFOST), 18–21 Sept 2012, pp 1–4
9. Tadesse E, Weihua S, Meiqin L (2014) Driver drowsiness detection through HMM based dynamic modeling. In: 2014 IEEE International conference on robotics and automation (ICRA), 31 May 2014–7 June 2014, pp 4003–4008
10. Chin-Teng L, Che-Jui C, Bor-Shyh L, Shao-Hang H, Chih-Feng C, Wang IJ (2010) A real-time wireless brain-computer interface system for drowsiness detection. IEEE Trans Biomed Circuits Syst 4(4):214–222
11. Kurian D, Johnson JPL, Radhakrishnan K, Balakrishnan AA (2014) Drowsiness detection using photoplethysmography signal. In: 2014 Fourth international conference on advances in computing and communications, 27–29 Aug 2014, pp 73–76

12. You C-W, Lane ND, Chen F, Wang R, Chen Z, Bao TJ, Montesde-Oca M, Cheng Y, Lin M, Torresani L, Campbell AT (2013) CarSafe app: alerting drowsy and distracted drivers using dual cameras on smartphones. In: The 11th international conference on mobile systems, applications, and services, 25–28 June 2013
13. Sahayadhas A, Sundaraj K, Murugappan M (2012) Detecting driver drowsiness based on sensors: a review. Sensors 12:16937–16953
14. OpenCV. http://opencv.org/
15. Pulli K, Baksheev A, Kornyakov K, Eruhimov V (2012) Real-time computer vision with OpenCV. Commun ACM 55(6):61–69
16. Yang X, Cheng KT (2012) LDB: an ultra-fast feature for scalable augmented reality on mobile devices. In: IEEE international symposium on mixed and augmented reality (ISMAR), pp 49–57
17. Spinsante S, Gambi E (2012) Home automation systems control by head tracking in AAL applications. In: Proceedings of 2012 IEEE first AESS European conference on satellite telecommunications (ESTEL), Rome (IT), pp 1–6
18. Montanini L, Cippitelli E, Gambi E, Spinsante S (2014) Real time message composition through head movements on portable android devices. In: Proceedings IEEE international conference on consumer electronics (ICCE), Las Vegas, 10–13 Jan 2014
19. Viola P, Jones MJ (2004) Robust real-time face detection. Int J Comput Vision 57(2):137–154
20. Lienhart R, Maydt J (2002) An extended set of Haar-like features for rapid object detection. In: IEEE ICIP 2002, vol 1, Sep 2002, pp 900–903
21. Ballard DH (1981) Generalizing the Hough transform to detect arbitrary shapes. Pattern Recogn 13(2), 111–122

Distributed Speech and Speaker Identification System for Personalized Domotic Control

Giorgio Biagetti, Paolo Crippa, Laura Falaschetti, Simone Orcioni and Claudio Turchetti

Abstract This paper presents a combined speech recognition/speaker identification system that can be efficiently used for personalized domotic control. The proposed system works as a distributed framework and it is designed to identify a speaker in home environments in order to provide user access to customized options. Human speech signals contain both language and speaker dependent information. Using this information the system realizes a personalized control in home environments and this approach can also be applied in more generic scenarios such as car customization settings. The system was optimized with the aim to allow an immediate use only with the addition of small and cheap audio front-ends that will capture commands spoken by the user. Meanwhile a remote server performs the speech recognition as well as user identification and combines these informations to provides user specific settings which are sent back to the desired actuator at home.

1 Introduction

Biometrics refers to the automatic identification of a person based on some of his/her unique physiological or behavioural characteristics.

G. Biagetti (✉) · P. Crippa · L. Falaschetti · S. Orcioni · C. Turchetti
Dipartimento di Ingegneria dell'Informazione, Università Politecnica delle Marche, via Brecce Bianche, 12, 60131 Ancona, Italy
e-mail: g.biagetti@univpm.it

P. Crippa
e-mail: p.crippa@univpm.it

L. Falaschetti
e-mail: l.falaschetti@univpm.it

S. Orcioni
e-mail: s.orcioni@univpm.it

C. Turchetti
e-mail: c.turchetti@univpm.it

M. Conti et al. (eds.), *Mobile Networks for Biometric Data Analysis*,
Lecture Notes in Electrical Engineering 392, DOI 10.1007/978-3-319-39700-9_13

In biometric person identification systems, speaker identification can be considered as the less invasive method as the voice is the more natural signal to produce and the simplest to acquire [4, 17], so biometric systems based on human speech are increasingly being used as a means for the recognition of people.

Speaker recognition is the process of automatically recognizing who is speaking on the basis of individual information included in speech waves and can be classified into two fundamental modes of operations: identification and verification [18, 21, 23].

Speaker identification is the process of determining which speaker from a given pool provides a given utterance. Speaker identification is a one to N (many) match where the voice is compared against N templates [8, 13, 22].

Speaker verification, on the other hand, is a one to one match where one speaker's voice is matched to one template [7].

Both tasks fall into the general problem of statistical pattern recognition, in which a given pattern is to be assigned to one of a set of different categories [16].

The system we propose in this paper is classified as *text-independent speaker identification system* since its task is to identify the person who speaks regardless of what is saying. The system uses fruitfully both speaker identification and speech recognition.

Speech recognition is the process of identifying what words have been spoken, not identifying who spoke them. Speech recognition can be used in combination with speaker identification to make it more powerful.

This paper aims at describing a distributed speech recognition/speaker identification system for user contextualization, that is a combination of two parallel tasks: a speech recognition [5] and a speaker identification [6] task, that work together in a distributed system to determine the commands spoken and who spoke them to perform a personal settings customization in user's home. This paper focus on the accuracy of a text independent speech recognition/speaker identification system using Mel frequency cepstral coefficients (MFCCs) and Gaussian mixture model (GMM) together with Figueiredo's [12] and expectation maximization (EM) algorithms. The goal of speaker identification is to determine which one of a group of known speakers best matches the input voice samples: voice samples were taken, MFCCs were extracted, and these coefficients were statistically analyzed by GMM in order to build each profile. The purpose of the speech recognition is to identify the utterance spoken by the user, discarding those that do not belong to the acceptable grammar.

2 Speech Recognition

The main issues in *speech recognition* systems are related to: activation, speech capture, elimination of spoken text unrelated to the commands, user feedback, interpretation and execution of commands, and continuous listening.

On the one hand capturing the speech signal is difficult due to the fact that the microphone is always active, and it is therefore necessary to distinguish the voice from the noise, recognize *distant speech*, and detect pronunciation errors.

On the other hand the interpretation of commands is also difficult due to the fact that the user may not adhere perfectly to the used grammar and due to the risk of confusing conversation with normal controls.

To solve these problems, the proposed system makes use of two reasonably simple techniques: (i) an adaptive automatic voice activity detection (VAD) thresholding technique that enables the distributed recognition system (DSR) only when is needed, and (ii) a mechanism that can easily be adapted to each particular application and which is able to effectively reject non-command utterances, background noise, and unwanted sounds by exploiting the generation of a *garbage model* [2, 14], which includes suitably placed decoy words [19] that will help the automatic speech recognition (ASR) identify out-of-vocabulary words, thus enabling it to discard non-command utterances.

3 Speaker Identification

The main operational tasks to be carried out in the *speaker identification* process are the feature extraction, the speaker modeling and the speaker classification.

Feature extraction was performed by the recognition system front-end and captures the speaker-specific characteristics, reducing the model complexity as well. The usually adopted model for speech parametrization is the source-filter model which leads to the extraction of parameters such as linear predictive coding, MFCCs, perceptual linear prediction coefficients, etc. [9, 10, 20]. Among these, MFCCs have demonstrated to be the most successful due to their particular robustness to the environment and flexibility [20]. Here we use MFCCs extracted from the spectrum. This is because in our application speaker identification is combined to further speech processing, speech recognition, to identify what is being said. So, to limit computation in a possible application, it makes sense to use the same features for speech and speaker identification.

Since an utterance from a speaker consists of a random sequence of frames, the speaker modeling has been realized using a statistical model, such as the GMM.

As far as the classification of voice samples is concerned, the optimal Bayesian classifier guarantees a minimum classification error by identifying the speaker model which exhibits the maximum GMM a posteriori probability [15].

4 System Architecture

A distributed recognition system is a system with a client-server architecture. The fundamental idea of such a system is to distribute the voice recognition processing between a client front-end (FE), where the speech features are extracted, and a remote back-end (BE), where the compressed stream is received by a server that finalizes the speech recognition/speaker identification process.

4.1 Front-End

The feature extraction algorithm adopted in this voice control framework for both speech recognition and speaker identification is based on the standard ETSI ES 201-212 [11], specifically defined for DSR. The standard specification describes the FE algorithm for the computation of the MFCCs, from speech waveforms sampled at 8 kHz rate. The feature vectors consist of 12 cepstral coefficients and a log-energy coefficient. In addition 26 dynamic features, the delta and delta-delta cepstral coefficients, are computed at the server-side, so that the final acoustic vector has 39 components. The MFCCs coefficients are extracted from frames of 25 ms generated every 10 ms, thus two consecutive 25 ms frames overlap by 15 ms.

Figure 1 shows a block diagram of the standard ETSI MFCC features extraction.

At the input of the processing chain the VAD block drops all non speech segments from the input audio records, exploiting the energy acceleration associated with voice onset. Hence buffering is required for storing overlapping regions among frames. Before computing the cepstral coefficients, a noise reduction is performed and subsequently a waveform processing is applied to the clear signal. Finally the blind equalization of the resulting features represents the last processing stage in the FE terminal.

4.2 Back-End

The recognition process is performed at the server-side by the BE which acts on the features stream for speech recognition and speaker identification.

The speech recognition process is performed by making extensive use of libraries of the Carnegie Mellon University Sphinx framework [24] and consists of the following steps: (i) identifying the segments that potentially contain commands, processing them to reduce noise, reverberation end extraneous components, and finally compressing them; (ii) processing the segments previously identified in order to extract the command phrases that may be present; (iii) interpreting the command phrases and executing the corresponding operations.

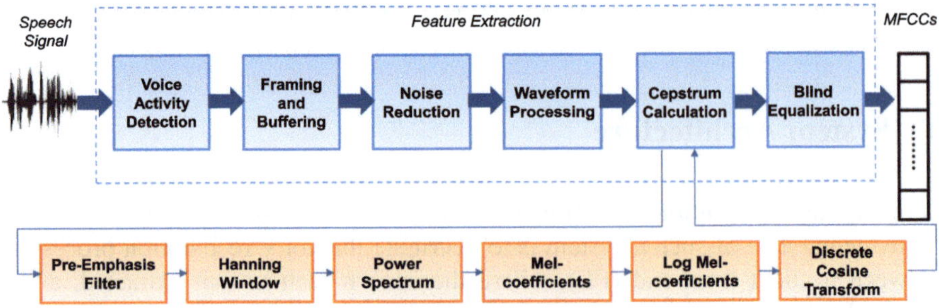

Fig. 1 Feature extraction scheme according to ETSI standard

At the end of the speech recognition process the BE returns to the FE the text corresponding to the spoken command, if this has been recognized as correct and belonging to the allowed grammar.

The speaker identification process is performed in two different phases namely: training and testing. In the training phase, each registered speaker has to provide samples of their speech so that the system can build or train a reference model for that speaker. In the testing phase, the input speech is matched with stored reference model(s) and a recognition decision is made. The training phase is performed only once at the system initialization (and eventually will be updated), while the testing phase is performed on BE at every spoken command and it is based on MFCCs and GMM. Then the model parameters are estimated on the basis of the maximum similarity score making use of the EM algorithm. A GMM can be considered as a parametric probability density function represented as a weighted sum of Gaussian component densities. GMMs are commonly used as a parametric model of the probability distribution of continuous measurements or features in biometric systems, such as the vocal-tract related spectral features in a speaker identification system. GMM parameters are estimated from data using the iterative improved version of the EM algorithm proposed by Figueiredo et al. in [12]. The Figueiredo's algorithm tries to overcome some major weaknesses of the standard EM algorithm. The standard EM algorithm requires to set the number of components and the number will be fixed during the estimation process. The Figueiredo's algorithm adjusts the number of components during estimation by annihilating the components that are not supported by the data. This leads to the other EM failure point: the boundary of the parameter space. The Figueiredo's algorithm avoids the boundary when it annihilates components that are becoming singular and also allows to start with an arbitrarily large number of components, which tackles the initialization issue with the standard EM algorithm. The initial guesses for component means can be distributed into the whole space occupied by training samples, even setting one component for every single training sample [12].

At the end of the speaker identification process, the BE returns both the text corresponding to the spoken command (in case it was recognized as an allowed command) and a tag specifying the speaker who spoke it. A device installed at the user side, typically the FE itself, decides the action that should be carried out. In particular, the command-speaker link allows the command to be executed with pre-loaded user settings so as to realize a customization of the home environment control.

Figure 2 shows the overall system workflow.

4.3 Communication Scheme

The system comprises a custom set of API that interacts with a recognizer in response to input from the network in HTTP protocol. The program is a Common Gateway Interface (CGI), currently used with the Apache web server. It provides a

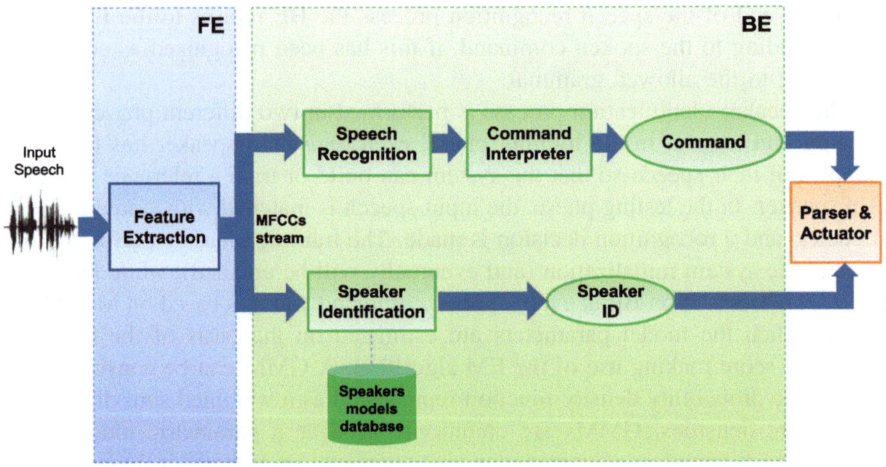

Fig. 2 Overall system workflow

data management via MySQL server, a multi-user access automatically handled by Apache, an authentication challenge response mechanism for user granting access, and on-demand recognition. More in detail, it provides this tools by the rigid definition of low level HTTP API, that is the basis of high level API implemented in various languages to provide more structured methods.

Table 1 summarizes the implemented low level API.

Figure 3 reports a schematization of the dialogue between the various entities that comprise the system. As it can be seen, data are exchanged between the audio FE and the recognition BE. After a command is spotted and recognised, another exchange takes place between the FE and the actuator. The upper portions of the exchanges deal with authentication and are performed only once at the system startup.

Table 1 HTTP VoiceService API

API function	Method	Description
Login	POST	Obtains an access token (one-time password) for authenticated requests
Logout	POST	Invalidates the current session on the server
Echo	POST	Tests the connection with the server
Batch adaptation	POST	Sends a file to the server in tar or compressed tar format
Recognize frames	POST with query string	Sends to the server a stream of frames for speech recognition/speaker identification
Recognize utterance	POST with query string	Sends the server a whole utterance for recognition
Get model info	GET	Obtains useful information on the acoustic model used

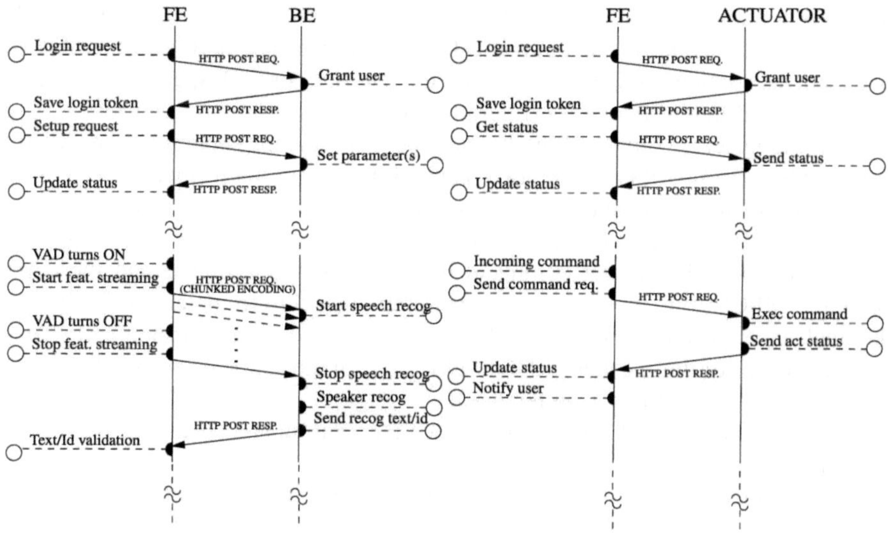

Fig. 3 DSR-dialogue

5 Results

The experiments were carried out using speech data produced by six different Italian speakers, four males (A, B, C, D) and two females (E, F), and setting-up a simple system composed with an inexpensive and power-efficient Raspberry Pi board equipped with a standard USB microphone. To test our system, the microphone was placed in an office room and a 30 min session with people doing their ordinary daily tasks was recorded, having asked them to speak commands from a given list. All recordings are mono, 8 kHz, 16-bit.

The speech recognition process includes also an hidden Markov model-based recogniser, Sphinx 4 [24], running on the back-end with a fine-tuned configuration optimized for the Italian language. The acoustic model was trained for a generic large-vocabulary continuous speech recognition task [1], while the language model is grammar-based with a hand-crafted grammar suitable to control, in our specific application, a lighting system. A garbage model has been implemented [2] to test the system effectiveness in spotting commands during a typical usage session: a phone loop-based generic word model [3], essentially able to capture any out-of-grammar (OOG) sequence of phones.

The speaker identification process requires a preliminary training stage. This phase needs to generate the speakers models used to match the single command with the ID of the speaker. To evaluate the performances, we used two different data sets for training the system: the first one contains records belonging to the grammar of commands (180 commands for a duration of about 10 min), the second one contains continuous speech of various content (i.e. for each speaker one chapter belonging to different books for a duration of about 20 min has been read).

The testing set is composed by the individual commands spoken by the user (60 commands). During this phase two parallel tasks are performed: one for the recognition of the command and the other for matching the person who delivered the command. The speech recognition system, through the use of the garbage model, identifies the correct command belonging to the grammar and the corresponding MFCCs are analyzed by the identifier algorithm to match the speaker. In the final stage a parsing program analyzes the command-speaker matching to select the preferences of the speaker and thus allows to run the action corresponding to the command according to user preferences.

Figure 4 and Tables 2 and 3 show speech recognition and speaker identification performances.

The parameters listed in the tables are calculated according to these formulas:

$$\begin{aligned} \text{Sensitivity} &= \frac{\text{TP}}{\text{TP} + \text{FN}}, \\ \text{Precision} &= \frac{\text{TP}}{\text{TP} + \text{FP}}, \end{aligned} \tag{1}$$

where, for the *speech recognition* task, TP are the true positives (elements that belong to the grammar and are recognized as belonging to it), FN the false negatives (elements that belong to the grammar and are discarded), FP the false positives (elements that belong to garbage and are recognized as belonging to grammar). For *speaker identification* task, instead, TP are the true positives (the diagonal elements

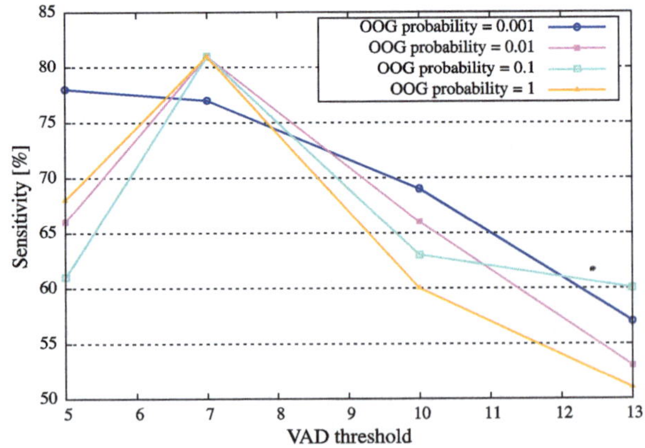

Fig. 4 Speech recognition performance changing the value of the OOG probability and VAD threshold

Table 2 Speaker identification performance using both training and testing data from the command grammar

Speaker	Gender	Sensitivity (%)	Precision (%)
		Sequence length = 200 frames	
A	M	100.00	100.00
B	M	100.00	100.00
C	M	100.00	100.00
D	M	100.00	100.00
E	F	100.00	100.00
F	F	100.00	100.00
		Sequence length = 100 frames	
A	M	95.45	95.45
B	M	100.00	100.00
C	M	100.00	100.00
D	M	95.45	95.45
E	F	92.59	100.00
F	F	100.00	97.06
		Sequence length = 70 frames	
A	M	93.55	90.62
B	M	100.00	100.00
C	M	100.00	100.00
D	M	96.77	96.77
E	F	89.47	97.14
F	F	97.87	93.88
		Sequence length = 30 frames	
A	M	85.14	81.82
B	M	98.75	92.94
C	M	97.56	96.39
D	M	86.30	86.30
E	F	76.67	90.79
F	F	90.99	87.07
		Sequence length = 1 frame	
A	M	48.10	45.99
B	M	71.83	65.86
C	M	77.89	64.95
D	M	54.92	51.14
E	F	47.54	60.14
F	F	58.35	66.47

of the confusion matrix), FN the false negatives (the sum of the other elements on the same row of the confusion matrix), FP the false positives (the sum of the other elements on the same column of the confusion matrix).

Table 3 Speaker identification performance using training data from the continuous speech and testing data from the command grammar

Speaker	Gender	Sensitivity (%)	Precision (%)
		Sequence length = 200 frames	
A	M	100.00	100.00
B	M	100.00	100.00
E	F	100.00	100.00
		Sequence length = 100 frames	
A	M	100.00	92.86
B	M	100.00	91.67
E	F	82.73	100.00
		Sequence length = 70 frames	
A	M	95.83	95.83
B	M	100.00	84.78
E	F	75.00	100.00
		Sequence length = 30 frames	
A	M	95.16	80.82
B	M	93.94	83.78
E	F	67.14	96.97
		Sequence length = 1 frame	
A	M	69.27	56.26
B	M	73.84	71.86
E	F	54.07	63.31

Figures 5 and 6 show the speaker identification performance, represented by the sensitivity, as a function of the sequence length (number of frames). As it can be easily seen from both the figures, the increase of the number of frames corresponds to a higher identification rate.

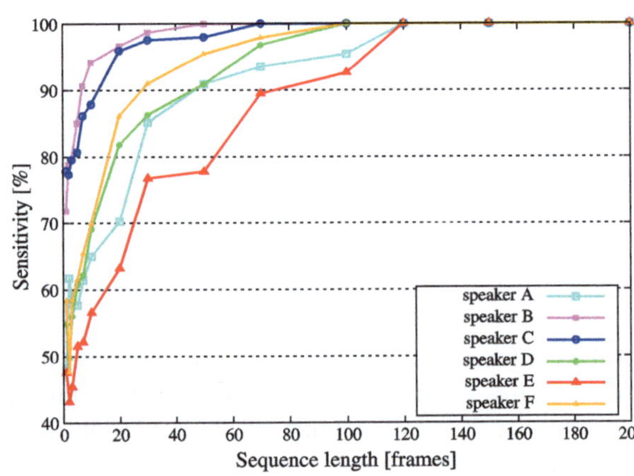

Fig. 5 Speaker identification performance as a function of the sequence length using both training data and testing data from the command grammar

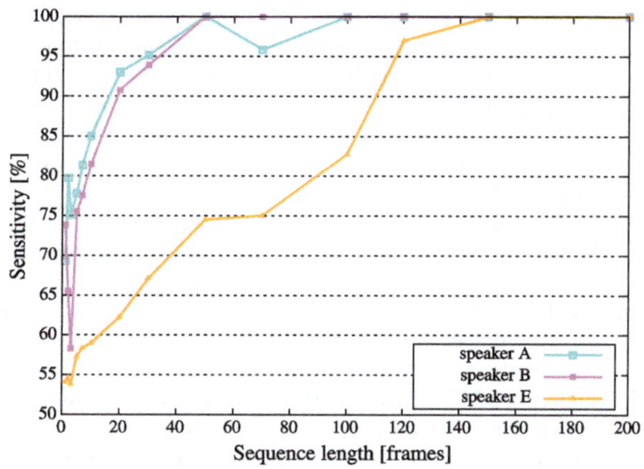

Fig. 6 Speaker identification performance as a function of the sequence length using training data from the continuous speech and testing data from the command grammar

6 Conclusions

This paper presents a combined speech recognition and speaker identification system using MFCC method. This is motivated by the fact that MFCCs have demonstrated to be particularly suitable for both speech recognition and speaker identification. The experimental results show good performance in terms of speech recognition and speaker identification. Moreover, the system is extremely versatile and can be used in a large number of applications.

References

1. Alessandrini M, Biagetti G, Curzi A, Turchetti C (2011) Semi-automatic acoustic model generation from large unsynchronized audio and text chunks. In: 12th Annual conference of the international speech communication association (Interspeech 2011). Florence, Italy, pp 1681–1684
2. Alessandrini M, Biagetti G, Curzi A, Turchetti C (2013) A garbage model generation technique for embedded speech recognisers. In: Signal processing: algorithms, architectures, arrangements, and applications (SPA 2013). Poznan, Poland, pp 318–322
3. Bazzi I, Glass JR (2000) Modeling out-of-vocabulary words for robust speech recognition. In: 6th International conference on spoken language processing (ICSLP 2000/INTERSPEECH 2000). Beijing, China, pp 401–404
4. Bhardwaj S, Srivastava S, Hanmandlu M, Gupta J (2013) GFM-based methods for speaker identification. IEEE Trans Cybern 43(3):1047–1058
5. Biagetti G, Crippa P, Curzi A, Falaschetti L, Orcioni S, Turchetti C (2015) Distributed speech recognition for lighting system control. In: Intelligent decision technologies. Springer, Berlin, pp 101–111
6. Biagetti G, Crippa P, Curzi A, Orcioni S, Turchetti C (2015) Speaker identification with short sequences of speech frames. In: International conference on pattern recognition applications and methods (ICPRAM 2015). Lisbon, Portugal, pp 178–185

7. Bimbot F, Bonastre JF, Fredouille C, Gravier G, Magrin-Chagnolleau I, Meignier S, Merlin T, Ortega-Garca J, Petrovska-Delacrétaz D, Reynolds DA (2004) A tutorial on text-independent speaker verification. EURASIP J Appl Sig Process 2004:430–451
8. Campbell JPJ (1997) Speaker recognition: a tutorial. Proc IEEE 85(9):1437–1462
9. Chen K, Wang L, Chi H (1997) Methods of combining multiple classifiers with different features and their applications to text-independent speaker identification. Int J Pattern Recognit Artif Intell 11(03):417–445
10. Dobrowolski A, Majda E (2011) Cepstral analysis in the speakers recognition systems. In: Signal processing algorithms, architectures, arrangements, and applications conference proceedings (SPA 2011). Poznan, Poland, pp 1–6
11. ETSI ES 202 050 V1.1.5 (2007) Speech processing, transmission and quality aspects (STQ); distributed speech recognition; advanced front-end feature extraction algorithm; compression algorithms
12. Figueiredo MAT, Jain A (2002) Unsupervised learning of finite mixture models. IEEE Trans Pattern Anal Mach Intell 24(3):381–396
13. Gish H, Schmidt M (1994) Text-independent speaker identification. IEEE Signal Process Mag 11(4):18–32
14. Hirota S, Hayasaka N, Iiguni Y (2012) Experimental evaluation of structure of garbage model generated from in-vocabulary words. In: 2012 International symposium on communications and information technologies (ISCIT 2012). Gold Coast, Australia, pp 87–92
15. Huang X, Acero A, Hon HW (2001) Spoken language processing: a guide to theory, algorithm, and system development. Prentice Hall PTR, Upper Saddle River
16. Jain A, Duin RPW, Mao J (2000) Statistical pattern recognition: a review. IEEE Trans Pattern Anal Mach Intell 22(1):4–37
17. Jain A, Ross A, Prabhakar S (2004) An introduction to biometric recognition. IEEE Trans Circuits Syst Video Technol 14(1):4–20
18. Kinnunen T, Li H (2010) An overview of text-independent speaker recognition: from features to supervectors. Speech Commun 52(1):12–40
19. Levit M, Chang S, Buntschuh B (2009) Garbage modeling with decoys for a sequential recognition scenario. In: IEEE workshop on automatic speech recognition understanding (ASRU 2009). Merano, Italy, pp 468–473
20. Reynolds D (1994) Experimental evaluation of features for robust speaker identification. IEEE Trans Speech Audio Process 2(4):639–643
21. Reynolds D (2002) An overview of automatic speaker recognition technology. In: 2002 IEEE international conference on acoustics, speech, and signal processing (ICASSP 2002). Orlando, Florida, vol 4, pp IV–4072–IV–4075
22. Reynolds D, Rose R (1995) Robust text-independent speaker identification using Gaussian mixture speaker models. IEEE Trans Speech Audio Process 3(1):72–83
23. Togneri R, Pullella D (2011) An overview of speaker identification: accuracy and robustness issues. IEEE Circuits Syst Mag 11(2):23–61
24. Walker W, Lamere P, Kwok P, Raj B, Singh R, Gouvea E, Wolf P, Woelfel J (2007) Sphinx-4: a flexible open source framework for speech recognition

Part IV
Biometric Data Analysis of Cardiac Signals

Surface EMG Fatigue Analysis by Means of Homomorphic Deconvolution

Giorgio Biagetti, Paolo Crippa, Simone Orcioni and Claudio Turchetti

Abstract In this paper we use homomorphic deconvolution to obtain the power spectrum of the motor unit action potential (MUAP) from the surface electromyography (sEMG) signal. This spectrum is then used to extract the parameters of a time-domain model of the MUAP itself, in particular its amplitude and time scale. The analysis of the extracted parameters leads to the estimation of cadence and muscle fatigue. The methodology is tested with a sEMG signal recorded during biceps curl exercises.

1 Introduction

The shapes and firing rates of motor unit action potentials (MUAPs) in electromyography (EMG) signals provide an important source of information in rehabilitation and biomedical applications [10, 15, 19]. In the past years, studies on single MUAP were mainly conducted using intramuscular EMG signals, because of the high selectivity of this detection technique, that can possibly target a single fiber.

However, surface EMG, being a non-invasive method, can still be used to quantify [9] and classify [17] the global muscular activity. This is of great importance in clinical diagnosis and biomedical applications as well as in recreational, sportive, and training activities [3, 4]. The main problem in using of surface EMG for discriminating the contributions of each single active motor unit is its low spatial selectivity [22]. Thus, the development of appropriate algorithms and methods for analyzing the surface EMG signal, in order to accurately understand the nature and characteristics of the signal itself for various applications, is a challenging task.

An interesting application of surface EMG signal processing can be muscle fatigue estimation. Spectral parameters derived from the EMG signal, such as mean or median frequency, can be used for this purpose as shown in [13, 14, 18]. This is

G. Biagetti · P. Crippa · S. Orcioni (✉) · C. Turchetti
Dipartimento di Ingegneria dell'Informazione, Università Politecnica delle Marche, via Brecce Bianche, 12, 60131 Ancona, Italy
e-mail: s.orcioni@univpm.it

M. Conti et al. (eds.), *Mobile Networks for Biometric Data Analysis*,
Lecture Notes in Electrical Engineering 392, DOI 10.1007/978-3-319-39700-9_14

due to the fact that, at least during a sustained isometric contraction, the low frequency band tends to show an increase in amplitude, while a corresponding relative decrease in the amplitude of the higher frequencies can be observed. This is called EMG spectrum compression [20], and is due to the fact that MUAP pulses tend to lengthen as the muscle becomes more fatigued.

However, for dynamic or cyclic movements, the EMG signal is inherently non-stationary. The overall spectrum, and consequently the parameters derived from it, have thus little physical significance. Time-frequency analysis techniques need to be used to deal with this time variability, and many signal processing methods suitable for application in EMG-based muscle fatigue evaluation have thus been proposed [1, 2, 7]. Despite these attempts, correlating the surface EMG signal with the shape of the underlying MUAP is still a hard task to accomplish.

In this paper, a novel approach to address the issue of discriminating the contributions of MUAP potential in the overall EMG signal is presented. This technique can also be applied to investigate muscle fatigue, or to calculate the pace of repetitive, cyclic movements. The method is based on the identification of the MUAP from the surface EMG signal, employing a power cepstrum-based deconvolution technique.

The cepstrum concept was first introduced in [6] for the analysis of seismic signals. Over time, different flavours, such as the real, power, and complex cepstrums, were introduced. A rather comprehensive survey on cepstral analysis of signals can be found in [16]. The real and the power cepstrum are quite similar, and both do not maintain the phase information of the signal, while the complex cepstrum does. The term homomorphic deconvolution is often used as a synonym for cepstrum-based deconvolution, although strictly speaking it should only be used for complex cepstrum-based deconvolution.

With these considerations in mind, we used homomorphic deconvolution to obtain the MUAP spectrum, from which the parameters of a time-domain model of the MUAP itself (i.e. amplitude and time scale), can be estimated. Finally, in order to validate the presented technique, a surface EMG signal was recorded during a fitness exercise using a wireless electromyograph, the MUAP parameters were estimated from it and compared with corresponding results reported in [3].

2 Surface EMG Signal Model

A surface EMG signal originates from the superposition of MUAPs, coming from different muscle fibers. According to [15] the signal of a single fiber can be modeled as the summation of delayed versions of a MUAP, also called MUAP train (MUAPT). Each MUAPT has its amplitude κ and time scale α. If we represent the MUAP with a function $f(t)$, we can write

$$\mathrm{MUAPT}_j(t) = \sum_i \frac{\kappa_j}{\alpha_j} f\left(\frac{t-\theta_{ij}}{\alpha_j}\right). \tag{1}$$

The parameters θ_{ij} represents the time shifts of each MUAP. The summation of different MUAPTs originates the surface EMG signal

$$s(t) = \sum_j \mathrm{MUAPT}_j = \sum_j \sum_i \frac{\kappa_j}{\alpha_j} f\left(\frac{t - \theta_{ij}}{\alpha_j}\right). \tag{2}$$

If in the previous equation we assume that $f(\cdot)$ has the same temporal scale in all MUAPTs ($\alpha_j = \alpha$), then (2) becomes

$$s(t) = \sum_j \sum_i \frac{\kappa_j}{\alpha} f\left(\frac{t - \theta_{ij}}{\alpha}\right). \tag{3}$$

At this stage we slice the signal with a sliding window. The reason is twofold. Firstly, in digital signal processing we always need to process a finite span of the signal. Secondly, as said in the introduction, both muscular fatigue and execution of movements (i.e., a non-isometric scenario) can make the signal non-stationary. This means that the parameters in (3) can be considered as functions of time (i.e. $\alpha = \alpha(t)$). Applying a sliding window $w(t - \theta_k)$ allows us to assume the MUAPT parameters as constant within the considered span. Thus the windowed version of $s(t)$ will be

$$s_k(t) = w(t - \theta_k) \sum_j \sum_i \frac{\kappa_j}{\alpha_k} f\left(\frac{t - \theta_{ij}}{\alpha_k}\right). \tag{4}$$

By posing

$$g_k(t) = \frac{1}{\alpha_k} f(t/\alpha_k), \quad u(t) = \sum_j \sum_i \kappa_j \delta(t - \theta_{ij})$$

Equation (4) can be written as

$$s_k(t) = w_k(t)[h_k(t) \circledast u(t)] \tag{5}$$

where $\circledast$ is the symbol of convolution and $w_k(t) = w(t - \theta_k)$. The function $u(t)$ is a train of delta functions. The mean frequency of this function is also called the firing frequency of the MUAPT.

By performing the Fourier transform of (5) we have

$$S_k(f) = W_k(f) \circledast [H(f)U(f)] = W_k(f) \circledast [F(\alpha_k f)U(f)] \tag{6}$$

where $S_k(f)$, $W_k(f)$, $F(f)$, and $U(f)$ are the Fourier transforms of $s_k(t)$, $w_k(t)$, $f(t)$, and $u(t)$, respectively. In particular $U(f) = \sum_j \sum_i \kappa_j e^{-j2\pi f \theta_{ij}}$.

To minimize the effect of $W(f)$ on $S(f)$, we have chosen a Blackman–Nuttall window to provide a sharp peak in frequency, improving the resolution of the spectrum, and low side ripple, reducing artifacts. In this way we can consider

$$S_k(f) \simeq F(\alpha_k f)U(f) \tag{7}$$

so that the effect of $U(f)$ can be removed from $|S(f)|$ by performing power cepstrum deconvolution.

3 Homomorphic Deconvolution

Considering the discrete time case, the power cepstrum $c_s(n)$ of the signal $s(n)$ can be written as

$$c_s(n) \triangleq \text{IFFT } \log|\text{FFT}\, s(n)|^2. \tag{8}$$

If the signal $s(n)$ can be written as the result of the convolution of two signals, like the output of a system is the convolution of an input with its impulse response,

$$s(n) = h(n) \circledast u(n) \tag{9}$$

its frequency representation will be

$$S(f) = H(f)U(f). \tag{10}$$

As a result, the power cepstrum of this signal can be written as

$$c_s(n) = \text{IFFT}\left\{\log|H(f)|^2 + \log|U(f)|^2\right\}. \tag{11}$$

If the spectra $H(f)$ or $U(f)$ have periodic components, these components will result in peaks in $c_s(n)$ that can be filtered. The filtering of $c_s(n)$, occurring in a domain which is the inverse of frequency, is called liftering.

By liftering the power cepstrum we can separate the contribution of the train of deltas from that of the MUAP function. It can be found [8, 21] that the firing frequency of a MUAPT can reach at maximum 50 Hz. As a consequence, in this work we decided to perform a low pass liftering with a cut-off frequency of 20 ms.

The liftered cepstrum, $\tilde{c}_s(n)$, can be written as

$$\tilde{c}_s(n) = lp(n)c_s(n) \tag{12}$$

with

$$lp(n) = \begin{cases} 1 & n \leq n_{\text{cutoff}} \\ 0 & n > n_{\text{cutoff}} \end{cases} \tag{13}$$

where n_{cutoff} = 20 ms$\cdot F_S$, F_S being the sampling frequency.

After liftering we have to come back to the frequency domain with

$$|\tilde{S}(f)|^2 = |e^{\text{FFT}\tilde{c}_s(n)}|. \tag{14}$$

As said, the liftering is intended to remove the $U(f)$ part from the spectrum of $S(f)$, so that $\tilde{S}(f) \simeq F(\alpha f)$.

4 MUAP Model

The model for the MUAP is based on Hermite-Rodriguez (HR) series expansion and was proposed in [12] as a convenient expansion of signals with essentially compact support, like EMG ones. The basis functions are defined as follows

$$\text{HR}_n(A, \alpha, t) = A \frac{1}{\sqrt{2^n n!}} \text{H}_n(t/\alpha) \frac{1}{\sqrt{\pi}\alpha} e^{(-t/\alpha)^2} \tag{15}$$

where n is the order of the function, H_n is the Hermite polynomial, the parameter A directly affects the amplitude of the MUAP, while α varies its time scale.

According to the theory proposed in the previous section, by comparing (3) to (15) it is possible to identify a basis function for the MUAP as

$$\text{f}_n(\xi) = \frac{A_k}{\sqrt{\pi 2^n n!}} \text{H}_n(\xi) e^{-\xi^2}. \tag{16}$$

The MUAP, in general, can be expressed as a series of these basis functions, nevertheless, in this work only one term in the series was used, while a model using a combination of the first two order can be found in [5].

Figure 1 shows different MUAPs and their spectra, obtained by using first order, second order Hermite-Rodriguez polynomials, and a combination of the two. As it can be observed, the first order HR polynomial has a biphasic shape while the second order one has a triphasic symmetrical shape. Also, the spectrum of the second order polynomial is almost symmetrical, while that of the first order is steeper at low frequencies.

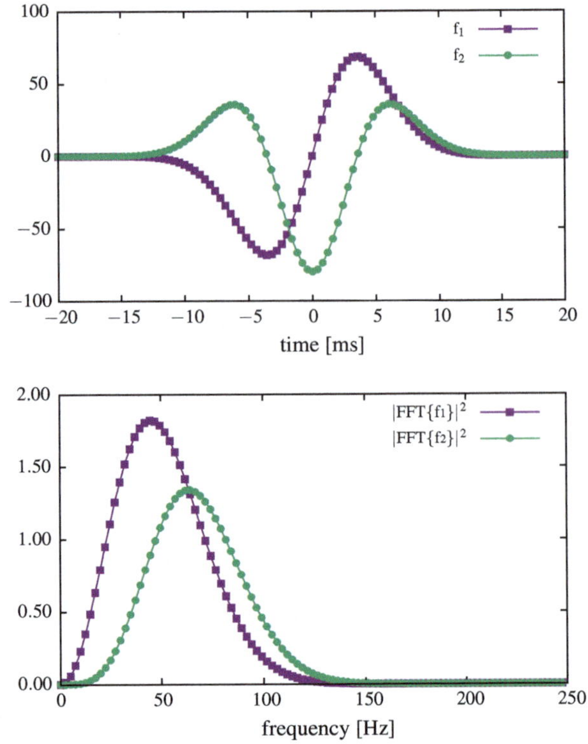

Fig. 1 Example of Hermite-Rodriguez polynomials obtained from (15) with $A = 1$ and $\alpha = 2$ ms. The power spectrum is computed over a 4 s window, in order to obtain a high resolution spectrum

5 Experimental Results

The data used to validate this methodology are that used in [3]. They were recorded during a fitness exercise consisting in performing biceps curls with dumbbells of different weight during a one minute session. The data was recorded following SENIAM recommendations [11], with the exception reported in [3].

They were asked to choose three dumbbell weights from the set of 2, 3, 5, and 7 kg and perform biceps curls, at an approximately constant pace of about 20 reps/min. They were then allowed to rest for several minutes before changing the weight.

Informed consent was obtained for all data, that were recorded using a wireless electromyograph system.

The top plot in Fig. 2 shows a repetition taken form the exercise of subject 4. The width of this time window is equal to 4.096 s, that sampled with $F_S = 2\,\text{kHz}$ corresponds to 8192 points.

The middle plot in Fig. 2 shows the signal after the application of the Blackman–Nuttall window. The cepstrum was calculated and then liftered,

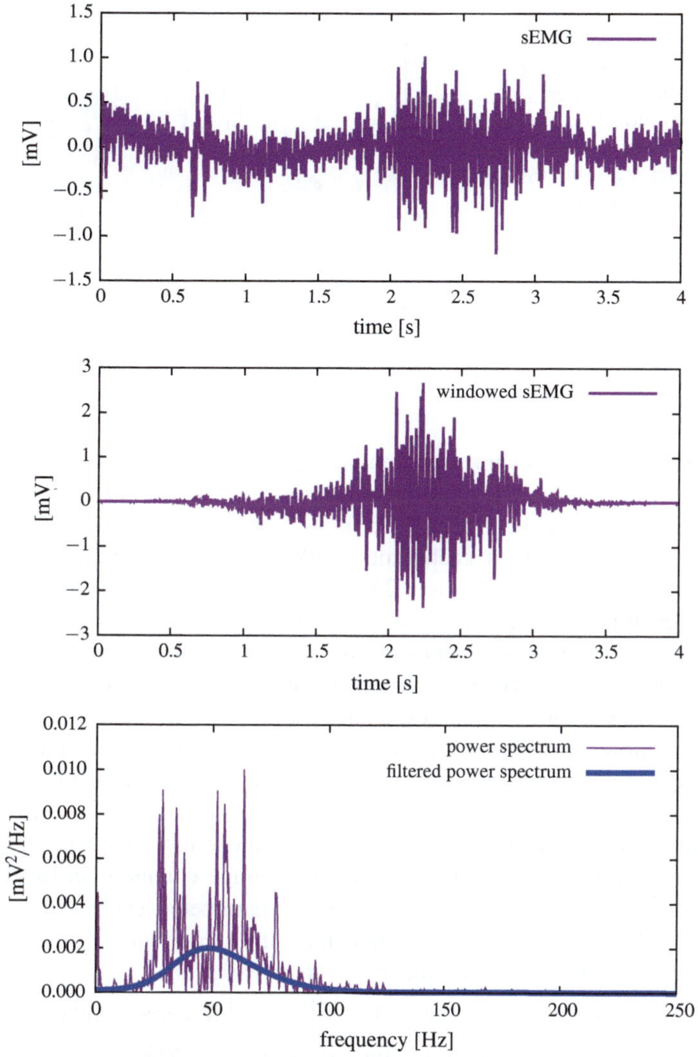

Fig. 2 A windowed of used EMG signal and the result of power cepstrum deconvolution

according to (12) and (13) respectively, with a cutoff quefrency of 20 ms, corresponding to $20\,\mathrm{ms} \cdot F_S = 40$ samples.

The filtered power spectrum, obtained with (14), is shown in the bottom plot of Fig. 2. The power spectrum of the original signal is added to the same plot for comparison. The convolution with the impulse train is responsible for the presence of many peaks in the spectrum of the signal. As can be seen from Fig. 2, all peaks have been removed by the nonlinear filtering performed by the cepstrum transform.

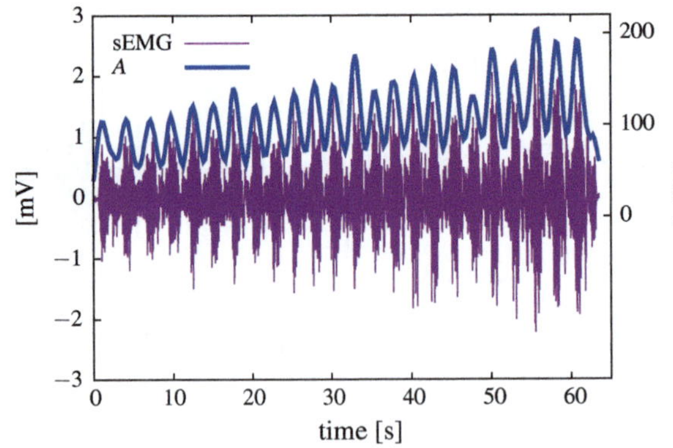

Fig. 3 Relationship between the EMG signal recorded from subject 4 during the 5 kg exercise and the parameter *A* of the EMG model f_1. The scales of sEMG signal and of the *A* parameter are at the *left* and at the *right* of the plot, respectively

The EMG spectrum obtained by homomorphic deconvolution was used to fit the parameters of the MUAP function. The parameters were estimated as described in Sect. 3 using the Levenberg-Marquardt nonlinear regression algorithm, as implemented in GNU Octave, in each time window. The signal was windowed with 4.096 s long windows, shifted by 0.256 s.

A first result is shown in Fig. 3, which relates the sEMG signal to the parameter *A*. The estimated *A* parameter and the original sEMG signal recorded during an exercise with a 5 kg dumbbell performed by subject 4 are shown. As can easily be seen, *A* follows with great accuracy the envelope of the EMG signal, as expected.

Figure 4 shows error-bar diagrams of the *A* parameter, as a function of time, estimated during an exercise with a dumbbell of a 5 kg, performed by subjects 1, 2 and 4. The plots in Fig. 4 show the values of the estimated parameter, with error bars equal to one standard deviation, and the linear regression of the estimated parameter, computed in order to evaluate the average change over time.

Figure 5 shows error-bar diagrams of the α parameter, estimated for the same signals as in the previous figure. All plots have the same scale in order to highlight the differences in the extracted values and their variations.

Figure 6 shows the normalized MSE $\varepsilon_k \Big/ \left|\tilde{G}_k(f)^2\right|$ in percent, relative to the same data as in the previous two figures. Even if the relative MSE does not follow well the envelope of the EMG signal (while the amplitude parameter does), its maximum values generally correspond to windows centered between repetitions, i.e. in regions with small or no signal, where we can expect the noise to dominate the recording. On the contrary, when the amplitude of the sEMG signal grows, the normalized MSE drops to low values.

While Figs. 4 and 5 show parameters estimated from different subjects performing the same exercise, Figs. 7 and 8 show results obtained from the same subject performing the exercise with different weights. In particular Fig. 7 shows

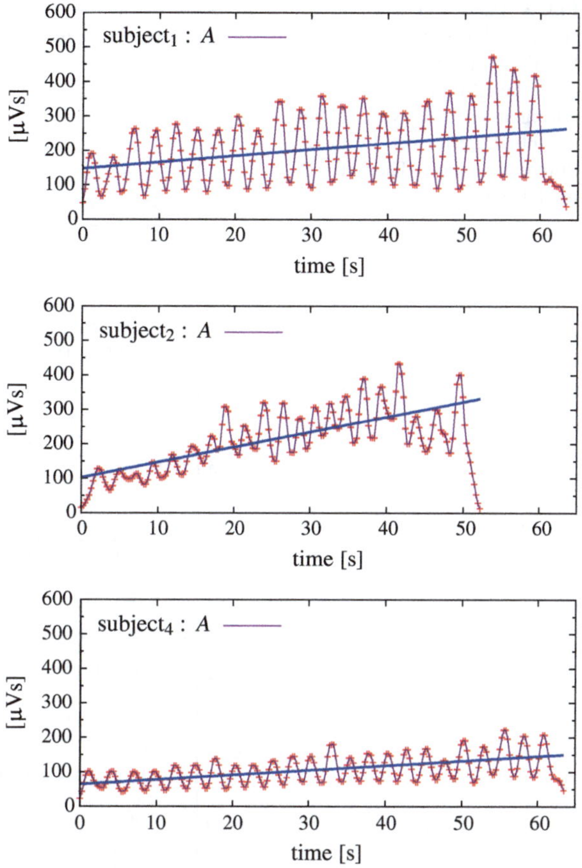

Fig. 4 Parameter A, with its standard deviations, as functions of time, estimated from the EMG recorded during an exercise with a 5 kg dumbbell, performed by three different subjects. The MUAP model is f_1

the error-bar diagram of the A parameter, estimated from a biceps curl exercise, performed by subject 3 with weights equal to 2, 3 and 5 kg, respectively. The estimation of the α parameter for the same signals is shown in Fig. 8.

5.1 *Fatigue and Cadence Estimation*

It is well known [13, 14, 18, 20] that fatigue can be related to the growth of the MUAP time scale, represented by the slope of linear regression of the α parameter in Fig. 5. From Fig. 4 we can see that, when a subject is fatigued, also the linear regression of the amplitude shows an appreciable slope. Nevertheless, the extent of amplitude variation, for the same exercise, is strongly subject dependent.

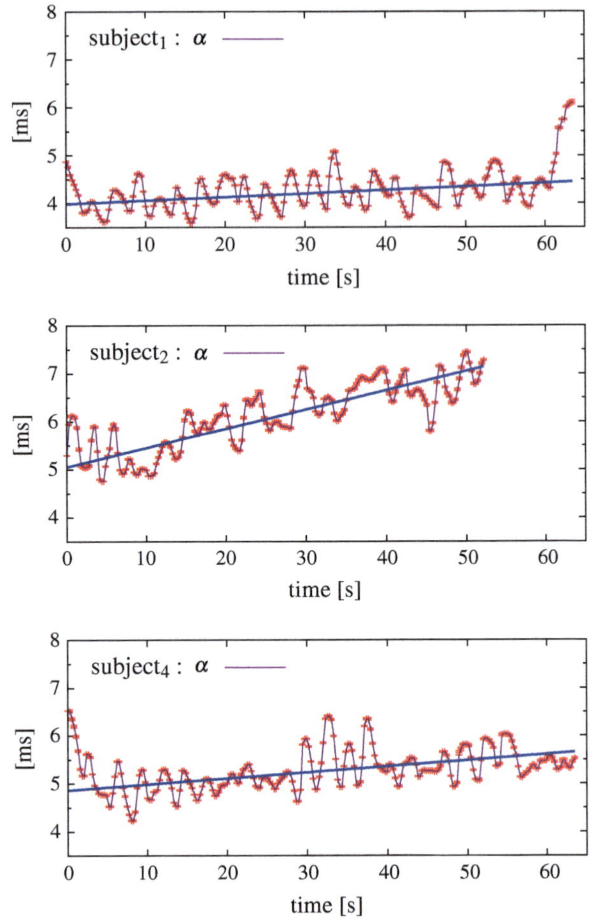

Fig. 5 Parameter α, with its standard deviations, as functions of time, estimated from the EMG recorded during an exercise with a 5 kg dumbbell, performed by three different subjects. The MUAP model is f_1

In Fig. 8 we can better note how using a bigger weight impacts the subject fatigue. Also the amplitude parameter, reported in Fig. 7, reflects the fatigue in the increase of mean value and range.

In order to investigate on the relation between fatigue and MUAP time scale α, estimated by the proposed technique, a numerical comparison with results obtained in [3] was carried out. From the linear regression of time scale parameter α, expressed as $q+mt$, a parameter representing the rate of variation of the line, $R = m/q$, was extracted. It represents the percentage variation per minute of the parameter itself. Table 1 shows the rate R of all available signals (in bold) compared to the rate of MFA variation as extracted in [3]. The meaning of the sign of the data is different for the two cases, since in the former it is a slope of a time, in the latter of a frequency. Thus opposite signs are indication of agreement.

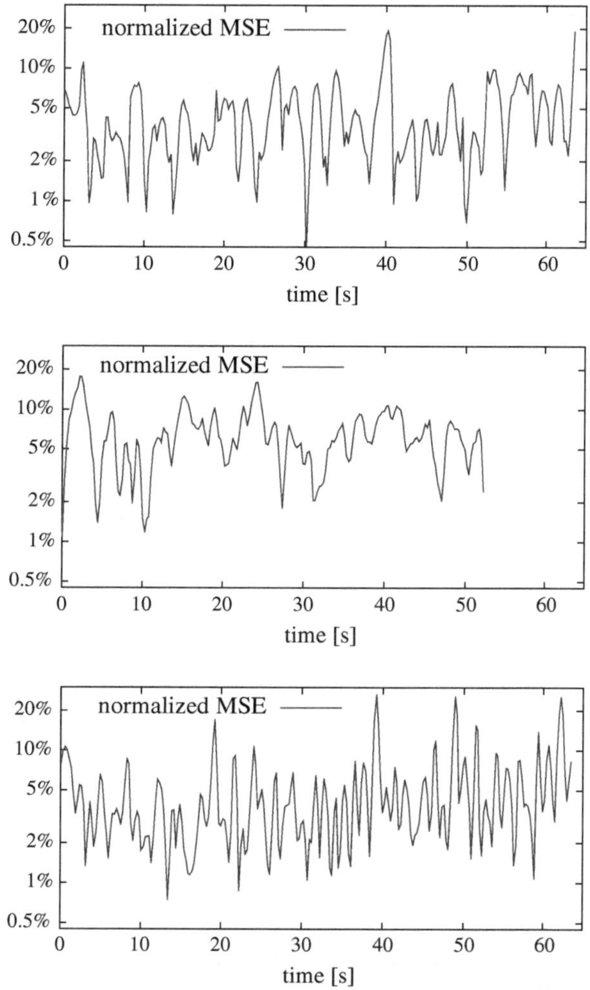

Fig. 6 Normalized MSE as functions of time, estimated from the EMG recorded during an exercise with a 5 kg dumbbell, performed by three different subjects. The MUAP model is f_1

In order to quantitative evaluate the agreement between the use of MFA and α in fatigue estimation, the Pearson correlation coefficient between the two data series was calculated and reported in Table 2 together with mean absolute error and standard deviation. The results in Table 2 show a good agreement between the two techniques, although they sometimes give slightly different results. However, estimation of fatigue using MFA (or MNF) may be affected by noise in the spectral representation of the sEMG signal introduced by the stochastic firing. Since this effect on the spectrum is removed by homomorphic deconvolution we can state that the estimate with the time scale α should be more accurate and realistic than that obtained with MFA or MNF.

Fig. 7 Parameter *A*, with its standard deviations, as functions of time, estimated from the EMG recorded during different exercises performed by subject 3. The MUAP model is f_1

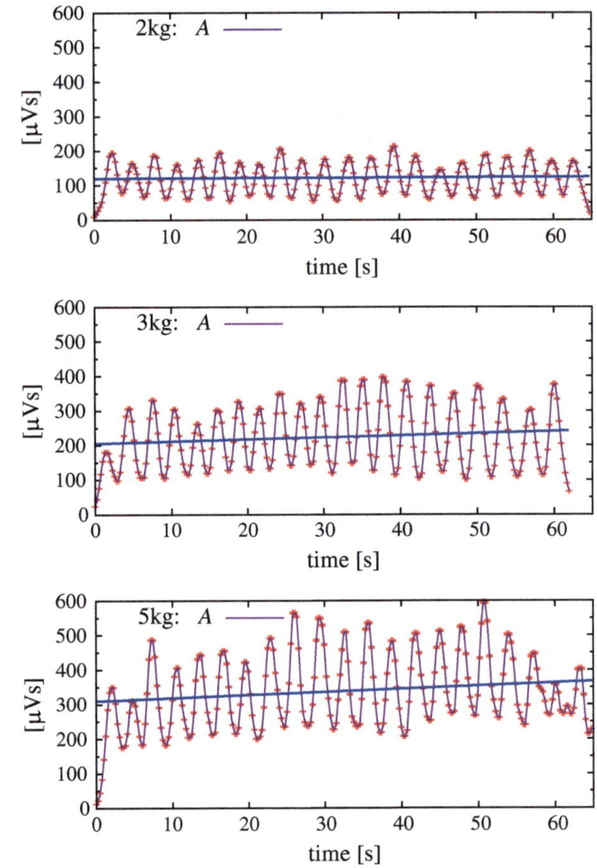

As for cadence estimation, from the observation of Figs. 4 and 8, we can deduce that the shape of *A* parameter can be used to estimate the cadence of the exercise.

The exercise repetition rate can thus be obtained with good accuracy by counting the crossings between *A* and its linear regression. Unfortunately, in a few cases the graph of the estimated parameter does not cross its linear regression, as also shown in Figs. 4 and 7, causing the missing of a repetition in the count. To mitigate this issue the linear regression can be estimated in a window shorter than the whole signal, as those used in our examples, giving more chances to each repetition to cross its regression line.

Table 3 shows the comparison with the results obtained with f_1 MUAP model and the correct results for the same signals, as published in [3]. As can be seen this method gives an optimum pace estimation. Table 4 compares the estimate obtained

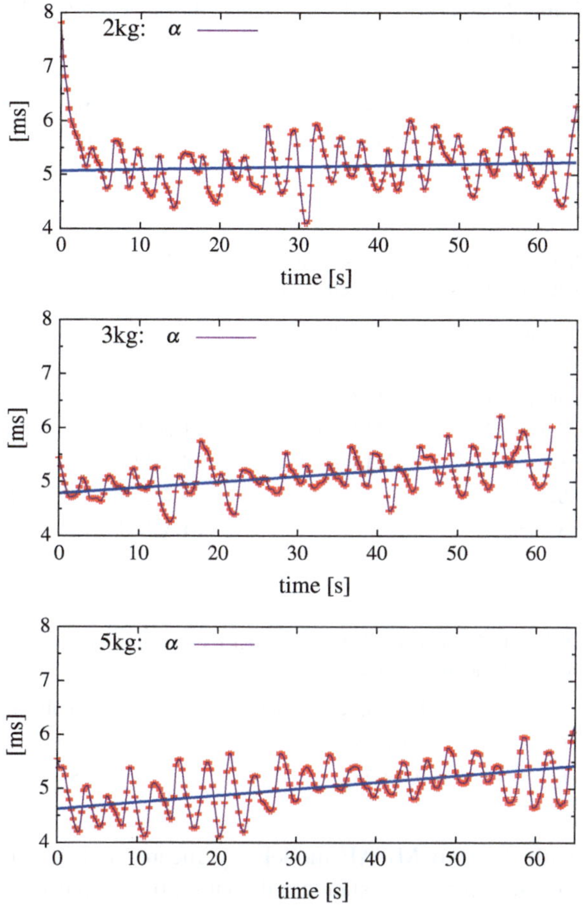

Fig. 8 Parameter α, with its standard deviations, as functions of time, estimated from the EMG recorded during different exercises performed by subject 3. The MUAP model is f_1

Table 1 Summary of the results of the fatigue test for all the subjects: estimated rate of α variation (f_1) versus estimated rate of MFA variation [3], (%/min)

Subject	Weight			
	2 kg	3 kg	5 kg	7 kg
Subject 1	**15.2**/−7.9	**11.4**/−4.2	**10.8**/−12.4	
Subject 2	**2.47**/−0.6	**−3.49**/−9.8	**47.6**/−30.3	
Subject 3	**2.98**/−1.2	**12.9**/−7.5	**15.9**/−19.1	
Subject 4		**8.23**/−7.4	**15.4**/−16.8	**36.5**/−23.7
Subject 5		**2.25**/−12.0	**−0.56**/−14.3	**23.0**/−34.3

Table 2 Comparison between fatigue estimation performed with f_1 and f_2 based MUAP models and fatigue estimated in [3]

	Pearson corr. coeff.	Mean abs. err.	Std. dev.
f_1	−0.7330	7.3289	5.5726
f_2	−0.7486	6.4144	4.3353

Table 3 Summary of the results of the cadence estimation for all the subjects: estimated mean cadence using the parameter *A* of the MUAP model f_1 versus [3], (reps/min)

Subject	Weight			
	2 kg	3 kg	5 kg	7 kg
Subject 1	**17.6**/17.7	**21.0**/20.5	**21.8**/23.4	
Subject 2	**23.8**/23.5	**21.5**/21.7	**19.4**/18.6	
Subject 3	**21.0**/20.9	**20.6**/20.1	**18.0**/19.3	
Subject 4		**24.2**/23.4	**23.1**/23.5	**21.2**/20.4
Subject 5		**21.3**/22.4	**21.8**/22.9	**21.4**/22.6

Table 4 Comparison between cadence estimation performed with f_1 and f_2 based MUAP models and real cadence, (reps/min)

	Pearson corr. coeff.	Mean abs. err.	Std. dev.
f_1	0.89728	0.7210	0.4721
f_2	0.89978	0.7287	0.4611

with the two MUAP models by means of Pearson correlation coefficient, mean absolute error, and standard deviation between estimation and real data. As it can be seen, there is not a substantial difference between the f_1 of f_2 models.

6 Conclusions

This work presents a technique for finding the MUAP parameters of a surface EMG signal by deconvolution. The deconvolution is obtained by passing the power cepstrum through a liftering stage. The filtered power spectrum so obtained is then used to fit the MUAP parameters. These parameters can be used for fatigue estimation and cadence detection. Indeed the amplitude (A) of the MUAP can be related to the energy of the signal, while the its time scale (α) can be related to the mean frequency of the spectrum. The results obtained by this method have been compared to previously published results computed with a totally different approach, and a good match was found, thus demonstrating the feasibility of the proposed method.

References

1. Bai F, Chew CM (2013) Muscle force estimation with surface EMG during dynamic muscle contractions: a wavelet and ANN based approach. In: 35th annual international conference of the IEEE engineering in medicine and biology society (EMBC), pp 4589–4592
2. Bai F, Lubecki T, Chew CM, Teo CL (2012) Novel time-frequency approach for muscle fatigue detection based on sEMG. In: Biomedical circuits and systems conference (BioCAS), 2012 IEEE, pp 364–367
3. Biagetti G, Crippa P, Curzi A, Orcioni S, Turchetti C (2015) Analysis of the EMG signal during cyclic movements using multicomponent AM–FM decomposition. IEEE J Biomed Health Inform 19(5):1672–1681. doi:10.1109/JBHI.2014.2356340
4. Biagetti G, Crippa P, Falaschetti L, Orcioni S, Turchetti C (2015) A rule based framework for smart training using sEMG signal. In: Neves-Silva R, Howlett RJ, Jain LC (eds) Intelligent decision technologies: 2015, smart innovation, systems and technologies series. Springer, Germany, pp 89–99
5. Biagetti G, Crippa P, Orcioni S, Turchetti C (2016) Homomorphic deconvolution for MUAP estimation from surface EMG signals. IEEE J Biomed Health Inform (in press):1. doi:10.1109/JBHI.2016.2530943
6. Bogert BP, Healy MJR, Tukey JW (1963) The quefrency analysis of time series for echoes: cepstrum, pseudo-autocovariance, cross-cepstrum, and saphe-cracking. In: Rosenblatt M (ed) Proceedings of the symposium on time series analysis. Wiley, New York, pp 209–243
7. Cifrek M, Medved V, Tonković S, Ostojić S (2009) Surface EMG based muscle fatigue evaluation in biomechanics. Clin Biomech 24(4):327–340
8. Conwit RA, Stashuk D, Tracy B, McHugh M, Brown WF, Metter EJ (1999) The relationship of motor unit size, firing rate and force. Clin Neurophysiol 110(7):1270–1275
9. De Luca C (1979) Physiology and mathematics of myoelectric signals. IEEE Trans Biomed Eng BME-26(6):313–325
10. Gazzoni M, Farina D, Merletti R (2004) A new method for the extraction and classification of single motor unit action potentials from surface EMG signals. J Neurosci Methods 136 (2):165–177. doi:10.1016/j.jneumeth.2004.01.002
11. Hermens HJ, Freriks B (1999) European recommendations for surface electromyography (CDROM). Roessingh Research and Development
12. Lo Conte LR, Merletti R, Sandri GV (1994) Hermite expansions of compact support waveforms: applications to myoelectric signals. IEEE Trans Biomed Eng 41(12):1147–1159. doi:10.1109/10.335863
13. Merletti R, Knaflitz M, De Luca CJ (1990) Myoelectric manifestations of fatigue in voluntary and electrically elicited contractions. J Appl Physiol 69(5):1810–1820
14. Merletti R, Lo Conte LR (1997) Surface EMG signal processing during isometric contractions. J Electromyogr Kinesiol 7(4):241–250
15. Merlo A, Farina D, Merletti R (2003) A fast and reliable technique for muscle activity detection from surface EMG signals. IEEE Trans Biomed Eng 50(3):316–323. doi:10.1109/TBME.2003.808829
16. Oppenheim AV, Schafer RW (2004) From frequency to quefrency: a history of the cepstrum. IEEE Signal Process Mag 21(5):95–106. doi:10.1109/MSP.2004.1328092
17. Ouyang G, Zhu X, Ju Z, Liu H (2014) Dynamical characteristics of surface EMG signals of hand grasps via recurrence plot. IEEE J Biomed Health Inform 18(1):257–265
18. Potvin J, Bent L (1997) A validation of techniques using surface EMG signals from dynamic contractions to quantify muscle fatigue during repetitive tasks. J Electromyogr Kinesiol 7 (2):131–139
19. Reaz MBI, Hussain MS, Mohd-Yasin F (2006) Techniques of EMG signal analysis: detection, processing, classification and applications. Biol Proced Online 8(1):11–35. doi:10.1251/bpo115

20. Sakurai T, Toda M, Sakurazawa S, Akita J, Kondo K, Nakamura Y (2010) Detection of muscle fatigue by the surface electromyogram and its application. In: IEEE/ACIS 9th international conference on computer and information science (ICIS), pp 43–47
21. Seki K, Narusawa M (1996) Firing rate modulation of human motor units in different muscles during isometric contraction with various forces. Brain Res 719(1):1–7
22. Stegeman DF, Blok JH, Hermens HJ, Roeleveld K (2000) Surface EMG models: properties and applications. J Electromyogr Kinesiol 10(5):313–326

sEMG-Based Evaluation of Muscle Recruitment Variability During Walking in Terms of Activation Length and Occurrence Frequency

A. Mengarelli, E. Maranesi, L. Burattini, S. Fioretti and F. Di Nardo

Abstract Surface electromyography (sEMG) is commonly used in gait analysis for detecting muscle activity in a non-invasive way, preserving the normal mobility of the subject. The purpose of the study was to assess the variability of sEMG signals acquired from lower-limb muscles during walking. To this aim, a statistical analysis of sEMG signals from a large number (hundreds) of strides per subject was performed in twenty-two healthy young caucasian volunteers. Tibialis Anterior, Gastrocnemius Lateralis, Rectus Femoris, Biceps Femoris and Vastus Lateralis were selected to represent both proximal and distal leg segments. Besides the muscular activation onset-offset instants, the study was aimed to analysed the occurrence frequency of muscle recruitment, a parameter seldom considered because of the low number of strides usually analysed in classic EMG studies. Findings illustrated that a single muscle showed a different number of activation intervals in different strides of the same walking. The number of times when muscle activates during a single gait cycle defined the modality of muscle recruitment, that in the present study was referred to as activation modality, i.e. *n*-activation modality consists of *n*-activation intervals for the considered muscle, during a single gait cycle. For each of the selected muscles, five activation modalities were detected. Each of these activation modalities is characterized by a different occurrence frequency and by different onset-offset activation instants. Concomitance of these results indicates a large variability in onset-offset muscular activation and occurrence frequency, which should be considered in discriminating pathological from physiological behaviour and for designing focused gait studies.

A. Mengarelli · E. Maranesi · L. Burattini · S. Fioretti · F. Di Nardo (✉)
Department of Information Engineering, Università Politecnica delle Marche, via Brecce Bianche, 60131 Ancona, Italy
e-mail: f.dinardo@univpm.it

M. Conti et al. (eds.), *Mobile Networks for Biometric Data Analysis*,
Lecture Notes in Electrical Engineering 392, DOI 10.1007/978-3-319-39700-9_15

1 Introduction

Gait analysis (GA) represents a widely used tool for clinical evaluation of subjects with impairments affecting walking mechanisms [1–4].

Surface electromyography (sEMG) is commonly used for the assessment of the activation patterns of muscles involved in gait [5–16]. This technique allows a non-invasive evaluation of muscles activity, preserving the essential requirement of a high degree of subject mobility, a fundamental requisite that allows the subject to walk in the most natural and physiological manner, without the influence of the acquisition system on gait kinematics. The sEMG signals vary from person to person and they also differ for the same motion even within the same person; this variability reflects the variability in muscles activation during walking [9–19]. In order to improve the interpretation of EMG signals in both physiological and pathological conditions, the assessment of this natural variability is valuable.

During walking most of the lower limb muscles are recruited, in both shank and thigh. In order to achieve a whole evaluation of the physiological gait characteristics, the muscles usually selected are Tibialis Anterior and Gastrocnemius Lateralis for the shank and Rectus Femoris, Biceps Femoris and Vastus Lateralis for the thigh.

Tibialis Anterior (TA) and Gastrocnemius Lateralis (GL) were analysed for their role on ankle joint control. During gait, TA acts as ankle dorsi-flexor, preventing slapping of the foot during initial stance, permitting the forefoot to clear the ground in initial stance and holding the ankle in position for initial contact [5]. GL acts as ankle plantar-flexor and knee flexor, restraining the forward rotation of the tibia on the talus during stance, providing ankle and knee stability and conserving energy by a minimization of the vertical oscillation of the body-centre of mass [8]. Moreover, these muscles provide a better sEMG signal than deeper shank muscles.

On the thigh there are two main muscle groups, the Quadriceps Femoris (QF), located in the front of the thigh and the Hamstrings, in the posterior side of thigh. For the QF group, Rectus Femoris (RF) and Vastus Lateralis (VL) were analysed. RF is a knee extensor and hip flexor and it helps the knee flexion by means of an active hip flexion during swing phase, producing a forward movement of femur that coupled with the tibia inertia, leading to the knee flexion during swing. Furthermore, RF can avoid a knee hyper-flexion acting directly as knee extensor during the late swing phase [5]. VL is a mono-articular muscle, it acts only as knee extensor and, with the BF, during the first period of gait cycle controls the knee joint in the critical heel strike phase, when this joint is loaded by a large part of body weight [5]. For the Hamstrings group, Biceps Femoris (BF) was chosen: it acts as knee flexor and hip extensor and concurs with the RF to the knee flexion by a direct action on the knee joint, during swing phase [5]. BF also attends to the stabilization of knee joint in the earliest part of gait cycle.

Thus, the aim of the present study was the quantitative assessment of the variability of the myoelectric activity of thigh and shank muscles in young healthy caucasian subjects during gait at self-selected speed and cadence, in terms of onset-offset muscular activation and frequency of occurrence. The goal of the study was pursued by performing a statistical analysis of sEMG signal acquired from TA, GL, RF, BF and VL.

Since this method depends on the accuracy of the process used to detect muscles onset/offset, this study is based on robust techniques for the detection of muscle activation intervals and specific tools for statistical gait analysis [20–22].

2 Materials and Methods

Twenty-two young healthy adult volunteers were recruited for this study. Mean (±SE) values were 23.9 ± 0.4 years for age, 173 ± 2 cm for height, 62.7 ± 2.7 kg for weight and 20.6 ± 0.4 kg m^{-2} for body mass index (BMI). Male/female ratio was 11/11. The EMG activity was recorded in both right and left lower limb of all subjects.

Exclusion criteria included history of neurological pathology, orthopaedic surgery within the previous year, acute or chronic knee pain or pathology, BMI $\geq$ 25 or abnormal gait. All participants signed informed consent.

2.1 Recording System: Signal Acquisition and Processing

Signals were acquired by means of a multichannel recording system for statistical gait analysis (Step32, DemItalia, Italy). Each subject was instrumented with foot-switches, goniometers and sEMG probes.

Three foot-switches were attached beneath the heel, the first and the fifth metatarsal head of each foot. A goniometer (accuracy 0.5°) was attached to the lateral side of each lower limb for measuring the knee joint angles in the sagittal plane [23]. Single differential sEMG probes with fixed (inter-electrode distance: 12 mm) and variable geometry (inter-electrode distance: 20 mm) were attached over TA, GL, RF, BF and VL of each lower limb, following the SENIAM recommendations [24]. Subjects were instructed to walk barefoot for around 5 min at their natural speed and cadence, following the path reported in Fig. 1. Natural pace was chosen because

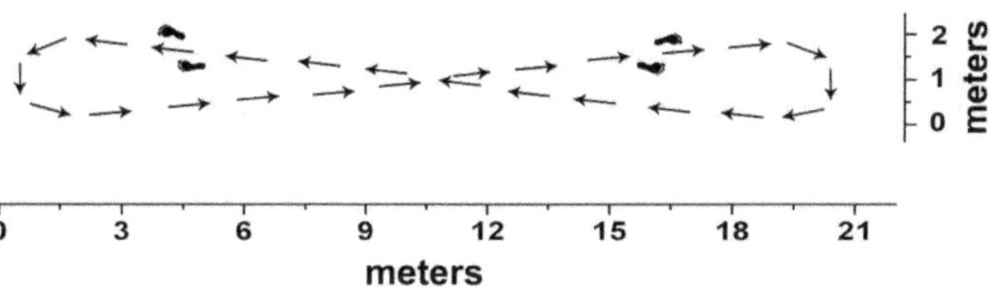

Fig. 1 Representation of the path walked by selected subjects during experimental trials

walking at a comfortable speed improves the repeatability of EMG data, while variability increases when subject are required to walk abnormally [18, 25].

Foot-switch signal was converted to four levels, corresponding to Heel contact (H), Flat foot contact (F), Push-off (P) and Swing (S) phases and processed to segment and classify the different gait cycles. During acceleration, deceleration and reversing the strides are different from those of steady state walking. Goniometric signals (low-pass filtered with cut-off frequency of 15 Hz) along with gait phases durations, were used by a multivariate statistical filter embedded in the Step32 system, to detect and discard outlier cycles. EMG signals were band-pass filtered (20–450 Hz) and then processed by a double-threshold statistical detector, embedded in the Step32 system, that provides the onset and offset time instants of muscle activity [20].

During gait a muscle activates a number of times which is variable from cycle to cycle [9–16, 26]. Thus, muscle on/off instants should be averaged considering each single modality of activation by itself. With modality of activation we mean the number of times when a muscle activates during a single gait cycle: n-activation modality consists of n activation intervals for the considered muscle, during a single gait cycle.

In the present study mean activation intervals (normalized respect to the gait cycle) for each modality of activation were achieved by means of the Step32 system, according to the following steps. First, muscle activations relative to each gait cycle were identified. Then, for all the gait cycles corresponding to straight line walking, muscle activations were grouped according to the number of activations detected, i.e. relatively to the modalities of activations detected [20]. Finally, the on/off time instants were averaged, for each specific modality of activation observed and relative standard deviation and standard error were computed. In the present study only gait cycles consisting of the sequence of H-F-P-S were considered.

2.2 Statistics

All data are reported as mean $\pm$ standard error (SE). The Lilliefors test was used to evaluate the hypothesis that each data vector had a normal distribution. Comparisons among normally distributed samples were performed with two-tailed, non-paired Student's t-test (two populations) or one-way ANOVA (more than two populations) followed by multiple comparison test. Kruskal-Wallis test, followed by multiple comparison test, was used to compare not normally distributed samples. Statistical significance was set at 5 %, i.e. p-value < 0.05.

3 Results

For each subject a mean (±SE) of 331 ± 21 strides has been considered; stride characteristics and gait phases duration are reported in Table 1.

The analysis of myoelectric signal in each subject shows a different number of activation intervals in different strides of the same walk, for all considered muscles. Variability of sEMG signals across subjects were represented in Fig. 2, for all the considered muscles.

For TA (Fig. 3) the most recurrent modality of activation occurs in 39.6 ± 1.5 % of total strides and consists of three activations: the first at the beginning of gait cycle, the second in the transition between P- and S-phase and the third during S-phase. The second most recurrent modality for TA is the 4-activation modality (25.4 ± 1.9 %), with a further activation, respect to the 3-activation modality, in the transition between F- and P-phase.

These two modalities occur with a non-significantly different frequency ($p > 0.05$). The 2-activation modality (23.8 ± 2.9 %) has an active interval at the beginning (H-phase) and at the end of gait cycle (late S-phase). The 5-activation modality recurs in the lowest percentage of strides (8.6 ± 1.1 %), with at least one active interval in every gait phase.

For GL (Fig. 4) the 2-activation modality shows the first activation in the flat foot contact and the second in the mid-swing; though this modality is the most recurrent one (38.4 ± 1.5 % of total strides), its occurrence frequency is not significantly higher (p-value for the evaluation of statistical significance was >0.05) than the 1- and 3-activation modalities. The 1-activation modality (27.9 ± 2.9 %) presents a single activation during the stance phase and the 3-activation modality (25.2 ± 1.9 %) shows the first activation in the F-phase, the second in the P-phase and the third at the end of swing. The 4-activation modality occurs in 7.4 ± 0.9 % of total strides and presents two activations during both stance and swing. The 1-activation modality for TA and the 5-activation modality for GL are not considered in the present study, because they were detected in a non-significant ($p > 0.05$) percentage of strides.

For RF (Fig. 5) the 3-activation modality (most recurrent) shows a significant ($p < 0.05$) higher occurrence (53.1 ± 2.8 % of total strides), compared to the other modalities. The first activation occurs at the beginning of gait cycle, the second at the end of P-phase and the third at the end of the swing. The 4 activation-modality

Table 1 Stride parameters and gait phases duration. S-S and D-S duration and gait phases values are reported as percentage of gait cycle ±SE

Cadence (cycle/min)	Cycle duration (s)	S-S[a] duration (%)	D-S[b] duration (%)	H-phase duration (%)	F-phase duration (%)	P-phase duration (%)	S-phase duration (%)
56.7 ± 0.5	1.1 ± 0.1	43.3 ± 0.4	13.3 ± 0.6	6.0 ± 0.3	26.7 ± 0.9	23.7 ± 0.7	43.5 ± 0.4

[a]Single support duration
[b]Double support duration

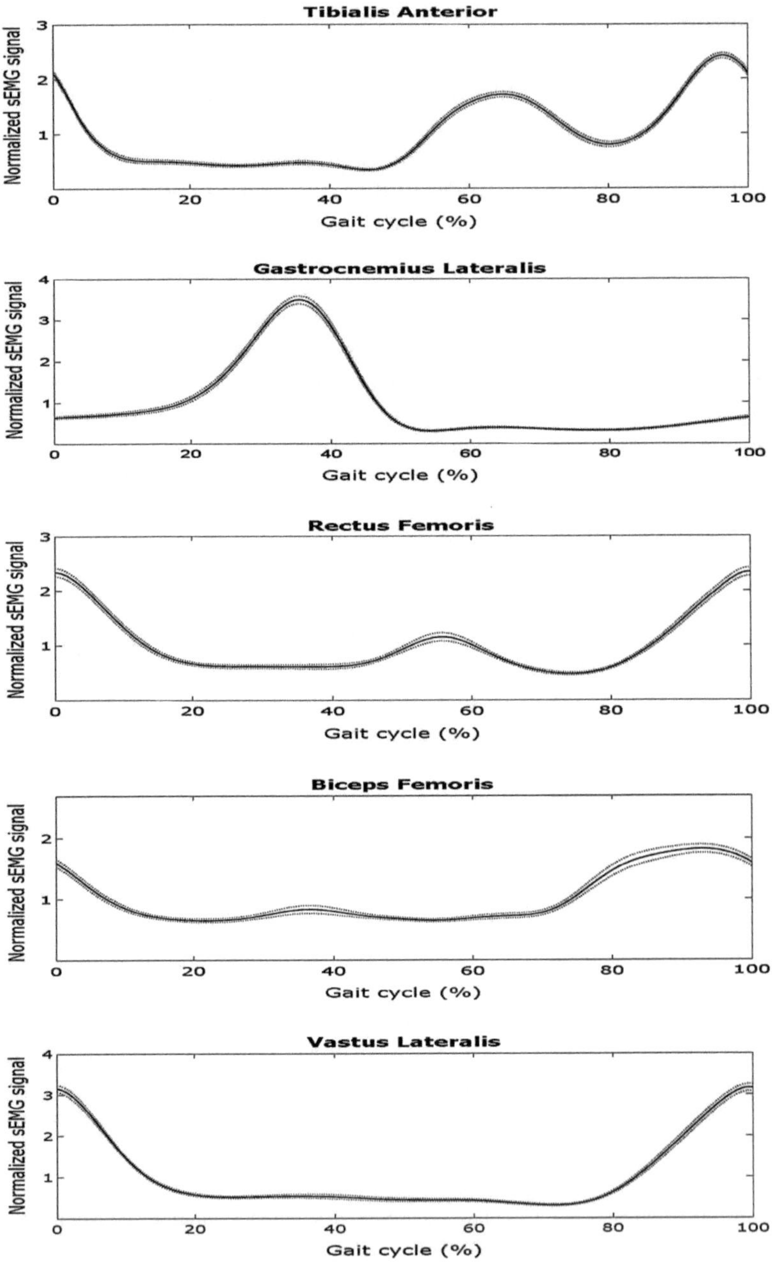

Fig. 2 sEMG patterning of all muscles, averaged (±SE) across all subjects. Only for this figure, sEMG signals were rectified, low pass filtered (5 Hz), and normalized in amplitude by mean value, for representative reasons

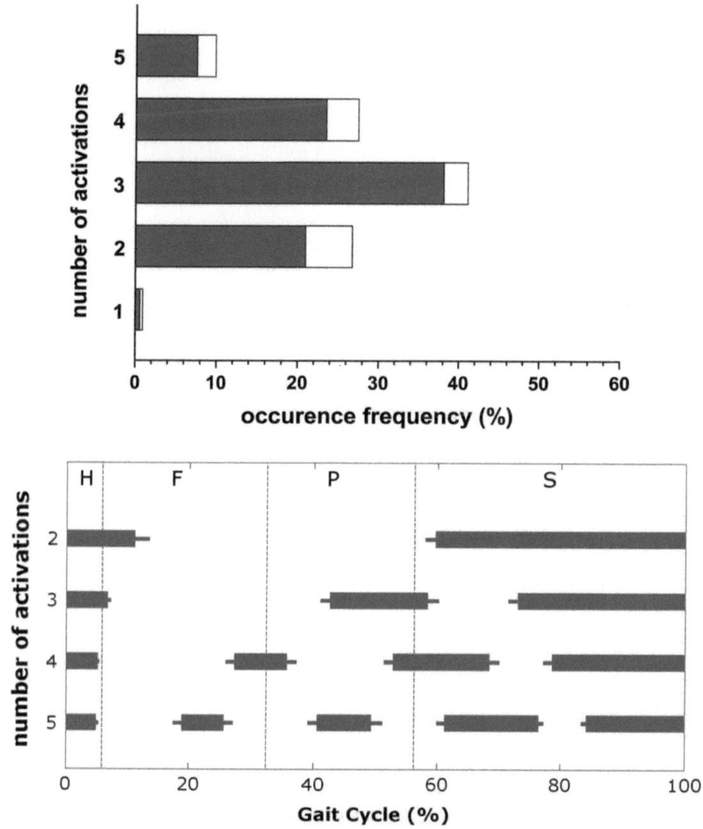

Fig. 3 Occurrence frequency (mean with confidence interval in *white*, *top panel*) and activation intervals (mean + SE, *horizontal bars*, *bottom panel*) versus percentage of gait cycle, of TA activation modalities. Heel contact (H), flat-foot contact (F), push-off (P) and swing (S) phases are delimited by *dashed vertical lines*. For both panel mean and SE values refers to the whole population

(22.4 ± 1.7 %) has the first and fourth activations in the earliest and latest part of gait cycle, the second in the transition between F- and P-phase and the third in the early swing. In the 2-activation modality (17.8 ± 2.4 %) the first activation lasts for the H- and F-phase and the second occurs in the latest part of the swing, while the 5-activation modality (5.3 ± 0.8 % of total strides) shows at least one active interval in every gait phase.

For BF (Fig. 6) the 3-activation modality results the most recurrent one: it occurs in 39.7 ± 2.1 % of total strides, with the first activation in the H-phase, the second in the mid push-off phase and the third in swing, whereas the activity of the 2-activation modality (36.3 ± 3.2 %) lies at the beginning and at the end of gait cycle. These two most recurrent modalities are not different ($p > 0.05$) in terms of occurrence frequency, but their recurrence is significantly higher ($p < 0.05$)

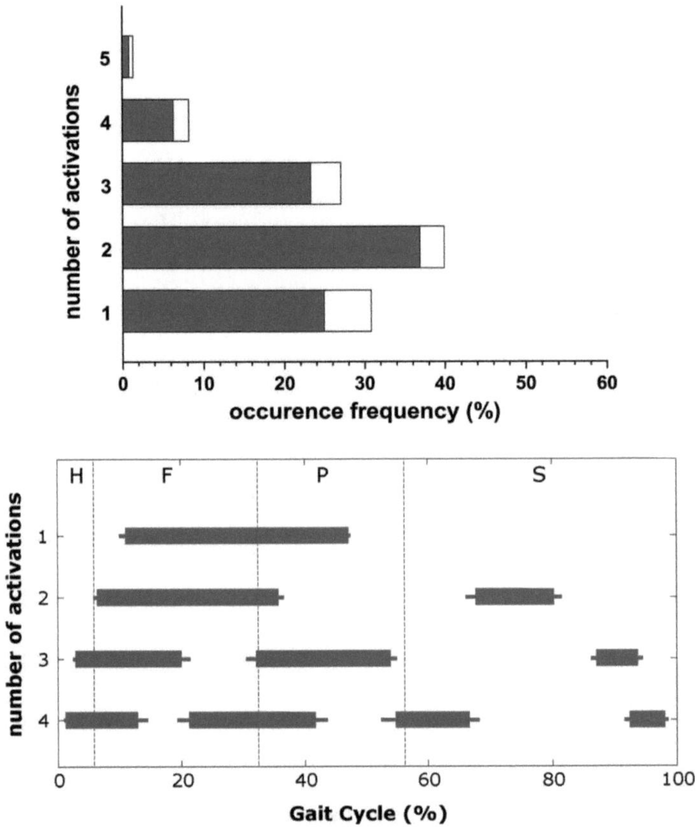

Fig. 4 Occurrence frequency (mean with confidence interval in *white*, *top panel*) and activation intervals (mean + SE, *horizontal bars*, *bottom panel*) versus percentage of gait cycle, of GL activation modalities. Heel contact (H), flat-foot contact (F), push-off (P) and swing (S) phases are delimited by *dashed vertical lines*. For both panel mean and SE values refers to the whole population

compared to all other modalities. The 4-activation modality appears in 16.5 $\pm$ 1.4 % of total strides and presents the second activation at the end of the F-phase and the third at the beginning of the swing. The 5-activation modality (3.3 $\pm$ 0.5 %) shows at least one active interval in every gait phase.

For the VL (Fig. 7) the 2-activation modality appears in 48.9 $\pm$ 3.6 % of total strides (most recurrent), with the first interval in the H- and F-phase and the second in the latest part of the swing. The 3-activation modality (35.5 $\pm$ 2.3 %) shows a further activation at the beginning of push-off phase, whereas the second interval of the 4-activation modality (12.0 $\pm$ 1.6 %) lies in the second half of F-phase and the third in the mid P-phase. The 5-activation modality for VL (2.5 $\pm$ 0.6 %) is similar to those of RF and BF, with at least one active interval in every gait phase. The occurrence frequency of the most recurrent modality for VL results significantly

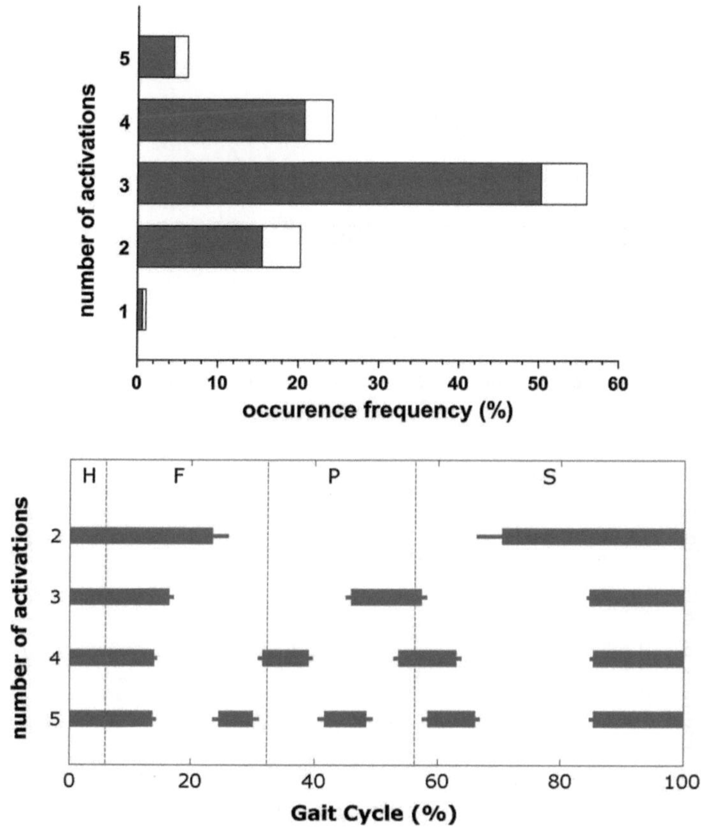

Fig. 5 Occurrence frequency (mean with confidence interval in *white*, *top panel*) and activation intervals (mean + SE, *horizontal bars*, *bottom panel*) versus percentage of gait cycle, of RF activation modalities. Heel contact (H), flat-foot contact (F), push-off (P) and swing (S) phases are delimited by *dashed vertical lines*. For both panel mean and SE values refers to the whole population

higher ($p < 0.05$) compared to the other modalities, except for the 3-activation modality ($p > 0.05$). The 1-activation modality for RF, BF and VL occurs in a non-significant percentage of strides ($p > 0.05$), thus it was not considered in the present study.

4 Discussion

The surface electromyography is a painless and non-invasive technique widely used to detect muscle activity. In particular sEMG results suitable for assessing muscle activity of lower limb during walking, because it allows a high degree of subject

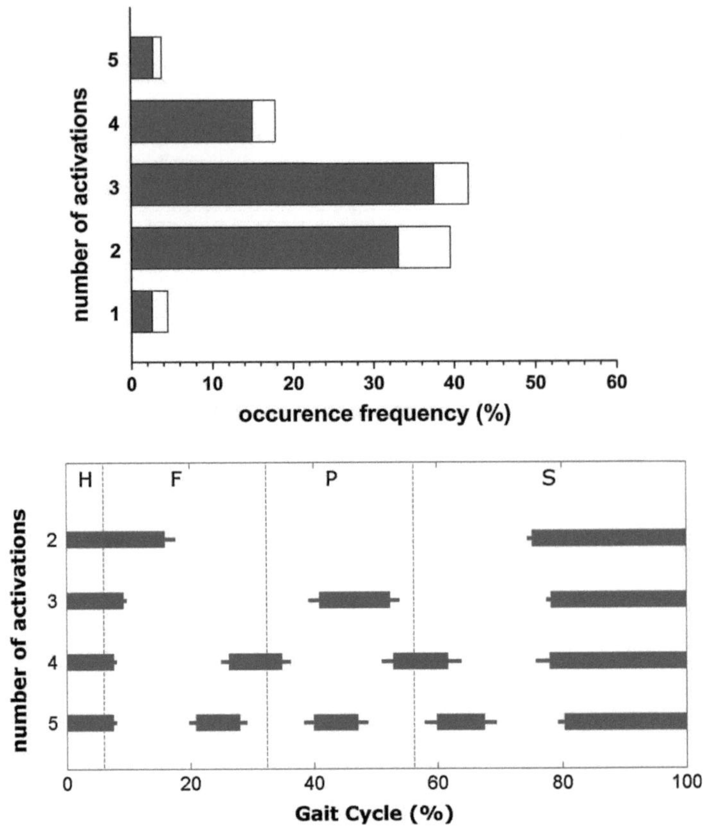

Fig. 6 Occurrence frequency (mean with confidence interval in *white*, *top panel*) and activation intervals (mean + SE, *horizontal bars*, *bottom panel*) versus percentage of gait cycle, of BF activation modalities. Heel contact (H), flat-foot contact (F), push-off (P) and swing (S) phases are delimited by *dashed vertical lines*. For both panel mean and SE values refers to the whole population

mobility. In the present study, the sEMG signal was acquired in 22 young healthy caucasian subjects during gait at self-selected speed and cadence, for the assessment of the activation modalities adopted by the most representative shank (TA and GL) and thigh muscles (RF, BF and VL). A statistical analysis of a large number (on average 331 $\pm$ 21) of strides for each subject was performed, in order to quantitatively represent the variability of sEMG activity. Besides the onset-offset muscular activation instants, the study was aimed to analyse the occurrence frequency of muscle recruitment, a parameter seldom considered because of the low number of strides analysed in classic EMG studies. A synthetic recap of main results is reported in Table 2.

The activation intervals follow roughly the typical pattern reported for each muscle during walking [4, 5]. However the statistical analysis highlights that both

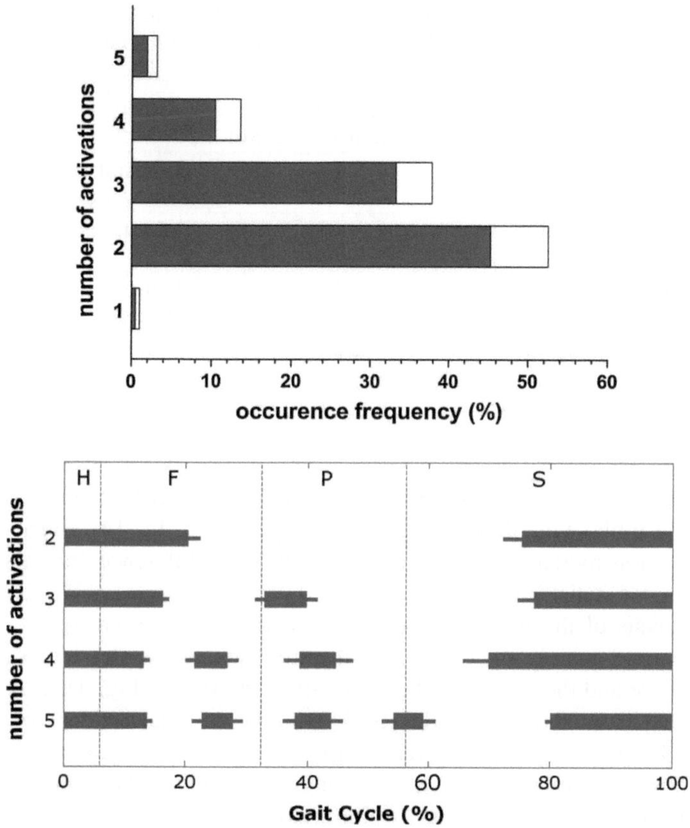

Fig. 7 Occurrence frequency (mean with confidence interval in *white*, *top panel*) and activation intervals (mean + SE, *horizontal bars*, *bottom panel*) versus percentage of gait cycle, of VL activation modalities. Heel contact (H), flat-foot contact (F), push-off (P) and swing (S) phases are delimited by *dashed vertical lines*. For both panel mean and SE values refers to the whole population

shank and thigh muscles adopt different modalities in the number of activations and in the timing of signal onset and offset, as reported previously [9–16]. This suggests that the activation modalities of each muscle should be considered with the respective frequency of occurrence, i.e. how frequently they were observed, in terms of percentage of total strides.

TA presents four main activation modalities (Fig. 3), with different occurrence frequencies. The analysis shows that the activity of TA during gait is centred mainly on three areas: the first at the beginning of gait cycle, the second in the mid-stance and the third in the swing phase. The activities in the first and third area are recognized as the typical TA activations for ankle dorsiflexion [4, 5]. Even though it occurs with a lower frequency (34.0 ± 2.2 % of total strides), the activity during mid-stance has been observed in previous studies [4, 9, 12–14, 16, 28] and could

Table 2 Most recurrent modality of activation for all considered muscles. For each activation modality the occurrence frequency and periods of gait cycle where the muscle is active are also reported

Muscle	Most recurrent modality	Occurrence frequency (% ± SE)	Activity periods
Tibialis anterior	3 activations	39.6 ± 1.5	Heel strike terminal stance late swing
Gastrocnemius lateralis	2 activations	38.4 ± 1.5	Mid-stance swing
Rectus femoris	3 activations	53.1 ± 2.8	Initial part of GC terminal stance late swing
Biceps femoris	3 activations	39.7 ± 2.1	Heel strike late stance late swing
Vastus lateralis	2 activations	48.9 ± 3.6	Initial part of GC late swing

likely be related to the role of TA as foot invertor muscle, to control balance and contralateral limb swing during single support period [9, 12, 14]. Even though the TA 3-activation modality is the most recurrent one, difference from 4-activation modality is not statistically significant ($p > 0.05$).

The analysis of the four main activation modalities detected for GL shows an activity focused mainly on two areas: the first lies across flat foot contact and push-off phase and the second in the final part of the swing (Fig. 4). The activity in the first area can show a single (66.3 ± 3.3 % of total strides) or a double (32.6 ± 2.1 %) activity and it is related to the GL task of restraining the forward progression of the tibia over the talus during the second rocker, hence controlling dorsiflexion [5, 8, 27]. In the 3-activation modality the pre-swing activity (Fig. 4) suggests a participation of GL in the knee flexion during this period, as reported by Winter [28]. However, in the remaining 73.7 ± 3.4 % GL is inactive during pre-swing, in according to what has been reported by Sutherland [4, 29] about the absence of plantar flexor muscles activity during this period. The analysis performed on the myoelectric signal of GL seems to suggest that these reports are not contradictory, because GL shows a pre-swing activity in the 25.2 ± 1.9 % of total strides but in the 73.7 ± 3.4 % GL seems to be silent during this period. The second area of GL activity is located in swing phase and this type of activation occurs in a high percentage of total strides (71.0 ± 2.6 %). Although this activity during swing is not of general consensus, it was reported [5, 9, 12–14, 16] that it can occur when GL acts as a foot invertor muscle, working in conjunction with TA for positioning the foot in preparation of the following heel strike. The GL 2-activation modality is the most recurrent one, even though differences from 1- and 3-activation modalities are not statistically significant ($p > 0.05$).

The RF activity appears focused on three areas: the first is located at the beginning of gait cycle, the second in the mid stance phase and at third at the end of gait cycle (Fig. 5). The RF activity at beginning of gait cycle seems to be related to participation of the muscle in the control of weight bearing, by means of an eccentric action, whereas the activity at the end of swing phase could assist knee extension and muscle tension for weight acceptance during the following loading response [30, 31]. A further possible interpretation of these RF activations could lie in the hypothesis of cross-talk from the surrounding vastii, as reported in a previous study [10]. The RF activity during mid stance phase is detected in the 3-, 4- and 5-activation modality (about 80 % of total strides) and seems to be present to help performing rapid knee extension and hip flexion [32]. A lack of activity during stance-to-swing transition was detected in the 2-activation modality: it is likely due to those strides where self-selected speed was temporary reduced, as reported by Nene [33]. Furthermore RF seems to adopt a favourite activation modality during gait: the most recurrent modality (3-activations) shows an occurrence frequency significantly ($p < 0.05$) higher compared to all other activation modalities.

BF activity covers three areas of gait cycle: the first and the second lie in the early and in the terminal part of stance phase respectively, the third covers the latest part of swing phase (Fig. 6). In 36.3 ± 3.2 % of total strides (2-activation modality) BF activity during the terminal stance is lacking (Fig. 6). BF does not seem to show a preferred activation modality: the two most recurrent ones (2- and 3-activation modality) are not significantly different ($p > 0.05$) in terms of occurrence frequency. Anyway all the modalities show the activity from the late swing to the early stance, that could reflects the primary function of BF, i.e. knee flexor and a hip extensor [5]. It's valuable noticing that bursts of activity from mid-stance to pre-swing slightly anticipate (10 % gait cycle) the RF activity to control hip and knee moments, when body weight is transferred on the contralateral foot [16].

VL presents three main activation modalities, which cover over 95 % of total strides (Fig. 7). The most recurrent modality (48.9 ± 3.6 %) consists of two intervals that form a single active period, from the terminal part of swing phase to the early part of stance: this type of activity is required to generate tension in terminal swing in preparation for weight bearing at initial contact, with the assistance of RF, and to control knee flexion during weight acceptance [4, 5, 34]. In the remaining 50.0 ± 2.9 % of total strides, VL shows a further activity during stance, between heel-off and foot-off, (Fig. 7): this activity can be characterized by a single (3-activation modality) or a double (4-activation modality) activation and is likely required to provide a stabilization of the knee joint, during the rapid flexion of the knee following the heel raise. As BF, also VL does not seem to show a favourite activation modality during gait: the 2-activation modality has no a significant higher occurrence frequency ($p > 0.05$) than the 3-activation modality, though both are significantly much recurrent ($p < 0.05$) than the third most recurrent modality (4-activation modality).

5 Conclusions

The statistical analysis performed on the sEMG signal from lower limb muscles (TA, GL, RF, BF and VL) of 22 young healthy subjects allows a non-invasive evaluation of the variability of muscle activity during gait at self-selected speed and cadence, preserving the subject natural mobility. Findings highlight that each considered muscle adopts different activation modalities within different strides of the same walking. This suggests the presence of a wide variability in the onset/offset of muscle activation interval and in occurrence frequencies, a parameter seldom considered in classic EMG studies. It is worth considering that, within this variability, not usually reported activities are also detected, suggesting a muscles capability of tuning their activity with respect to physiological requests arising during gait. Thus, the present analysis on the variability of physiological activity of lower limb muscles could be a useful reference for discriminating physiological and pathological behaviour and for designing more focused gait studies.

References

1. Frigo C, Crenna P (2009) Multichannel sEMG in clinical gait analysis: a review and state-of-the-art. Clin Biomech 24:236–245
2. Patikas D, Wolf SI, Schuster W et al (2007) Electromyographic patterns in children with cerebral palsy: do they change after surgery? Gait Posture 26:362–371
3. Romkes J, Hell AK, Brunner R (2006) Changes in muscle activity in children with hemiplegic cerebral palsy while walking with and without ankle–foot orthoses. Gait Posture 24:467–474
4. Sutherland DH (2001) The evolution of clinical gait analysis part l: kinesiological EMG. Gait Posture 14:61–70
5. Perry J (1992) Gait analysis: normal and pathological function. Slack Incorporated, Thorofare
6. Petersen TH, Farmer SF, Kliim-Due M et al (2013) Failure of normal development of central drive to ankle dorsiflexors relates to gait deficits in children with cerebral palsy. J Neurophysiol 109:625–639
7. Stewart C, Postans N, Schwartz MH et al (2007) An exploration of the function of the triceps surae during normal gait using functional electrical stimulation. Gait Posture 26:482–488
8. Sutherland DH, Cooper L, Daniel D (1980) The role of the ankle plantar flexors in normal walking. J Bone Joint Surg Am 62:354–363
9. Di Nardo F, Ghetti G, Fioretti S (2013) Assessment of the activation modalities of gastrocnemius lateralis and tibialis anterior during gait: a statistical analysis. J Electromyogr Kinesiol 23:1428–1433
10. Di Nardo F, Fioretti S (2012) Statistical analysis of surface electromyographic signal for the assessment of rectus femoris modalities of activation during gait. J Electromyogr Kinesiol 23:56–61
11. Di Nardo F, Fioretti S (2014) Emg-based analysis of treadmill and ground walking in distal leg muscles. IFMBE Proc 41:611–614. doi:10.1007/978-3-319-00846-2_151
12. Di Nardo F, Mengarelli A, Ghetti G et al (2014) Statistical analysis of EMG signal acquired from tibialis anterior during gait. IFMBE Proc 41:619–622. doi:10.1007/978-3-319-00846-2_153
13. Di Nardo F, Mengarelli A, Maranesi E et al (2014) Assessment of the ankle muscle co-contraction during normal gait: a surface electromyography study. J Electromyogr Kinesiol 25(2):347–354

14. Di Nardo F, Mengarelli A, Maranesi E et al (2014) Influence of gender on the myoelectric signal of shank muscles. In: MESA 2014—10th IEEE/ASME international conference on mechatronic and embedded systems and applications conference proceedings, Senigallia
15. Di Nardo F, Mengarelli A, Maranesi E et al (2015) Gender differences in the myoelectric activity of lower limb muscles in young healthy subjects during walking. Biomed Sig Process Control 19:14–22
16. Agostini V, Nascimbeni A, Gaffuri A et al (2010) Normative EMG activation patterns of school-age children during gait. Gait Posture 32:285–289
17. Arsenault AB, Winter DA, Marteniuk RG (1986) Is there a 'normal' profile of EMG activity in gait? Med Biol Eng Comput 24:337–343
18. Winter DA, Yack HJ (1987) EMG profiles during normal human walking: stride to stride and inter-subject variability. Electroencephalogr Clin Neurophysiol 67:402–411
19. Yang JF, Winter DA (1984) Electromyographic amplitude normalization methods: improving their sensitivity as diagnostic tools in gait analysis. Arch Phys Med Rehabil 65:517–521
20. Bonato P, D'Alessio T, Knaflitz M (1998) A statistical method for the measurement of muscle activation intervals from surface myoelectric signal during gait. IEEE Trans Biomed Eng 45:287–299
21. Balestra G, Knaflitz M, Molinari F (2002) Principles of statistical gait analysis. In: Proceedings of XIV ISEK conference, Vienna
22. Staude G, Flachenecker C, Daumer M et al (2001) Onset detection in surface electromyographic signals: a systematic comparison of methods. J Adv Signal Process 2:67–81
23. Maranesi E, Di Nardo F, Ghetti G et al (2014) A goniometer-based method for the assessment of gait parameters. In: 10th IEEE/ASME international conference on mechatronic and embedded systems and applications conference proceedings, Senigallia
24. Freriks B, Hermens HJ, Disselhorst-Klug C et al (2000) Development of recommendations for sEMG sensors and sensor placement procedures. J Electromyogr Kinesiol 10:361–374
25. Powers CM, Landel RF, Perry J (1996) Timing and intensity of vastus muscle activity during functional activities in subjects with and without patellofemoral pain. Phys Ther 76:946–955
26. Agostini V, Knaflitz M (2012) Statistical gait analysis. In: Acharya RU, Molinari F, Tamura T, Naidu DS, Suri JS (eds) Distributed diagnosis and home healthcare. American Scientific Publishers, Stevenson Ranch, pp 99–121
27. Sutherland DH, Olshen R, Cooper L et al (1980) The development of mature gait. J Bone Joint Surg Am 62:336–353
28. Winter DA (1990) Biomechanics and motor control of human movement, 2nd edn. Wiley, New York
29. Sutherland DH (1966) An electromyographic study of the plantar flexors of the ankle in normal walking on the level. J Bone Joint Surg Am 48:66–71
30. Shiavi R, Bugle HJ, Limbird T (1987) Electromyographic gait assessment, part 1: adult EMG profiles and walking speed. J Rehabil Res 24:13–23
31. Murray MP (1984) Kinematic & EMG patterns during slow, free, and fast walking. J Orthop Res 2:272–280
32. Conrad B, Meinck HM, Benecke R (1986) Motor patterns in human gait: adaptation to different modes of progression. In: Bles W, Brandt T (eds) Disorders of posture and gait. Elsevier Science Publishers BV, Amsterdam, pp 53–67
33. Nene A, Byrne C, Hermens H (2004) Is the rectus femoris really part of the quadriceps? Assessment of the rectus femoris function during gait in able-bodied adults. Gait Posture 10:1–13
34. Winter DA (1991) The biomechanics and motor control of human gait: normal elderly and pathological, 2nd edn. University of Waterloo Press, Waterloo

High Performance Dynamic Threshold Calibration for RR Interval Detection in a QRS Complex Using a Parallel Programing

Mohammed Rajab and Ralf Seepold

Abstract These days computer analysis of ECG (Electrocardiograms) signals is common. There are many real-time QRS recognition algorithms; one of these algorithms is Pan-Tompkins Algorithm. Which the Pan-Tompkins Algorithm can detect QRS complexes of ECG signals. The proposed algorithm is analysed the data stream of the heartbeat based on the digital analysis of the amplitude, the bandwidth, and the slope. In addition to that, the stress algorithm compares whether the current heartbeat is similar or different to the last heartbeat after detecting the ECG signals. This algorithm determines the stress detection for the patient on the real-time. In order to implement the new algorithm with higher performance, the parallel programming language CUDA is used. The algorithm determines stress at the same time by determining the RR interval. The algorithm uses a different function as beat detector and a beat classifier of stress.

1 Introduction

The electrocardiogram (ECG) provides electrical signals from the hearts during the cardiac cycle. The ECG analysis is an important tool in the measurement of cardiac diseases. The ECG signals have different features in a single cardiac cycle as 'P' wave, "Q" wave, 'R' wave, 'S' wave, and 'T' wave. The QRS complex is the important feature of an ECG. There are many algorithms used to detect the QRS on the real time using a microcontroller. Furthermore, there are many systems designed and implemented to perform signal processing tasks, for example, Hotlet tap analysis, and real-time patient monitoring. In 1985, the algorithm called (Pan and

M. Rajab (✉) · R. Seepold
HTWG Konstanz, Brauneggerstreet 55, 78462 Constance, Germany
e-mail: mrajab@htwg-konstanz.de
URL: http://uc-lab.in.htwg-konstanz.de

R. Seepold
e-mail: ralf.seepold@htwg-konstanz.de
URL: http://uc-lab.in.htwg-konstanz.de

M. Conti et al. (eds.), *Mobile Networks for Biometric Data Analysis*,
Lecture Notes in Electrical Engineering 392, DOI 10.1007/978-3-319-39700-9_16

Tompkins) was implemented. The QRS detection algorithm implemented by Pan and Tompkins [1] is one of the most widely used; the algorithm detects QRS complexes from ECG signal.

QRS complexes are an important part of the ECG signal processing system [1]. In this paper, the QRS will be used to measure the stress. Stress has become a problem affecting to a lot of people in their life [2]. Stress can be an illness directly (physiological effects) or indirectly (maladaptive health behaviors) [3]. There are some factors that put people under stress, like work overload, time pressure, complex tasks, high responsibility, job insecurity, and poor physical work conditions [4]. The aims of this paper are the definition of a real-time algorithm capable of pattern detection. Secondly, we will test the algorithm. Thirdly, we will implement the algorithm on a microcontroller. Furthermore, after detecting the QRS complexes, the stress algorithm will detect the RR interval from QRS complexes. Thus, the algorithm has to analysis all QRS complexes to take a decision if the patient under stress or not. Using paralleling possessors will lead to make an efficient decision, where this decision will help the patient to make a reaction for his/er stress statue. A weakness of ECG sensors are the sensibility to interferences caused by high-frequency noise, in the case, we have to rerecord this amplified [4]. The Pan and Tompkins algorithm uses a low-pass and a high-pass filter to delete the noise from the ECG signal. Thus, it will pass through derivative, then squaring and the window integration steps. Finally, the R-peaks are detected when a threshold technique is applied. R peaks from QRS complexes have been extensively investigated over the past two decades [5]. In this paper will investigate in the detection algorithm of the Pan and Tompkins using a parallel language. Furthermore, explain the new method of stress detection. Finally, The implementation description is given in the next section.

2 Background

Many QRS detection algorithms have been implemented. Some algorithms are based on the amplitude and derivative. Table 1 shows different suggestions for QRS detection from different researchers.

2.1 Pan & Tompkins QRS Detection Algorithm

The algorithm developed by Pan and Tompkins is one of the most popular algorithms for QRS detection. We will use this algorithm in our work to measure the stress. Figure 1 gives an overview of the algorithm. In a first step, the Pan and Tompkins algorithm passes the signal over the filters. It is passed through a low and

Table 1 Different ways to detect the QRS

Researcher name	QRS detection algorithms
These algorithms are based on the first derivative, it has to calculat be a given formula, furthermore by a fraction of the maximum slops in the first derivative, it can calculat the slope threshold:	
Fraden and Neuman [11]	Developed QRS detection, where the threshold is calculated as a fraction of the peak
Gustafson	QRS peak candidate when the first derivative and following to next three derivative values overreach, the threshold, and the next two sample points should have positive amplitude
These algorithms based on the first and second derivatives:	
Balda [11]	He proposed searching values were the threshold in the weighted collection of the first and second derivatives
Ahlstrom and Tompkins [11]	Suggested to use the absolute values of the first derivative and they are smoothed then added with the absolute values of the second derivative. Where it is used the primary and secondary threshold
There are some algorithms, which these algorithms are based on some steps of digital filters which are developed to remove components of QRS complex from ECG. Then the final steps for these algorithms are a square wave whose on-intervals correspond to the segments with QRS [12]	
There are some other algorithms are based on the length transformation and on syntactic methods by calculating the length and energy signals of ECG [11]	

high pass filter to reduce the effect of noise. According to the algorithm the low-pass filter Eq. (1) is:

$$y(n) = 2y(n-1) - y(n-2) + x(n) - 2x(n-6) + x(n-12) \quad (1)$$

The low-pass filter Eq. (2) is:

$$y(n) = y(n-1) - 1/32\,x(n) + x(n-16) - x(n-17) + 1/32\,x(n-23) \quad (2)$$

After these two steps of filtering, the ECG single is passed through differentiated to provide the QRS slop information by this equation:

$$y(n) = 1/8[2x(n) + x(n-1) - x(n-3) - 2x(n-4)] \quad (3)$$

Using Eq. (4), the ECG signal is squared to make the entire point positive with higher frequencies

$$y(n) = x^2(n) \quad (4)$$

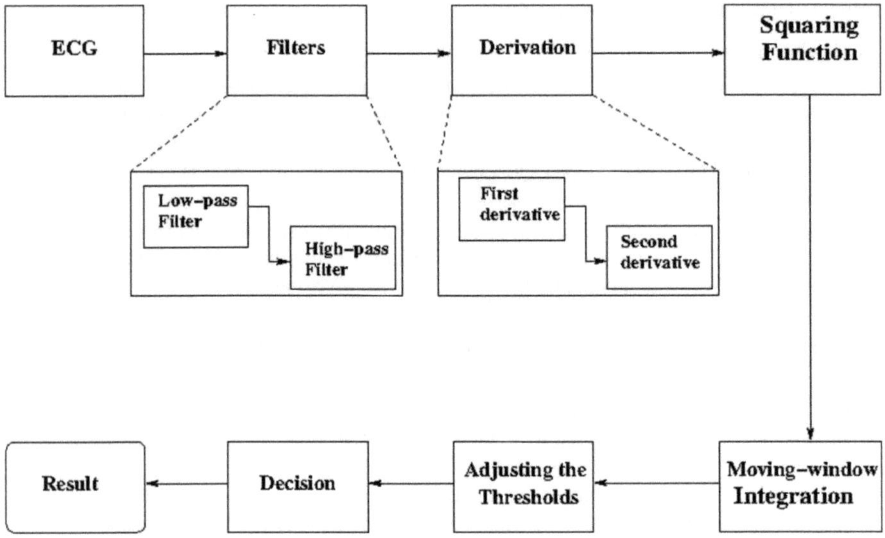

Fig. 1 Overview steps of the Pan and Tompkins algorithm

After that, the algorithm is using the window integration technique to take out waveform feature information using this Eq. (5):

$$y(n) = 1/N[x(n - (N - 1)) + x(n - (N - 2)) + \cdots + x(n)] \quad (5)$$

Where N is the size of sliding window. The window size depends on the sample rate. In the next step, the thresholds are adjusted. In the lower case the algorithm has to use search back technique to find a lost peak. Thus, the algorithm has to classify the peak as a signal peak whether it exceeds or as a noise peak. Finally, the integration is checked by the filtered signal.

On the other hand, some algorithms have improved the QRS detection, for example, the algorithm [6] improved the QRS complexes detection in the Arrhythmia ECG signals by using non-stationary effects, low signal-to-noise ratio, negative QRS polarities, low amplitudes, and ventricular.

3 The Architecture of Threshold Calibration for RR Interval Detection in a QRS Complex

The architecture of the stress detection algorithms is shown in Fig. 2. The idea of the algorithms is to use the QRS complex detection algorithms result to determine an R-to-R interval classification and to estimate the width for each beat.

First of all, the algorithm detects the QRS complex of ECG signal. Thus, we have to check some condition rules for our basic detection, if the R peak in the signal larger than the threshold, then the peak is QRS complex, otherwise the peak will be noise. Furthermore, if an interval is equal to 1.5 times the average of R-to-R interval, it is counted as R peak in case it is larger than half of the detection threshold, and the R peak pursued the preceding detection by at least 360 ms [7]. Then the peak is classified as a QRS complex. Depend on these two detection rules, it can estimate the QRS peak and noise peak. Also from the R-to-R interval, it can calculate the mean of the last eight R-to-R intervals.

By all the result from Pan and Tompkins algorithm, we can measure whether a person is under stress or not by analysis the last eight R-R intervals through comparing the present peak and the previous peak. If the QRS complex is equal to a fixed value 360 ms and the R-R interval has the same width compared to the next

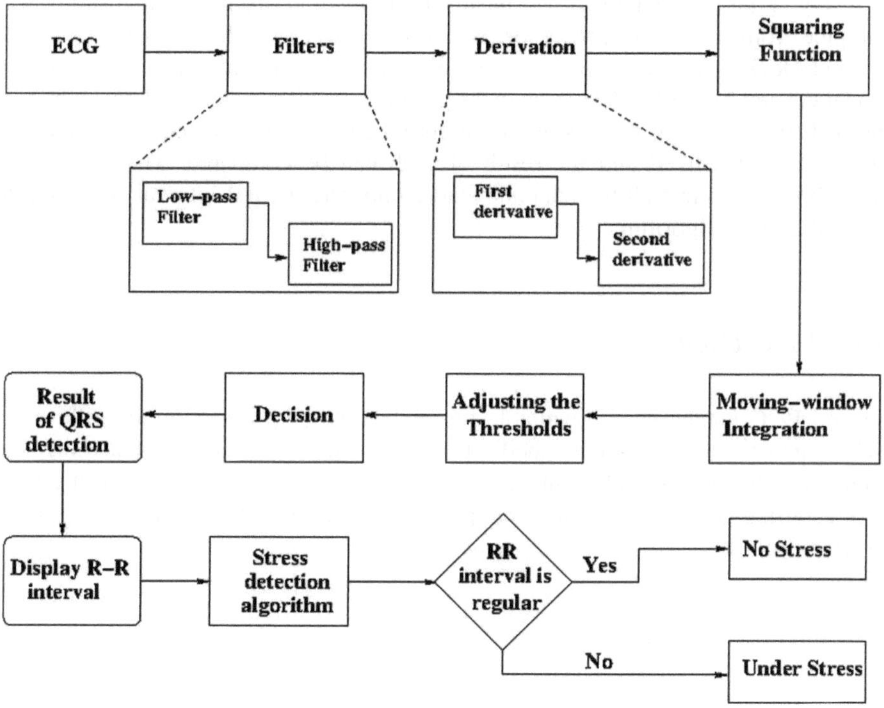

Fig. 2 Overview steps of the stress detection

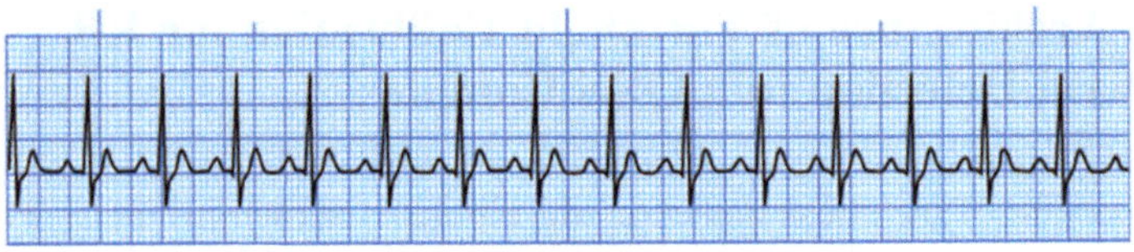

Fig. 3 The R-R interval rhythm is regular

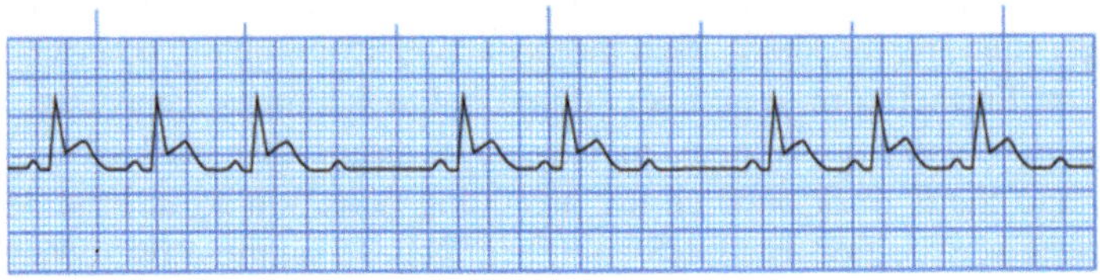

Fig. 4 The R-R interval rhythm is irregular

R-R interval then we can classify the person as a no stress state where the heart rhythm is regular as shown in Fig. 3.

If the R-R interval rhythm is irregular, as shown in Fig. 4, the person is considered under stress.

Using a parallel language, it means each method from our algorithm will compute from a different processor. By using CUDA (which stands for Compute Unified Device Architecture, is a parallel computing platform and programming model created by NVIDIA) we segment the data in several interval. This interval is dispatched to different threads and different processors that the sequential analysis of data is parallelized and the result of each can be compared. That means the algorithm will collect all the data and start comparing the RR interval according to stress detection algorithm.

3.1 Data Used

Several ECG databases are available for evaluating QRS detection algorithms for ECG signals. PhysioNet published 11 representative datasets for analyzing and comparing the proposed algorithms. The sets categorized in different subject groups and recording conditions, such as sampling rates (between 128 Hz and 1 kHz) and interferences [8].

Table 2 R-R interval classification

Classification name	Conditions
Two interval match	If the they are within 12.5 % of their mean [7]
Normal	If the present interval matches the previous interval (*no stress*)
Abnormal	If the present interval is not matching with the previous interval (*stress*)

3.2 R-R Interval Classification

To classify the R-R interval shown in the Table 2 [7], the following assumptions are made.

3.3 Pseudocode of Stress Detection Algorithm

```
Import the result of QRS detection from Pan and Tompkins algorithm
  For n=1 to the length of the peak do
   If The peak is QRS do
      Divide the data stream in different sigma
                Dispatch each sigma in different threads
Then Feed the stress detection algorithm

      Compare the RR interval with the previous RR interval from the last eight RR
      intervals
                If they have same width
                   Add to the buffer 1
                      Acount1 = +1
             else
                Add to the buffer 0
                   Account0=+1
           End if
           End if
          N++
          End for
    Wait all the result from each thread

    Compare the result
    End for
    Result = camper the account 1 with account 0
          If acount0> acount1
            The person UNDER STRESS
         else
           NO STRESS
        End if
    End function
```

4 Result and Future Work

The software has been implemented using a parallel programming language called CUDA to have higher performances. Firstly, the result of the Tompkins QRS detection algorithm will apply directly to the new stress algorithm with the same data taken from [9, 10]. The result shows a correlation between stress and the heartbeat. According to this project, in the future, we will implement this algorithm in the mobile application which will be easy for the patient to get the stress information from a mobile application. Furthermore, this prototyping will give the final data analysis and the final decision (if the person is under stress or there is no stress).

5 Conclusions

In this paper, we propose a software implementation of a stress detection algorithm based on the Pan and Tompkins QRS detection algorithm. Our current work is focused on explaining and enhancing the performance of the QRS detection by using the new algorithm architecture to detect stress. In this algorithm, we presented a different method, which each method has enhanced to get a faster decision, starting from an ECG signal data, then filtering, derivation, squaring, integration, and then from the output will use the QRS complex and the R-to-R interval. Finally, we have to pass these results into an analysis of the QRS complex and the R-to-R interval and to feedback whether a person is under stress or not using a parallel programming.

References

1. Pan J, Tompkins WJ (1985) A real-time QRS detection algorithm. IEEE Trans Biomed Eng 3:230–236
2. Bakker J, Pechenizkiy M, Sidorova N (2011) What's your current stress level? Detection of stress patterns from GSR sensor data. In: IEEE 11th international conference on data mining workshops (ICDMW). IEEE
3. Glanz K, Schwartz M (2008) Stress, coping, and health behavior. Health behavior and health education: theory, research, and practice, pp 211–236
4. Lascu, M, Dan L (2007) LabVIEW event detection using Pan-Tompkins algorithm. In: Proceedings of the 7th WSEAS international conference on signal, speech and image processing, Beijing, China
5. Elgendi M Fast T-wave detection with annotation of P and T waves in the MIT-BIH arrhythmia database
6. Elgendi M et al (2008) A robust QRS complex detection algorithm using dynamic thresholds. In: International symposium on computer science and its applications, CSA'08. IEEE (2008)
7. Hamilton P (2002) Open source ECG analysis. Computers in cardiology. IEEE
8. MIT-BIH Arrhythmia Database www.physionet.org
9. Scherz WD, Seepold R (2014) Physiologic parameter calculation supporting local and telemetric analysis. In: 6th European conference of the international federation for medical and biological engineering (MBEC)

10. Martínez Madrid N, Martínez Fernández J, Seepold R, Augusto JC (2013) Ambient assisted living (AAL) and smart homes. Autonomous Sensor Networks, Springer Series on Chemical Sensors and Biosensors, vol 13. ISSN: 1612-7617, pp 39–71
11. Pavlatos C et al (2005) Hardware implementation of Pan & Tompkins QRS detection algorithm. Proceedings of the EMBEC'05 conference
12. Okada M (1979) A digital filter for the QRS detection complex. IEEE Trans Biomed Eng 26:700–703

An Analog Front-End for Combined EMG/ECG Wireless Sensors

Giorgio Biagetti, Paolo Crippa, Simone Orcioni and Claudio Turchetti

Abstract In this work we describe a combined wireless sensor, able to capture either the electromyographic (EMG) or the electrocardiographic (ECG) signal. Since the two signals differ mainly because of their bandwidths, with the ECG being shifted towards lower frequencies, a simple and inexpensive circuit solution has been developed to allow an optional software-based bypass of the high-pass filtering action incorporated in the EMG signal amplifier, without sacrificing neither signal quality nor bandwidth in the much more demanding EMG path.

1 Introduction

Electromyographic (EMG) and electrocardiographic (ECG) signals are two important means by which a person's health status, fitness level, physical performance, etc., can be assessed [1–4]. These two kinds of signals both originate from the electrical potentials generated by contracting muscles and are both typically collected by means of electrodes contacting the surface of the skin, requiring a carefully designed, high-input-impedance, low-noise amplifier before they can be recorded.

Despite these similarities, devices used to record EMG and ECG signals are usually distinct [5], with perhaps a few exceptions [6]. The main reason for this is that the frequencies of interest for the ECG signals are typically confined to those below 100 Hz. For skeletal muscles EMG signals, on the other hand, there are

G. Biagetti (✉) · P. Crippa · S. Orcioni · C. Turchetti
Dipartimento di Ingegneria dell'Informazione, Università Politecnica delle Marche, via Brecce Bianche, 12, 60131 Ancona, Italy
e-mail: g.biagetti@univpm.it

P. Crippa
e-mail: p.crippa@univpm.it

S. Orcioni
e-mail: s.orcioni@univpm.it

C. Turchetti
e-mail: c.turchetti@univpm.it

M. Conti et al. (eds.), *Mobile Networks for Biometric Data Analysis*,
Lecture Notes in Electrical Engineering 392, DOI 10.1007/978-3-319-39700-9_17

significant components at least up to 500 Hz, whilst below 5 Hz there are many motion-induced artifacts that must be rejected by the amplifier, lest saturation of the gain stages should occur.

In this work we present a circuit solution for the input stage of a combined EMG/ECG wireless sensor, loosely based on the wireless EMG system described in [7]. The input stage of the EMG sensors has been altered so as to incorporate a feedback loop that actively rejects frequency components below 5 Hz, to obtain a clean and efficient EMG signal path. This change would have made the system unable to capture the ECG signal. But by sampling the rejected low-frequency components as they are being cancelled, and by proper digital combination of the two signals, it is possible to recover the complete bandwidth needed to record a (single-channel) ECG signal, as will be shown next.

2 System Architecture

The input signal is collected by means of AC-coupled electrodes, and after applying proper bias, is amplified by a high-gain, high-CMRR, low-offset instrumentation amplifier (IA). The schematic of the input stage is shown in Fig. 1. The signal is split into two paths after having been amplified. The upper path shown in the figure is the standard EMG signal path. It comprises an anti-aliasing low-pass filter (R_{lp} and C_{lp}) that, together with the high-pass filtering effect obtained by the DC-cancellation feedback loop (lower path, comprising the Miller integrator $R_{\mathrm{hp}}, C_{\mathrm{hp}}$), results in a 5 Hz–500 Hz band-pass filtered signal v_F, which is then digitized by the ADC. Additionally, the employed ADC has a programmable gain amplifier (PGA) in front of it, to further reduce quantization noise while acquiring extremely low-level EMG signals. The action of the DC-cancellation feedback loop is extremely important to avoid saturation of the PGA, but, as will be shown next, will impede direct measurement of signals containing low-frequency components.

A detailed analysis of the circuit will help understand how this problem can be circumvented without altering the circuit structure. In particular, neglecting

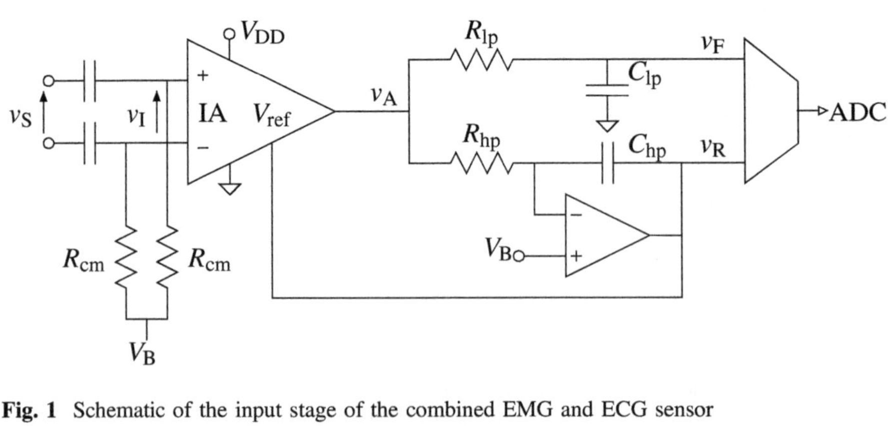

Fig. 1 Schematic of the input stage of the combined EMG and ECG sensor

higher-order effects and circuit non-idealities, the instrumentation amplifier gives an output voltage given by

$$V_A(s) = V_R(s) + GV_I(s) \tag{1}$$

where G is the fixed amplifier gain, and the feedback reference voltage $V_R(s)$ is obtained by integrating the amplifier output itself:

$$V_R(s) = \frac{-1}{sR_{hp}C_{hp}}(V_A(s) - V_B(s)) + V_B(s) \tag{2}$$

where $V_B = V_{DD}/2$ is a constant bias voltage. It thus results

$$V_A(s) = \frac{sR_{hp}C_{hp}}{1 + sR_{hp}C_{hp}} GV_I(s) + V_B(s) \tag{3}$$

i.e., the DC-cancellation feedback loop also results in a high-pass filtering action on v_A with a cut-off frequency $f_{hp} = 1/(2\pi R_{hp}C_{hp})$.

By substituting (3) into (2) it is possible to obtain

$$V_R(s) = \frac{-1}{1 + sR_{hp}C_{hp}} GV_I(s) + V_B(s) \tag{4}$$

from which it can be seen that v_R contains the missing 0 Hz to f_{hp} band of the original signal v_I.

The signal sent to the PGA/ADC combo in the EMG signal path is v_F, which is simply a low-pass filtered version of v_A:

$$V_F(s) = \frac{1}{1 + sR_{lp}C_{lp}} \frac{sR_{hp}C_{hp}}{1 + sR_{hp}C_{hp}} GV_I(s) + \frac{V_B(s)}{1 + sR_{lp}C_{lp}} \tag{5}$$

with $1/(2\pi R_{lp}C_{lp}) = f_{lp}$ being the cut-off frequency of the first-order filter employed. The circuit has been designed so that $f_{hp} \simeq 5$ Hz and $f_{lp} \simeq 500$ Hz, hence v_F is all that needs to be sampled to acquire the EMG signal.

The ECG signal extends to much lower frequencies, so that it cannot be recovered by sampling v_F alone. But if the signal v_R is also sampled, it is possible to reconstruct the missing low frequencies. Since this signal is already available because it is needed to cancel the offset even for EMG-only units, and many ADCs also have input multiplexers anyway, adding this capability to the circuit is essentially cost-free.

In order to rebuild an ECG signal, we need to bypass the high-pass filtering action in (5). This is indeed an easy task, as (5) can be rewritten as

$$V_F(s) = H_{lp}(s)H_{hp}(s)GV_I(s) + H_{lp}(s)V_B(s) \tag{6}$$

where

$$H_{\rm lp}(s) = \frac{1}{1 + s\,R_{\rm lp}C_{\rm lp}}$$

and

$$H_{\rm hp}(s) = \frac{s\,R_{\rm hp}C_{\rm hp}}{1 + s\,R_{\rm hp}C_{\rm hp}}.$$

By noting that $v_{\rm R}$ can be low-pass filtered to obtain

$$H_{\rm lp}(s)V_{\rm R}(s) = \frac{1}{1 + s\,R_{\rm lp}C_{\rm lp}}\frac{-1}{1 + s\,R_{\rm hp}C_{\rm hp}}GV_{\rm I}(s) + \frac{V_{\rm B}(s)}{1 + s\,R_{\rm lp}C_{\rm lp}} \tag{7}$$

it is apparent that it can be subtracted from $v_{\rm F}$ to get

$$V_{\rm X}(s) = V_{\rm F}(s) - H_{\rm lp}(s)V_{\rm R}(s) = \frac{1}{1 + s\,R_{\rm lp}C_{\rm lp}}GV_{\rm I}(s) = H_{\rm lp}(s)GV_{\rm I}(s) \tag{8}$$

which no longer contains any high-pass filtering, and it is thus more than suitable to record an ECG signal. Of course, no DC cancellation is performed in this case, but it is not usually an issue for ECG signals that, because of their larger amplitude compared to EMG signals, do not require setting the PGA to high gains. The AC coupling done at the input (between $v_{\rm S}$ and $v_{\rm I}$, with a cut-off of about 150 mHz) suffices.

Actually, in (7), the action of the filter $H_{\rm lp}(s)$ on the signal $V_{\rm R}(s)$ is negligible because the filter cutoff frequency $f_{\rm lp}$ is two decades above the bandwidth of the signal $V_{\rm R}(s)$ (limited to $f_{\rm hp}$). It can thus be altogether omitted, so that

$$V_{\rm X}(s) \simeq V_{\rm F}(s) - V_{\rm R}(s). \tag{9}$$

If the ADC had been able to sample the two channels $v_{\rm F}$ and $v_{\rm R}$ simultaneously, we would only have needed to digitally subtract one from the other to get an ECG signal out of the sensor. Unfortunately, cheap ADCs do not have simultaneous sampling capabilities, and the system must rely on a signal multiplexer. To compensate the delays, a digital interpolation filter will be used, as described next.

3 Digital Filtering

In the normal EMG mode, the ADC would sample only its input signal $v_{\rm F}$, at a sample rate of 2 kHz. These samples would then be gathered into packets comprising 32 ms worth of data and wirelessly transmitted to a coordinator node, for real-time display, analysis, and archival.

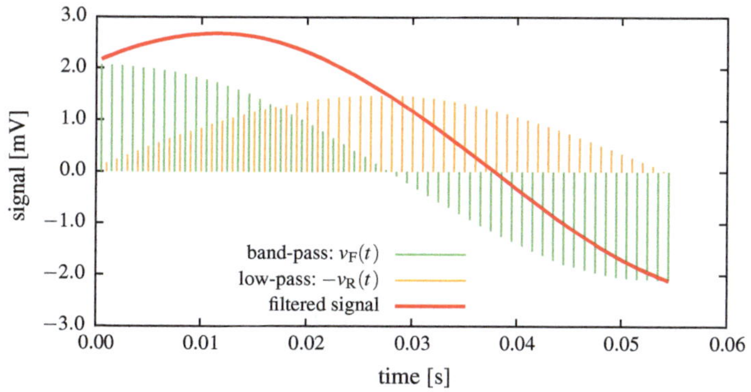

Fig. 2 Interpolation of the low-pass and band-pass signals by a single low-pass digital filter with DC gain of 2. The filter group delay has been compensated in the figure for illustration purposes

Since we aim at obtaining a sensor that can be put either in EMG or in ECG mode only by means of a software command, we minimized alterations to the underlying wireless protocol. The bit rate has thus been kept at the same value also for the ECG mode. In this mode of operation, the ADC would alternatively sample the two signals $v_F(t)$ and $v_R(t)$, each at $f_s = 1$ kHz, interleaving the two channels, so that samples are 0.5 ms apart. Samples pertaining to the v_R channel are then digitally complemented to account for the sign change in (9).

Having halved the sampling frequency, some aliasing near the upper portion of the bandwidth, f_{lp}, is to be expected, but this has been deemed to be acceptable because of the lower range of frequencies of interest in the ECG signal. It will be largely removed by a digital filtering performed at the receiver end.

Formally, the receiver should indeed extract the two streams related to v_F and $-v_R$, interpolate them to compensate for the phase shift between the sampling instants of the two streams, and finally add them together. In practice, a single interpolation digital filter acting directly on the interleaved data stream can be used to perform exactly all these tasks at once, as shown in Fig. 2.

4 Results

Figure 3 shows a sample recording obtained with the described system. From bottom to top, the interpolated low-pass signal $-v_R(t)$, the interpolated band-pass signal $v_F(t)$, and the rebuilt signal $v_X(t)$ are shown. The interpolation filter used was a fifth-order Butterworth filter with a corner frequency of 75 Hz. The middle signal is what we would have got using a standard band-limited EMG sensor alone. By adding the missing low frequencies, an accurate measure of an ECG signal can be obtained, as shown in the top trace.

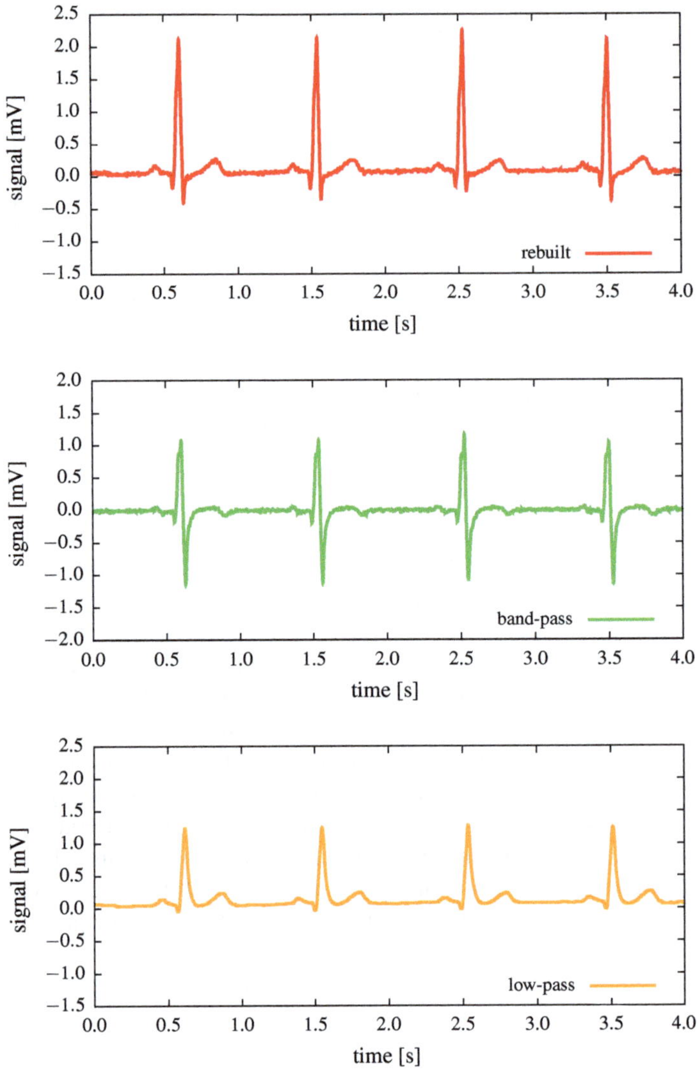

Fig. 3 Example of acquired ECG signal. From bottom: low-pass component $-v_R(t)$, band-pass component $v_F(t)$, rebuilt signal

To further validate the quality of the proposed recombination method, the frequency response of the system was also measured. To do so, a sinusoidal signal generator was set to output a signal with 2.5 V amplitude and varying frequency, between about 0.2 Hz and 500 Hz. This signal was then applied through a 60 dB attenuator to the input v_S of the system, thus obtaining a 2.5 mV input. The amplified signals v_F and $-v_R$ were then digitally acquired and recombined as

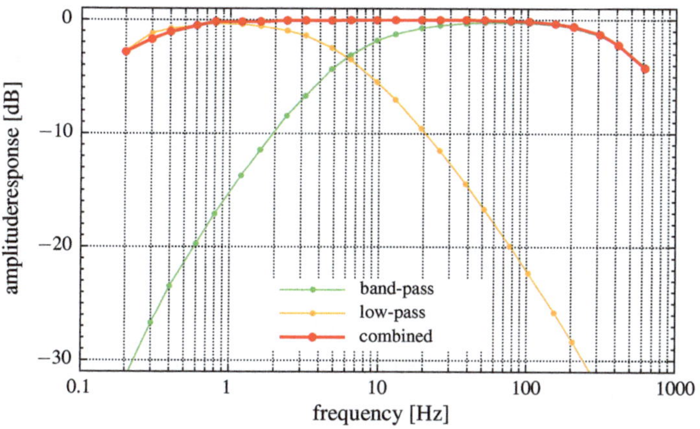

Fig. 4 Measured system frequency response

discussed earlier. To avoid acquisition artifacts and mitigate the effects of noise at these very low signal levels and frequencies, at least 5 s of signal was recorded for every frequency step.

The recorded tracks were then digitally processed to interpolate the samples and estimate their mean amplitude. To remove interpolation artifacts near the boundaries of the tracks, and occasional noise spikes occurring in a few of them, two averages were computed. A first average was computed from instantaneous amplitude estimates, together with their standard deviation. The final average was computed after having removed estimates diverging by more than twice their standard deviation from the original average. The interpolation filter used for this test was chosen with a corner frequency of 500 Hz so as not to interfere with the measurement.

The results are shown in Fig. 4. At extremely low frequencies, the AC-coupling of the input dictates the frequency response of the whole system. But within the ECG bandwidth of interest (typically between about 0.5 Hz and 50 Hz), an almost perfectly flat response was obtained, despite the crossover of the filters being just in the middle of the useful band. Higher frequencies are only of interest for the EMG mode of operation, which do not perform any signal recombination and is thus unaffected by the digital interpolation filter. In the ECG mode, this interpolation filter will most likely cut much of the band above about 100 Hz to further reduce noise.

The noise performance of the system was also measured. To this end, an acquisition was performed with the sensor inputs shorted together, and the raw signals $v_R(t)$ and $v_F(t)$ (i.e., before the reconstruction filter) analyzed. The resulting noise spectral density is shown in Fig. 5. The same figure also reports the analysis of the rebuilt signal, but there the effects of the digital interpolation filter, which has a cut-off frequency of 75 Hz, dominate the higher frequency range.

The effect of the Miller integrator on the noise introduced by the instrumentation amplifier is clearly apparent in the slope of the noise in the low-pass component,

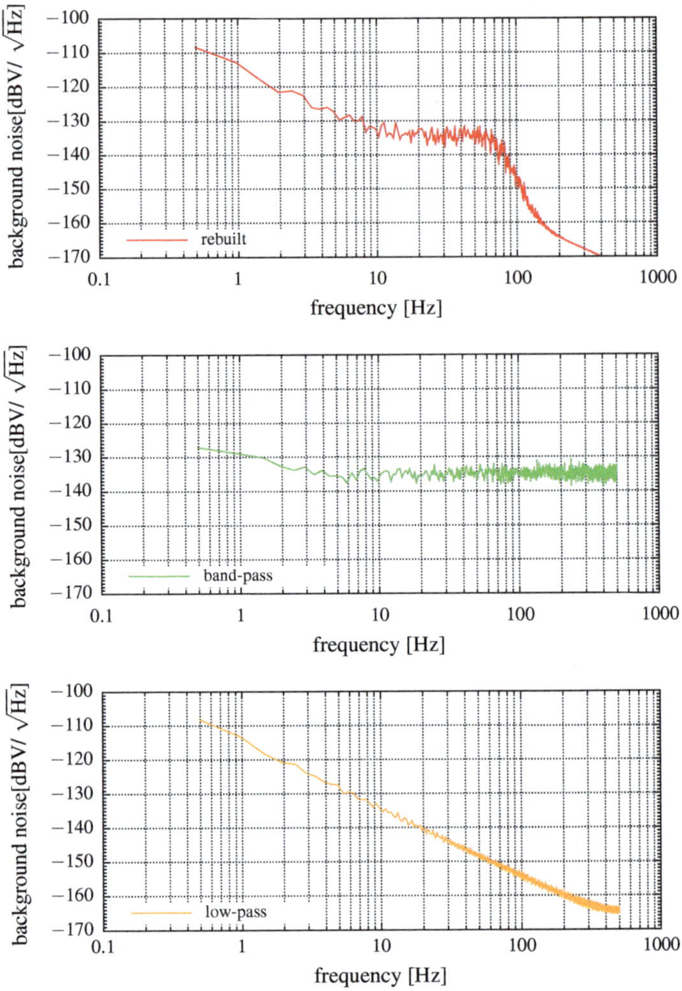

Fig. 5 Measured input-referred noise spectral density for the acquired signals $v_R(t)$ (low-pass component) and $v_F(t)$ (band-pass component), computed from the ADC output with the external electrodes shorted, together with that of the rebuilt signal

while the noise in the band-pass component is essentially white within the frequencies of interest. This results in an RMS noise of just 4.1 μV for $v_F(t)$, (as a reference, the input-referred ADC LSB is 10 μV), while it is twice as much for $v_R(t)$. This is not much of an issue since, as stated above, ECG signals usually have larger amplitudes than EMG signals. To better compare the noise densities, Fig. 6 reports the same three curves smoothed out by averaging frequency bins occurring

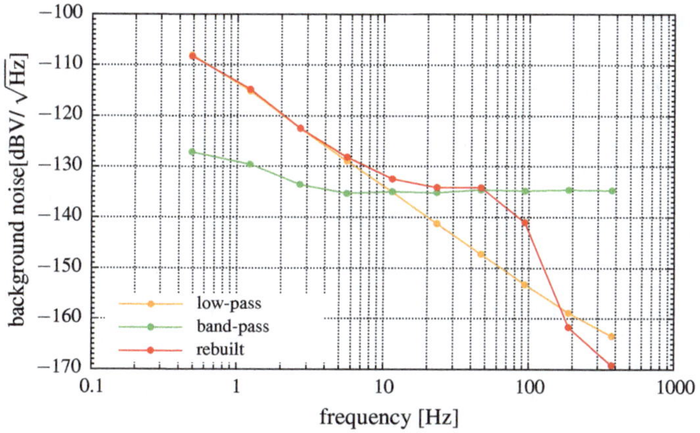

Fig. 6 Comparison of the noise spectral densities for the three signals considered. The densities have been smoothed by averaging their values in octave-wide bins

within the same octave. It is apparent that the noise in the ECG signal is dominated by the low-frequency components amplified by the Miller integrator, but it remains within acceptable limits.

5 Conclusions

In this paper an inexpensive circuit solution for adding the capability of acquiring either the ECG or the EMG signal using the same wireless sensor has been presented.

The ECG signal is acquired by simply modifying an EMG amplifier through an ad-hoc bypass of the high-pass limiting stage of the amplifier itself. This bypass is implemented almost entirely in software so as not to degrade the system capability of acquiring a clean EMG signal, and no additional components have been added in the EMG signal path, ensuring both an essentially cost-free feature extension of the sensor and no degradation of its noise performance in EMG mode. The noise performance in ECG mode are slightly worse but still well within acceptable limits.

Being software-based, the bypass can be included or excluded on command, so that the wireless sensor can be put in either mode, and quickly switched from one to the other, according to possibly changing needs. This allows further overall system cost savings, as it might reduce stocking costs and complexity.

Finally, the good quality of the recorded ECG signal, tested on a prototype of the system, and the very flat overall frequency response measured on the same, demonstrate the validity and feasibility of the proposed approach.

References

1. Biagetti G, Crippa P, Curzi A, Orcioni S, Turchetti C (2014) A multi-class ECG beat classifier based on the truncated KLT representation. In: Proceedings of the 2014 UKSim-AMSS 8th European modelling symposium on mathematical modelling and computer simulation (EMS2014), Pisa, Italy, pp. 93–98
2. Biagetti G, Crippa P, Curzi A, Orcioni S, Turchetti C (2015) Analysis of the EMG signal during cyclic movements using multicomponent AM-FM decomposition. IEEE J Biomed Health Inf 19(5):1672–1681
3. Biagetti G, Crippa P, Falaschetti L, Orcioni S, Turchetti C (2015) A rule based framework for smart training using sEMG signal. In: Neves-Silva R, Jain LC, Howlett RJ (eds) Intelligent decision technologies: 2015, smart innovation, systems and technologies series, vol 39. Springer, Switzerland, pp 88–99
4. Morici A, Biagetti G, Turchetti C (2009) 2.4 GHz wireless electromyograph system with statistically optimal automatic gain control: design and performance analysis. In: Proceedings of the 2009 international conference on bio-inspired systems and signal processing (BIOSIGNALS 2009), Porto, Portugal, pp 39–45
5. Pantelopoulos A, Bourbakis N (2008) A survey on wearable biosensor systems for health monitoring. In: 30th annual international conference of the IEEE engineering in medicine and biology society, (EMBS 2008), Vancouver, BC, pp 4887–4890
6. Burns A, Doheny EP, Greene BR, Foran T, Leahy D, O'Donovan K, McGrath MJ (2010) SHIMMER: an extensible platform for physiological signal capture. In: 2010 annual international conference of the IEEE engineering in medicine and biology society (EMBS 2010), Buenos Aires, Argentina, pp 3759–3762
7. Kobayashi H (2013) EMG/ECG acquisition system with online adjustable parameters using ZigBee wireless technology. Electron Commun Jpn 96(5):1–10

Stress Determent via QRS Complex Detection, Analysis and Pre-processing

Wilhelm Daniel Scherz, Juan Antonio Ortega, Ralf Seepold and Natividad Martínez Madrid

Abstract Stress is recognized as a predominant disease with raising costs for rehabilitation and treatment. Currently there several different approaches that can be used for determining and calculating the stress levels. Usually the methods for determining stress are divided in two categories. The first category do not require any special equipment for measuring the stress. This category useless the variation in the behaviour patterns that occur while stress. The core disadvantage for the category is their limitation to specific use case. The second category uses laboratories instruments and biological sensors. This category allow to measure stress precisely and proficiently but on the same time they are not mobile and transportable and do not support real-time feedback. This work presents a mobile system that provides the calculation of stress. For achieving this, the of a mobile ECG sensor is analysed, processed and visualised over a mobile system like a smartphone. This work also explains the used stress measurement algorithm. The result of this work is a portable system that can be used with a mobile system like a smartphone as visual interface for reporting the current stress level.

W.D. Scherz (✉) · R. Seepold
HTWG Konstanz, Brauneggerstr. 55, 78462 Konstanz, Germany
e-mail: wscherz@htwg-konstanz.de
URL: http://uc-lab.in.htwg-konstanz.de

R. Seepold
e-mail: ralf.seepold@htwg-konstanz.de
URL: http://uc-lab.in.htwg-konstanz.de

J.A. Ortega
Universidad de Sevilla, Avda. Reina Mercedes s/n, 41012 Seville, Spain
e-mail: jortega@us.es
URL: http://www.us.es

N.M. Madrid
Reutlingen University, Alteburgstr. 150, 72762 Reutlingen, Germany
e-mail: natividad.martinez@reutlingen-university.de
URL: http://iotlab.reutlingen-university.de

M. Conti et al. (eds.), *Mobile Networks for Biometric Data Analysis*,
Lecture Notes in Electrical Engineering 392, DOI 10.1007/978-3-319-39700-9_18

1 Introduction

Nowadays stress is regarded as a negative sensation and is recognized as a predominant disease [1] because of an overabundance of stress in modern life. Overabundance of high stress levels can cause erratic response to physical, emotional or mental demands [2, 3].

Back to the beginning of human history, stress has developed as a natural reaction that allowed primitive people to evaluate the situation and to act in the most effective way when they were confronted to dangerous situation. Stress triggers biological mechanism that reorganises the work of the organism depending on the priorities in order to reach high performance. That leads to the increase of physical capacities and performance in dangerous situation, in other words a fight or flight response.

Among the consequences of stress for modern society are the increasing amounts of people who face severe and irreversible long-term limitations like burnouts, cardiac problems like myocardial infarction, etc. This leads to increasing costs of treatment and healing for people that are suffering from high levels of stress.

By modern people stress usually appears as a result of the exposure of high demands and pressure in daily life, both mental and physical [4], e.g. constant decision taking demands or constant time pressure. Currently stress has become permanent and as result it leads to different disorders. A good example of the consequences of overabundance of stress caused by overwork are fatigue, sleep problems, etc. [5].

Stress can be created or induced in a laboratory using special exercises like the Trier tests [6], the Stroop color word test [7], etc.

Some symptoms of stress are shown in Fig. 1. Lifestyle, technologies and culture made significant changes in our lives but the reaction of our body to stress remains the same.

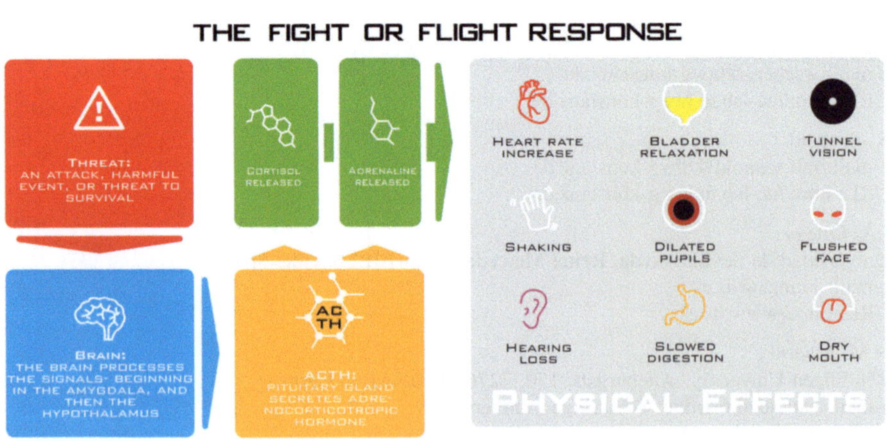

Fig. 1 Symptoms of stress and their physical responses

In case of danger and thread, processes of preparing the body for fight or flight are activated. Brain starts to release hormones like cortisol and adrenaline; this increases the performance of organs that are vital for surviving and saves resources by reducing the activity of organs that are not vital for surviving a dangerous situation. The following steps are important and useful for surviving and responding effectively to dangerous situations by increasing the blood flow to bring more oxygen to muscles (the heart rate increases); the pupils are dilated to give a better view; etc. The systems that are not necessary for surviving in a dangerous situation are genitourinary system, digestion, hearing, peripheral view, etc.

It is very important to define an objective way for determining stress because of the subjective perception of stress. This is one of the main challenges and one of the reasons why stress is sometimes underestimated, underrated. Some people demonstrate clear indications of stress, others do not notice when they pass the threshold of just ‘being busy’ to high stress level [8].

In our approach we use the electrical characteristics of the heart and the influences caused by stress. The system used for the detection incorporates a portable hard/software platform that can host different algorithms and sensors for calculation of the biological parameters. The platform also provides connectivity to a body area network and supports professional online analysis. The user of the system is constantly continually informed about his state enabling him to take decisions based on it. This research is based on our previous studies and models for stress measurements technologies [3, 9].

2 State of the Art

Most of the approaches that tries to calculate stress can be divided in three categories like approaches that do not require sensors, approaches with sensors and approaches with laboratory environment.

The approach that do not require sensors uses known behaviour patterns. They analyse the difference between normal behaviour and the behaviour of a stressed person. An example of this are [10, 11], they examine the difference between typing in normal situation and under stress. As a disadvantage of this approach can be mentioned difficulty of its applicability in multiple environments. Stress detection in this case is context-based and not human-centred.

The second kind of approaches uses different external sensors. For example [12], they measure the stress while driving. The sensors used for monitoring are an electrocardiogram (ECG), an electromyogram (EMG) that records the electrical activity of the muscles, a SC (Skin conductivity), a breath sensor and a video camera pointing at the driver. Because of the big amount of data the measurements are analysed offline, after recording. The main disadvantage is the huge amount of collected data, the need for offline analysis and the reduced degree of freedom and comfort caused by the different senor wires.

The third approach uses laboratory equipment. The results provided using this method are the most precise of all. For determining stress with this method the amount of cortisol, adrenaline and other hormones in the body and the amount of saliva are measured giving a direct hint to the stress level [13]. The biggest problems by this method are that the results are not available in real-time, the method is invasive and the equipment is expensive and not mobile.

The developed approach considers the main advantages and disadvantages from the methods researched. Our approach uses a low-cost ECG that is compact, wearable, and non-invasive. Another important quality is the real time capabilities that make it usable in different contexts.

3 System Architecture

Figure 2 describes the architecture of the system that is used in the approach to collect biometric data. As mentioned before, a low cost ECG sensor is used to capture continually the biological data.

The ECG sensor computes the signal obtained from the electrodes that are placed on the skin of the test person. The output from the ECG sensor is an analogue signal that contains the electrical characteristics of the heart. The output of the ECG is digitalised by an analogue digital converter from the microcontroller that will process the data. The processed data is then delivered to a digital IO interface (digital stream) like a Bluetooth or a serial interface. After this the obtained data is analysed, the heart rate (HR) and the RR Interval (the RR interval is defined as the interval between two R-peaks. See Fig. 3) derived from a QRS complex[1] are extracted. Later on the heart rate and the RR intervals are used to detecting stress. As visualisation and storage of the data and the results a smart-phone is used, as well as for the future analysis of the data could be used an external memory card or network.

4 Stress-Detection Method

Our method for stress detection provides a real time analysis using an ECG sensor. The signal obtained from the ECG is normally used for diagnostics proposes [14] but it also has unique data that can be used for example for identification of different subjects using the electrical characteristics of the heart [15].

As mentioned before in Fig. 1, stress has a direct impact on the heart rate (HR). The approach that we used for this work, analyses the ECG signal and calculates the

[1]QRS complex: Name for the combination of three of the graphical deflections can be seen on a typical electrocardiogram (ECG).

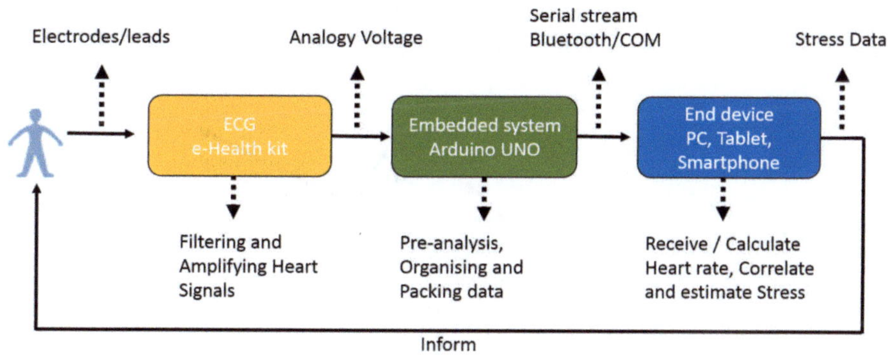

Fig. 2 System architecture for collecting, processing and visualising of biological data

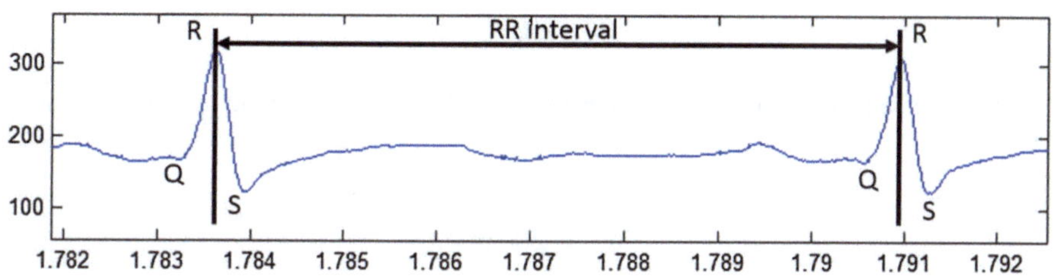

Fig. 3 Definition of an RR interval in a QRS segment

HR and RR intervals from a QRS complex obtained from ECG. The RR interval that was calculated before (see Fig. 3) is used for calculating the Heart Rate Variability (HRV). The HRV contains useful data about the stress. The RR interval represents the duration between two heartbeats.

The HRV is the variability between two heart beats. The difference between the heartbeats is modulated by the breathing rhythms. In [16] the relation between the heartbeats and respiration is described. Knowing the characteristic, that the HRV is modulated by the respiratory sinus, we can assume that the variance between beats behaves equal and periodically, in case that a person is not stressed. In case that a person is stressed the variation becomes bigger and irregular.

Figure 3 is normalised in the Y-axis over a range from 0 to 350 mv (1) and the X-axis shows the time in 10^2 ms (2).

$$0 \leq Y \geq 350, \tag{1}$$

$$x \geq 0, \tag{2}$$

$$RR_{interval} = R_i - R_{i-1}, \tag{3}$$

To enable an efficient detection of R-peak, we define that the current value R_j has to be higher than 200 mV (4).

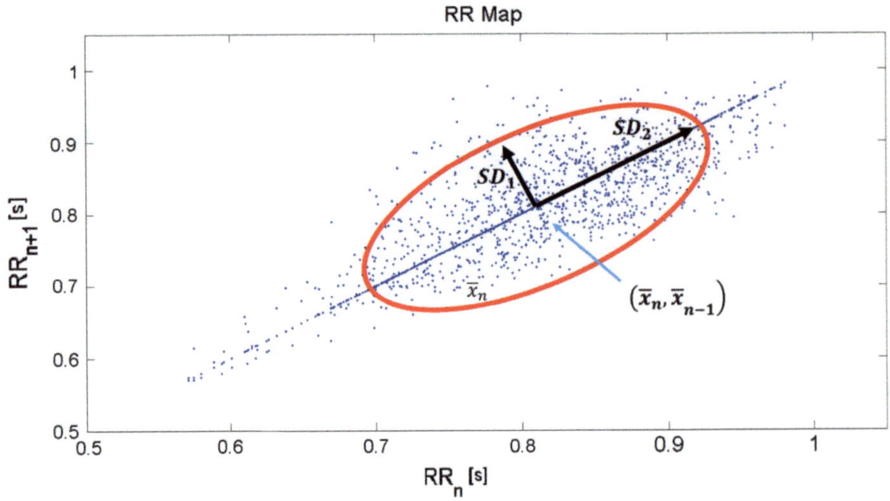

Fig. 4 Poincaré plot that shows variance between the RR intervals and the self-similarity of the values

$$R_j > 200\,\text{mV}, \tag{4}$$

Also the current R_j value has to be a maxima (5). Only if this condition is fulfilled, the R_j is an R-peak from a QRS complex.

$$(R_{j-1} < R_j) \wedge (R_j > R_{j+1}) \overset{Then}{\longrightarrow} R_i = R_j, \tag{5}$$

When the conditions (4) and (5) are fulfilled an R-peak (5) has been successfully detected. After two consecutive R-peaks has been detected, the duration of the RR interval can be calculated by calculating the time between the two R-peaks (3).

After the RR intervals are calculated, we analyse the HRV and we auto correlate the HRV values. When the results of the autocorrelation are visualised in a two-dimensional plot like in Fig. 4 we obtain a so called correlation plot. The X and Y axis are defined as (6) and (7). A concentration of v with low spreading around the axis indicates a low stress. If the values are widely spread, it means that the person is under stress.

$$X = \{RR_i\}, \tag{6}$$

$$Y = \{RR_{i+1}\}, \tag{7}$$

The total variability can be expressed as the product of SD_1 (8) and SD_2 (9).

$$SD_1 = \sqrt{var(x_1)} \quad where\, x_1 = \frac{x_n - x_{n+1}}{\sqrt{2}}, \tag{8}$$

$$SD_2 = \sqrt{var(x_2)} \quad where\, x_2 = \frac{x_n + x_{n+1}}{\sqrt{2}}, \tag{9}$$

Current/individual variance of the RR interval is calculated using the standard derivation like in (10).

$$\sigma = \left(\frac{1}{n-1}\sum_{i=0}^{n}(x_i - \bar{x})^2\right)^{\frac{1}{2}}, \tag{10}$$

5 Experiment

As method for inducing stress we decided to use the Stress Test (TSST) [6] because it is the most classical method for inducing stress in laboratory conditions. As second method for inducing stress we used a driving simulator (video game) that generates variable levels of stress by increasing the difficulty level over the time.

We adapted the TSST. The TSST that we used consists as the normal TSST of three phases. The anticipation period is where the person has to prepare a random topic that has to be presented. The second phase, the person has to do a presentation in English language (English is not the mother language of the person). The last phase is the cool down, here we tell the person it was only a test. Every phase takes 5 min.

The driving simulator employed in the experiment consists of two parts: a point reward system and an increasing difficult level system. The person receives a reward for fast, complex driving scenarios and the duration of the manoeuvre. With the increment of points the difficulty increases also. If he crashes or makes mistakes he loses all the point he has collected until now.

All the participants of the test were aged between 23 and 28, none of them were smokers or had alcohol problems. The methods used in this work assume that none of the candidates suffered from cardiac problems or mental anomalies and used pacemaker.

6 Results

The collected data from experiment was visualized using the proposed concept for stress detection. Heir we compare the results of two persons that where summited to the tests. As special characteristics, the first person has a lower heart rate but is stressed. The second person has a higher heart rate but is not stressed. Both datasets have the same length.

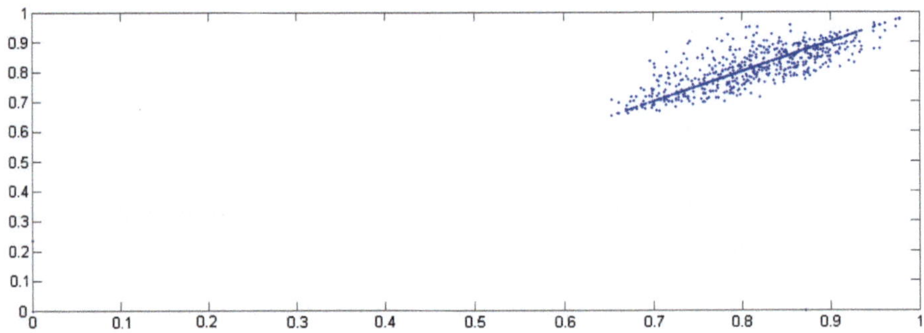

Fig. 5 Candidate under stress and with low pulse

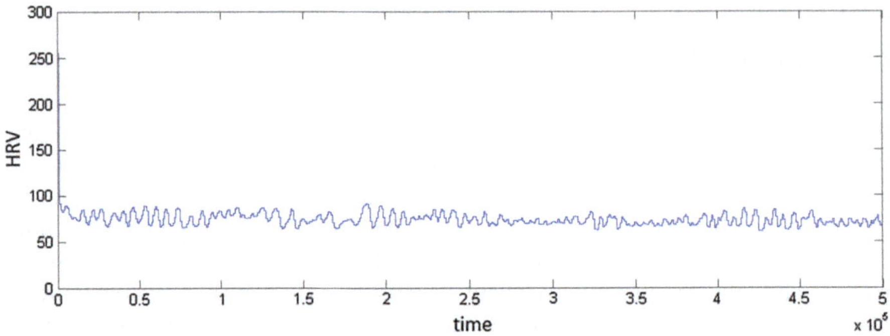

Fig. 6 Heart rate variance (HRV) visualisation a not uniform signal of a stressed candidate

In Figs. 5 and 7 the autocorrelation of the heart rate intervals is shown. In Figs. 6 and 8 the heart rate variance as a function of time is shown.

In Fig. 4 the auto correlation in stress was introduced. Clear visible is that the values are wide spread. The reason for the spreading of the values is the variation of the breathing and heart rate in stressful situation. It is observable that most of the values are between 0.7 and 0.9 s for the RR interval, this means that the interval for the heart rate is between 66.7 and 85.7 bpm.

The Fig. 6 shows the irregularities in the heartbeat. It can be interpreted as indicator for stress because normally, except of illness, the signals behave regularly (cyclical) and similar to the breath sinus.

The Fig. 7 shows the values of a person who has a higher heart rate but is not under stress. Contrarily the person is very calm and relaxed. The RR interval is between 130 and 85 bpm. It is obvious that the values are close to each other and are not spread.

The Fig. 8 shows that the variation in the heartbeats is lower. In this case the variation between the heartbeats is lower because the person is calm and breaths regularly.

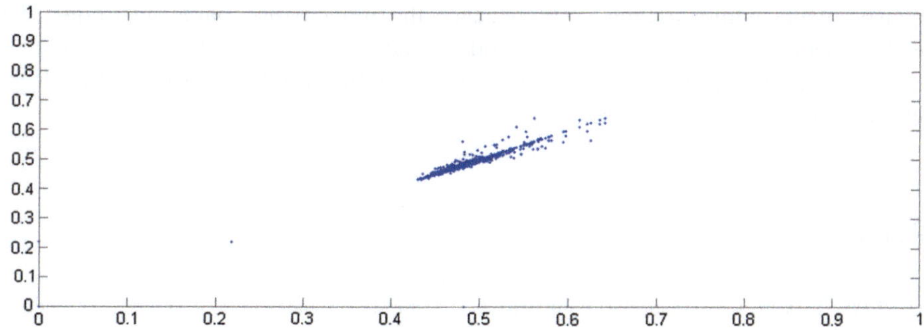

Fig. 7 Candidate with higher heart but not in stress

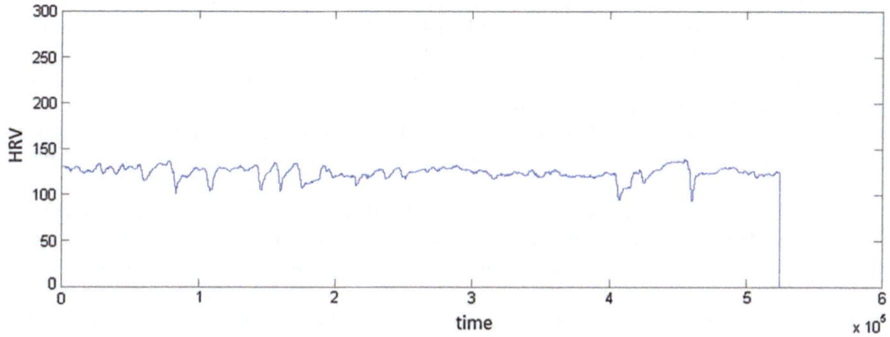

Fig. 8 HRV visualisation of not stressed candidates

7 Conclusions

The concept presented in this work shows a possible algorithm for estimating the stress using the correlation and HRV. The system is designed to be real time capable of monitoring stress, it is also cheap and wearable. The results of this experiment shows that the stress can be detected with independent of biological conditions lower or higher heart rate.

The current prototype can be connected to different systems like PCs or smartphones for calculation and visualisation proposes. An important characteristic of the algorithm for stress detection is a small footprint. The algorithm can also be deployed in a small prototyping board. For this work an Arduino Uno[2] was used as deployment target. While another test that was conducted the test person was wearing the system while sport activities. We have big problems in making

[2]http://arduino.cc.

measurements while making sport because the movements make electrodes and leads generate too much interference and artefacts. These errors have to be filtered and corrected later; otherwise, has to be used another kind of sensor that delivers data that is free from interferences.

References

1. WHO (2000) Cross-national comparisons of the prevalences and correlates of mental disorders. Bull World Health Organ, pp 413–426
2. Martínez Fernández J, Augusto JC, Seepold R, Martínez Madrid N (2010) Why traders need ambient intelligence. Springer, Heidelberg
3. Martínez Fernández J, Augusto JC, Trombino G, Seepold R, Martínez Madrid N (2013) Self-aware trader: a new approach to safer trading. J Univ Comput Sci
4. Kidd T, Carvalho LA, Steptoe A (2014) The relationship between cortisol responses to laboratory stress and cortisol profiles in daily life. Biol Psychol 25(02):34–40
5. Torbjörn Å, John A, Mats L, Nicola O, Göran K (2014) Do sleep, stress, and illness explain daily variations in fatigue? J Psychosom Res 20(01):280–285
6. Kirschbaum C, Pirke KM, Hellhammer DH (193) The 'Trier Social Stress Test'—a tool for investigating pyschobiological stress responses in a laboratory settings. Neuropychobiologie 28:78–81
7. Stroop JR (1935) Studies of interference in serial verbal reactions. J Exp Psychol 18:643–662
8. Martinez Madrid N, Martinez Fernandes J, Seepold R, Augusto JC (2013) Ambient assisted living (AAL) and smart homes. In: Springer series on chemical sensors and biosensors. vol 13, pp 39–71
9. Fernandez JM, Augusto JC, Seepold R, Madrid NM (2012) A sensor technology survey for a stress aware trading process. IEEE Trans Syst Man Cybernet Part C Appl Rev 42(6):809–824
10. Gunawardhane SD, De Silva PM, Kulathunga DS, Arunatileka SM (2013) Non invasive human stress detection using key stroke dynamics and pattern variations. In: International conference on advances in ICT for emerging regions (ICTer), Colombo
11. Vizer L, Zhou L, Sears A (2009) Automated stress detection using keystroke and linguistic features: an exploratory study. Int J Human-Comput Stud 67(10):870–886
12. Healey JA, Picard RW (2005) Detecting stress during real-world driving tasks using physiological sensors. IEEE Trans Intell Transp Syst 6(2)
13. Juliane H, Melanie S (2012) The physiological response to Trier Social Stress Test relates to subjective measures of stress during but not before or after the test. Psychoneuroendocrinology 37(1)
14. Dubin D (2000) Rapid interpretation of EKG's. COVER Publishing Co., Tampa, Florida
15. Israel SA, Irvineb JM, Chengb A, Wiederholdc MDD, Wiederholdd BK (2004) ECG to identify individuals. Pattern Recogn 21(05):133–142
16. Hirsch JA, Bishop B (1981) Respiratory sinus arrhythmia in humans: how breathing pattern modulates heart rate. Am J Physiol Heart Circulatory Physiol (New York)

Health Monitoring in Sport Through Wearable Sensors: A Novel Approach Based on Heart-Rate Variability

E. Maranesi, M. Morettini, A. Agostinelli, C. Giuliani, F. Di Nardo and L. Burattini

Abstract Sudden cardiac death (SCD) is one of the leading cause of death during sport activities. Heart rate (HR) and HR variability (HRV) provide a measure of how the organism adapts to physical fatigue, and can be monitored by commercial wearable sensors. Still, HR and HRV, widely used to optimize a training session, were rarely used to evaluate the athlete's health-status, even though widely known to provide indexes of risk for SCD. This work, developed in collaboration with Bio-Medical Engineering Development Srl, aims to provide a contribution to the problem of preventive identification of athletes at increased risk of SCD, by developing and testing a low-cost, large-scale procedure for HR and HRV monitoring from signals obtained using comfortable wearable sensors. To this aim a new protocol for the acquisition of the tachogram was proposed. It included recordings of the signals during resting, exercise and recovery phases, to allow evaluation of prevention as well as performance indexes. The procedure was tested on 10 sedentary subjects (SS) and 10 amateur athletes (AA). Compared to SS, AA showed a better health-status, quantified in a lower resting HR (63 bpm vs. 73 bpm; $P < 0.005$) and a higher resting HRV (29 ms vs. 23 ms; $P < 0.05$), and a better performance level, quantifies in a lower recovery time (130 ms vs. 174 ms; $P < 0.05$). Thus, the proposed procedure allows evaluation of both the health-status and the performance level of an athlete, and represents a valuable tool to contrast SCD in sport.

E. Maranesi (✉) · A. Agostinelli · C. Giuliani · F. Di Nardo · L. Burattini
Department of Information Engineering, Università Politecnica delle Marche,
Via Brecce Bianche, 60131 Ancona, Italy
e-mail: e.maranesi@univpm.it

E. Maranesi · M. Morettini · A. Agostinelli · L. Burattini
Bio-Medical Engineering Development Srl, Via Brecce Bianche,
60131 Ancona, Italy

M. Morettini
Interuniversity Centre of Bioengineering of the Human Neuromusculoskeletal System,
University of Rome "Foro Italico", Piazza Lauro de Bosis, 15, 00135 Rome, Italy

M. Conti et al. (eds.), *Mobile Networks for Biometric Data Analysis*,
Lecture Notes in Electrical Engineering 392, DOI 10.1007/978-3-319-39700-9_19

1 Introduction

Worldwide, 60 % of premature deaths in apparently-healthy subjects under 35 are due to sudden cardiac death (SCD), often occurring during physical activity. Cardiovascular evaluations, however, are only required for subjects practicing sport at competitive levels, typically checked once or twice a year with specific tests which too often turned out to be insufficient to asses cardiovascular risk. Instead, many amateur players usually undergo an annual medical examination which does not include cardiovascular tests. Consequently, cardiovascular monitoring is desirable among athletes of any level in order to reduce the number of cases of SCD during sport.

Heart rate (HR) and HR variability (HRV) provide a measure of how the organism reacts and adapts itself to stress, physical fatigue and metabolic-request changes. HR is the mean number of cardiac beats per minute and, at rest, its normal value is around 70–75 bpm, values that decreases to about 60 bpm for trained people. Typically, the time interval between two consecutive beats, called cardiac period (CP), is not constant but varies continuously generating HRV, defined as the fluctuation degree of HR around its mean value. HRV represents the control effect performed by the autonomic nervous system on cardiovascular system, and can be quantified using both time-domain and frequency-domain parameters.

Both HR and HRV can be easily monitored by means of commercial wearable sensors, that are usually made available with software applications finalized to estimate the performance level of the athlete and to make his/her training sessions more efficient [1–3]. Still, HR and HRV have been rarely [4] used to evaluate the health conditions of an athlete, even though they are widely known to provide indexes of cardiovascular risk [5] and SCD [6], as also recognized by the European Society of Cardiology and the North American Society of Pacing and Electrophysiology. Different tools have been developed by the world medical community in order to prevent SCD in sport but till now, three unsolved issues, namely (1) economic feasibility; (2) applicability in mass screening and (3) comfort of the measurements, have limited their application. This work, developed in collaboration with Bio-Medical Engineering Development Srl (www.bmed-bioengineering.com), has the ambition to provide a contribution to the unsolved issue of preventive identification of athletes at increased risk of major cardiac events and SCD. More specifically, the aim of the present work was to develop and test a low-cost, large-scale procedure for HR and HRV monitoring from signals obtained using comfortable wearable sensors, finalized to evaluate the health status of an athlete besides his/her performance level. To this aim a new protocol for the acquisition of the cardiac-period signals was proposed. It included recordings of the signals during all phases of a training, that are resting, exercise and recovery, for the evaluation of well-known as well as innovative health and performance HR and HRV parameters. Eventually, the entire procedure was tested on 10 sedentary subjects and 10 amateur athletes.

2 Materials and Methods

2.1 Data Acquisition Protocol and Experimental Set-up

Our new tachogram (i.e. CP signal) acquisition protocol included three phases: (1) a 5-min resting phase (*Resting*), (2) a time-variable exercise phase (*Exercise*) and (3) a 5-min recovery phase (*Recovery*). The data acquisition during the 5-min resting phase occurred preferably in supine or seated position before the beginning of the physical activity. During the recording the subjects were require to stay still and quite. The exercise phase consisted in a running session performed at a variable intensity and duration dependent on subject's physical training level, and ended when the fatigue status occurred. Eventually, the data acquisition during the recovery phase occurred similarly to that during resting. The three phases are shown in Fig. 1.

The cardiac-period or HR signals were acquired using the professional sensor BioHarness 3 by Zephyr (Annapolis, Maryland, USA) [7]. The device, even though not particularly cheap, was chosen because able to simultaneously record and memorize other biosignals (such as the ECG signal and the respiratory one) beside the one of interest in the present paper.

Eventually, a software application, designed and implemented in Matlab by B. M.E.D. SRL was used in order to automatically recognize the three phases described above, and to compute the health and performance HR and HRV indexes described below.

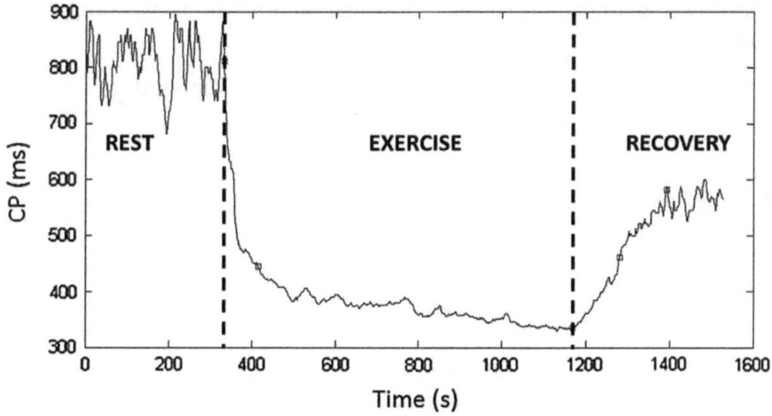

Fig. 1 Example of cardiac-period (CP) signal acquired according to our protocol in which the three phases, rest, exercise and recovery are easily identified

2.2 *Definitions of Indexes*

2.2.1 Prevention Indexes

The prevention indexes are computed using the CP-signal during the resting and exercise phases, and can be classified either as time-domain parameters and frequency domain parameters, depending on the analysis domain.

Time-domain prevention indexes:

- **mHR (bpm)**. Mean heart rate, computed during resting phase.
- **HRmax (bpm)**. Maximum heart rate reached during exercise.
- **SDNN (ms)**. Standard deviation of all CPs (also called NN intervals) which reflects global HRV, computed during resting phase.
- **RMSSD (ms)**. The root mean square of the sum of the squares of differences between adjacent NN intervals, computed during resting phase.

Frequency-domain prevention indexes:

- **HF Power (ms^2)**. Index is obtained by integrating the CP-power spectrum in the 0.15–0.4 Hz frequency range. It mainly reflects the vagal activity. It is computed during resting phase.
- **LF Power (ms^2)**. Index is obtained by integrating the CP-power spectrum in the 0.04–0.15 Hz frequency range. It reflects both parts of the autonomic nervous system, sympathetic and vagal. It is computed during resting phase.
- **LF/HF ratio**. Index that quantifies the interaction (balance) between sympathetic and vagal systems. Normally it has a value between 0.6 and 10. If it is close to 0 means a vagal prevalence or sympathetic inhibition, if it is close to 10 means a sympathetic prevalence or vagal inhibition. It is computed during resting phase.

The normality ranges for the prevention indexes are reported in Table 1.

Table 1 Range of normality of the computed parameters [8]

Parameter	Min	Max
mHR (bpm)	52	76
HRmax (bpm)	60	220-age
SDNN (ms)	32	93
SDSD (ms)	19	75
RMSSD (ms)	19	75
HF Power (ms^2)	82	3630
LF Power (ms^2)	193	1009
LF/HF ratio	1.1	11.6

2.2.2 Performance Indexes

The performance indexes are computed using the CP-signal during the exercise, recovery and resting phases. In the following definitions the re-established HRV index represents the instant, during the recovery phase, in correspondence of which the CP matches or overcome the mean CP minus the CP standard deviation.

- **ΔCP (ms)**. Delta of increase between the average cardiac period during exercise and the mean CP in the recovery phase.
- **ΔCPstab (ms)**. Delta of increase between the average CP during exercise and the CP of the re-establish HRV index.
- **T tot** (s). Time between the end of exercise and the end of the recording.
- **T stab** (s). Time between the end of the exercise and the re-establish HRV instant.
- **%CPrec**. Percentage value of CP evaluated in the recovery phase compared with the mean CP detected at rest.
- **d tot, d stab**. Derivatives that quantifies the recovery times, calculated from the end of the exercise.
- **T stim (s)**. Estimated time for a complete recovery, i.e. the time that the subject would needed for his/her mean CP in the recovery phase to become equal to that in the resting phase.

The normality ranges for the performance indexes are not available being them proposed here for the first time.

2.3 Subjects

Twenty healthy (without any known cardiac or metabolic disease, and not taking drugs) volunteers were selected for this study: 10 sedentary subjects (SS) and 10 amateur athletes (AA). A subjects was classified as sedentary if satisfied both these characteristics: no constant sport activity for a year and a sedentary life style. Instead, the AA group included subjects who regularly (at least twice a week) practiced a sport such as running, soccer, cycling. Their anthropometric data together with HR upper limit (i.e. the HR that should never be overcome by an athlete, estimated as 220-age) are reported in Tables 2 and 3 for the SS group and the AA group, respectively. All subjects involved in the study gave oral and written consent.

Table 2 Anthropometric data of sedentary subjects (SS)

SS	Age (year)	Weight (kg)	Height (cm)	HR upper limit (bpm)
#1	25	66	179	195
#2	26	72	165	194
#3	25	77	183	195
#4	37	71	167	183
#5	25	80	179	195
#6	28	83	178	192
#7	31	53	167	189
#8	36	70	171	184
#9	25	78	183	195
#10	25	74	183	195
Median	26	73	179	194.5
[25th–75th percentile]	[25–31]	[70–78]	[167–183]	[189–195]

Table 3 Anthropometric data of amateur athletes (AA)

AA	Age (year)	Weight (kg)	Height (cm)	HR upper limit (bpm)
#1	43	67	168	177
#2	25	70	178	195
#3	36	72	182	184
#4	24	50	160	196
#5	27	52	159	193
#6	36	89	191	184
#7	29	75	190	191
#8	20	76	185	200
#9	18	57	165	202
#10	31	68	169	189
Median	28	69	173	192
[25th–75th percentile]	[24–36]	[57–75]	[165–185]	[184–196]

2.4 *Statistics*

Normality of subjects parameters was performed using the Lilliefors test. Not normal distributions were described using the 50th (median) and [25th–75th] percentiles. The Wilcoxon Rank Sum Test was applied to compare parameters distributions between SS and AA. Statistical level was set at 0.05 in all cases.

3 Results

The two SS and AA groups were clinically equivalent (not statistically different in terms of age, weight and height).

3.1 Prevention Indexes

Tables 4 and 5 report the prevention indexes for SS and AA groups, respectively. Three SS show too high mHR, while HRmax-range is overcame by 1 SS and 3 AA. Globally, at rest the AA are significantly more bradycardiac than the SS (63 bpm vs. 73 bpm, $P < 0.005$). Instead, HRmax during the exercise is comparable in the two groups (SS: 182 bpm, AA: 179 bpm). SDNN is out of range in 2 SS (one over and one under), while RMSSD is too low in 4 SS. Overall, the SDNN was comparable in the two groups (SS: 54 ms, AA: 61 ms) whereas RMSSD was lower in the SS than in the AA (23 ms vs. 29 ms; $P < 0.05$). Eventually, HF value is borderline in 1 SS, LF is over range in 3 SS and 4 AA, and LF/HF is borderline in one SS and over the maximum in another SS. Only the HF value is significantly

Table 4 Prevention indexes computed during the rest phase

SS	mHR* (bpm)	HRmax (bpm)	SDNN (ms)	RMSSD* (ms)	HFpower* (ms^2)	LFpower (ms^2)	LF/HF
#1	73	189	68	23	262	956	3.6
#2	71	181	34	13[b]	109	271	2.5
#3	71	182	29[b]	16[b]	184	317	1.7
#4	66	169	94[a]	27	84	680	8.1
#5	76	189	57	25	339	1404[a]	4.1
#6	73	192	50	24	316	670	2.1
#7	85[a]	161	64	27	81[b]	1465[a]	18.1[a]
#8	65	211[a]	61	23	205	1434[a]	7.0
#9	79[a]	179	34	18[b]	383	399	1.0[b]
#10	91[a]	179	34	14[b]	211	435	2.1
Median [25th–75th percentile]	73 [71–79]	182 [179–189]	54 [34–64]	23 [16–25]	208 [109–316]	675 [399–1404]	3.1 [2.1–7.0]

Median values are also reported for SS group
*$P < 0.05$ when comparing median values a parameter in the SS versus AA groups
[a]Values overcoming the upper limit of the range
[b]Values undergoing the lower limit of the range

Table 5 Prevention indexes computed during the rest phase

AA	mHR* (bpm)	HRmax (bpm)	SDNN (ms)	RMSSD* (ms)	HFpower* (ms^2)	LFpower (ms^2)	LF/HF
#1	65	177	48	27	486	956	1.5
#2	52	166	68	44	813	271	1.9
#3	60	199[a]	64	33	1172	317	1.3
#4	70	238[a]	54	24	84	8361[a]	7.3
#5	63	175	83	36	715	1404[a]	3.0
#6	63	206[a]	62	31	1290	670	1.5
#7	63	190	59	20	580	1465[a]	2.8
#8	70	171	33	19	195	1434[a]	3.4
#9	62	176	65	52	631	399	2.7
#10	71	180	35	20	470	435	1.2
Median [25th–75th percentile]	63 [62–70]	179 [175–199]	61 [48–65]	29 [20–36]	606 [470–813]	813 [399–1434]	2.3 [1.5–3]

Median values are also reported for AA group
*$P < 0.05$ when comparing median values a parameter in the SS versus AA groups
[a]Values overcoming the upper limit of the range

different between SS and AA; more specifically it is lower in the former than in the latter group (208 ms^2 vs. 606 ms^2; $P < 0.01$).

3.2 *Performance Indexes*

Tables 6 and 7 report the performance indexes for SS and AA groups, respectively. Only two parameters are statistically different when comparing the two groups, which are the ΔCP and dstab. More specifically, ΔCP is significantly lower in the SS than AA (130 ms vs. 174 ms; $P < 0.05$).

4 Discussion

This study aims to provide a contribution to the problem of SCD during physical activity. It proposes an innovative low-cost, large-scale procedure for HR and HRV monitoring from signals obtained using comfortable wearable sensors finalized to evaluate the health status of an athlete besides his/her performance level. According to the procedure, the tachogram is acquired during three recording phase, namely resting, exercise and recovery, and is used to evaluate prevention indexes as well as performance indexes. More specifically, prevention indexes are computed during the initial resting phase and the exercise phase, while the performance indexes are computed during the exercise and recovery phases.

Table 6 Performance indexes computed during the exercise and the recovery phase

SS	ΔCP* (ms)	ΔCPstab (ms)	Ttot (s)	Tstab (%)	% CPrec	dtot	dstab*	Tstim(s)
#1	109	75	202	51	68	0.8	1.6	94
#2	128	63	178	47	65	1.0	1.3	94
#3	226	119	180	56	70	1.4	2.0	78
#4	130	190	297	40	96	0.5	1.5	120
#5	163	141	387	72	65	0.7	1.8	209
#6	132	95	264	64	73	0.9	1.4	97
#7	70	88	264	27	76	0.1	1.4	189
#8	91	77	434	83	68	0.6	1.0	206
#9	151	97	216	49	61	0.8	2.0	136
#10	83	50	236	67	75	0.6	0.8	77
Median [25th–75th percentile]	130 [91–163]	92 [75–119]	250 [202–297]	54 [47–67]	69 [65–75]	0.8 [0.6–0.9]	1.5 [1.3–1.8]	109 [94–189]

Median values are also reported for SS group
*$P < 0.05$ when comparing median values a parameter in the SS versus AA groups

Table 7 Performance indexes computed during the exercise and the recovery phase

AA	ΔCP* (ms)	ΔCPstab (ms)	Ttot (s)	Tstab (%)	% CPrec	dtot	dstab*	Tstim(s)
#1	181	87	128	44	81	1.7	1.6	30
#2	161	117	381	75	73	0.4	1.6	141
#3	162	134	315	43	71	0.4	3.0	128
#4	83	54	293	14	67	0.4	2.8	160
#5	208	142	294	65	65	0.9	2.3	161
#6	174	27	146	42	84	0.9	0.8	29
#7	189	81	160	38	70	1.7	2.2	67
#8	167	117	271	60	78	1.0	1.8	75
#9	201	117	158	41	65	1.4	2.9	87
#10	206	117	194	53	62	1.4	1.9	118
Median [25th–75th percentile]	174 [161–201]	117 [81–117]	233 [158–294]	44 [41–60]	71 [65–78]	1.0 [0.4–1.4]	2.1 [1.6–2.8]	103 [67–141]

Median values are also reported for AA group
*$P < 0.05$ when comparing median values a parameter in the SS versus AA groups

It is well know that an intense training forces the entire body to adapt to this condition through the development of morphological and functional changes [9–11]. These adjustments mean that the heart of an athlete appears different from that of a sedentary subject, and is coined as “athlete’s heart” [12]. Typically, SCD occurs in athletes when pathological hypertrophic enlargement of the heart went undetected or was incorrectly attributed to the benign athletic adaptations. Other alternative causes of SCD are episodes of isolated arrhythmias which degenerated into lethal ventricular fibrillation and asystole, and undetected asymptomatic cardiac congenital defects of the vessels, chambers, or valves of the heart.

The HR and HRV of a subject are the effect of the control of the nervous system on the cardio-circulatory system in order to optimize the adaptation of the organism to the different conditions (for example of exercise) [5, 9–16]. The autonomic nervous system, specifically, acts on the heart muscle by means of its two components, which play in a synergic way: the sympathetic nervous system, which intervenes in situations of stress and emergency and causes an acceleration of the HR; and the parasympathetic nervous system, or vagal system, whose activation is instead aimed at saving energy and causes a slowing of the HR. Because of the synergic action of the sympathetic and parasympathetic nervous system, the time between the two consecutive heart beats is not constant but changes continuously. This change, which causes a continuous variation of the HR, is not only physiological, but also a sign of good health [17]. A net decrease of HRV is universally considered a predictor for the risk of SCD. If the study of HR allows an assessment of cardiac function, the study of HRV allows an indirect assessment of the capabilities of the autonomic nervous system [18].

Aerobic training, such that considered in the present study, causes important changes in the load of the autonomic nervous system of the heart, characterized by a reduction in sympathetic tone with a predominance of vagal tone. The main consequence of this new regulation of the autonomic nervous system of the heart is the reduction in heart rate at rest. Indeed, our results show that during the rest phase the AA are significantly more bradycardic (lower mHR) than the SS (Tables 4 and 5). However, training does not substantially change the maximum heart rate (i.e. HRmax) during exercise (Tables 4 and 5). Another consequence of the vagal predominance is the increase of the HRV indexes that main reflect its activity [19], such as the HF. Indeed, AA are found to be characterized by a significantly higher HF power (Tables 4 and 5). Eventually, the AA are also characterized by significantly a higher RMSSD (Tables 4 and 5), which being a measure of HRV, indicate a better health status of the AA with respect to the SS.

The total absence of physical activity in our SS constituted by young (medially 26 years old) subject causes 30 % of them to have values of mHR higher than normal, and 40 % of them to have values of RMSSD lower than normal. Such results indicate that an amateur physical activity induce a better health status (lower resting HR and higher resting HRV). Only 2 SS (20 %) were simultaneously showing mHR and RMSSD values out of range, indicating that these two parameters should be combined rather than used in a mutually exclusive way to provide a complete view of the health status of a subject.

Beside evaluating the prevention indexes, this study also evaluated the performance indexes. As expected, generally, such indexes quantified and highlighted a best athletic performance of AA than SS. Although all parameters are indicative of a better recovery of the trained subjects, only two of them, ΔCP and dstab, reached a statistical significance, likely because of the low number of subjects involved in the two groups. Both these parameters indicate a faster recovery of the CP in the AA, ΔCP in terms of recovered HR, and dstab in terms of HR-recovering speed.

5 Conclusions

The proposed innovative low-cost, large-scale procedure for HR and HRV monitoring from signals obtained using comfortable wearable sensors allows evaluation of both the health status and the performance level of an athlete. Consequently, it represents a valuable tool to contrast the problem of SCD in sport.

References

1. Kiviniemi AM, Hautala AJ, Kinnunen H et al (2010) Daily exercise prescription on the basis of HR variability among men and women. Med Sci Sports Exer 42(7):1355–1363
2. Kiviniemi AM, Hautala AJ, Kinnunen H et al (2007) Endurance training guided individually by daily heart rate variability measurements. Eur J Appl Physiol 101(6):743–751
3. Stanley J, Peake JM, Buchheit M (2013) Cardiac parasympathetic reactivation following exercise: implications for training prescription. Sports Med 43(12):1259–1277
4. Buchheit M, Simon C, Viola AU et al (2004) Heart rate variability in sportive elderly: relationship with daily physical activity. Med Sci Sports Exerc 36(4):601–605
5. Akselrod S, Gordon D, Ubel FA et al (1981) J Power spectrum analysis of heart rate fluctuation: a quantitative probe of beat-to-beat cardiovascular control. Science 213:220–222
6. Task Force of The European Society of Cardiology and The North American Society of Pacing and Electrophysiology (1996) Heart rate variability guidelines, standards of measurement, physiological interpretation, and clinical use. Eur Heart J 17:354–381
7. BioHarness 3 User Manual, http://zephyranywhere.com/media/pdf/BH_MAN_P-BioHarness3-User-Manual-FCC_20120912_V01.pdf
8. Nunan D, Sandercock GRH, Brodie DA (2010) A quantitative systematic review of normal values for short-term heart rate variability in healthy adults pacing. Clin Electrophysiol 33 (11):1407–1417
9. Robinson BF, Epstein SE, Beiser GD et al (1966) Control of heart rate by the autonomic nervous system. Studies in man on the interrelation between baroreceptor mechanisms and exercise. Circ Res 19:400–411
10. Katona PG, McLean M, Dighton DH et al (1982) Sympathetic and parasympathetic cardiac control in athletes and nonathletes at rest. J Appl Physiol 52:1652–1657
11. Lewis SF, Nylander E, Gad P, Areskog NH (1980) Non-autonomic component in bradycardia of endurance trained men at rest and during exercise. Acta Physiol Scand 109:297–305
12. Maron BJ, Pelliccia A (2006) The heart of trained athletes: cardiac remodeling and the risks of sports, including sudden death. Circulation 114(15):1633–1644
13. Frank H. Netter MD. Atlante di anatomia umana. Second edition, Masson Innova Pharma
14. La Rovere MT, Bigger JT Jr, Marcus FI et al (1998) Baroreflex sensitivity and heart rate variability in prediction of total cardiac mortality after infarction. ATRAMI (autonomic tone and reflexes after myocardial infarction). Lancet 351:478–484
15. Rossini P (2008–2009) Analisi della regolazione neurovegetativa del cuore durante l'esercizio fisico prolungato. Istituto superiore di educazione fisica pareggiato di Bologna

16. Grasso R, Schena F, Cevese A (1994) Cardiovascular reflex regulation and exercise. A spectral analysis approach. Funct Neurol 21–26
17. Kleiger RE, Miller JP, Bigger JT Jr et al (1987) Decreased heart ratevariability and its association with increased mortality after acutemyocardial infarction. Am J Cardiol 59:256–262
18. Akselrod S, Gordon D, Ubel FA et al (1981) Power spectrum analysis of heart rate fluctuation: a quantitative probe of beat-to-beat cardiovascular control. Science 213:220–222
19. Hottenrott K, Hoos O, Esperer HD (2006) Heart rate variability and physical exercise. Current status. Herz 31(6):544–552

A New Segmented-Beat Modulation Algorithm for Maternal ECG Estimation from Abdominal Recordings

A. Agostinelli, C. Giuliani, S. Fioretti, F. Di Nardo and L. Burattini

Abstract The noninvasive fetal electrocardiogram (fECG) provides precious information about the physiological fetus state. It is extracted from abdominal recordings, obtained positioning surface electrodes on the maternal abdomen, by subtraction of the maternal ECG (mECG), often roughly estimated by simply concatenating a maternal-beat template. Aim of the present study is to propose a new algorithm for the mECG estimation based on a segmented-beat modulation method (SBMM) that adjusts the template length to the maternal physiological heart-rate variability (HRV) and reduces the level of noise. According to the SBMM, each maternal cardiac cycle (CC) is segmented into two segments, QRS and TUP, respectively independent and proportional to preceding RR interval. The estimated mECG is the concatenation of the template-beat, obtained as the median of the maternal beat after modulation and demodulation of TUP segment. The algorithm was applied to two (ARec1 and ARec2) 4-channel abdominal recordings obtained from pregnant women. ARec1 and ARec2 were both 60 s long and characterized by similar heart rate (HR: 80 bpm and 82 bpm) but different HRV (42 ms vs. 139 ms). Results indicate that the error in the mECG estimation is always small (<2.5 μV) but increases with HRV (ARec1: 0.87–1.65 μV; ARec2: 1.98–2.37 μV). In conclusion, the proposed algorithm based on the SBMM allows a clean mECG estimation from abdominal recordings thanks to a modulation procedure introduced to track physiological variation in the maternal heart rhythm.

1 Introduction

Congenital heart defects are among the most common birth defects and the leading causes of birth death [1]. Moreover, several pathologies and complications, even not directly linked to the fetal heart (such as fetal hypoxia, defined a deprivation of

---

A. Agostinelli · C. Giuliani · S. Fioretti · F. Di Nardo · L. Burattini (✉)
Department of Information Engineering, Università Politecnica delle Marche,
Via Brecce Bianche 12, 60131 Ancona, Italy
e-mail: l.burattini@univpm.it

M. Conti et al. (eds.), *Mobile Networks for Biometric Data Analysis*,
Lecture Notes in Electrical Engineering 392, DOI 10.1007/978-3-319-39700-9_20

an adequate supply oxygen that, if prolonged, can lead to irreversible neurological diseases [2]) show abnormalities in the cardiac activity as side effect. Premature diagnosis of fetal cardiac defects and abnormalities during pregnancy may allow the treatment of the pathologies in the early stages of the fetus development and may prevent a permanent disease or, in the worst cases, a fatal outcome [3].

Fetal electrocardiography may provide precious information about the fetal well-being and the physiological state of the fetus. Sometimes, the fetal monitoring is noninvasively performed using abdominal recordings obtained positioning electrodes on the abdomen of a pregnant woman, which behaves as a conductor. The abdominal recording is a signal representing the summation of three main components [4, 5]: (1) fetal ECG (fECG) or signal of interest; (2) maternal ECG (mECG) or primarily interference; and (3) noise, composed by measurement interferences and physiological interferences of different nature from cardiac. From a morphological point of view, the mECG and the fECG are signals that contain the same basic ECG waves: the P wave, associated to atrial depolarization; the QRS complex, associated to ventricular depolarization; the T wave and eventually the U wave, associated to ventricular repolarization. However, the fECG amplitude is significantly smaller (0.06–0.07 mV) than that the mECG (0.10–0.15 mV), and the frequency bands are largely overlapped (medially, 1.8–45 Hz for fECG and 1.3–32.5 for mECG). Consequently, traditional linear filters (with fixed frequency passing bands) are not efficient to separate the two ECG components. Instead, the mECG has to be accurately estimated to be then subtracted from the abdominal recording in order to get the fECG. Several algorithms to filter out the mECG signal have been developed over the past 30 years [4–6]. The most recent methods are based on innovative techniques of signal processing, such as adaptive filtering and signals decomposition [5, 6], while the traditional approaches based on maternal template subtraction [4] continue to occupy a very important role. To morphologically reconstruct the mECG, the adaptive filter either requires a maternal reference channel (generally thoracic) or several linearly independent channels. Still, the purely mathematical basis of the signal decomposition techniques does not consider the several physiological characteristics of the involved signals. Instead, the template-based approaches are single-channel methods that rely on the computation of one template representing the most frequent beat kind. These types of approaches provide a fECG signal with not sufficient accuracy mostly because of poor mECG estimation due to a fixed length of the template. Eventually, the noise (whose amplitude is very variable, from 0 to a few tenths of mV), is characterized by a very-wide frequency band [4] (medially, from 0 to hundreds of Hz). Most noise components are eliminated by applying a passband filter with cut-off frequencies of 0.5 and 45 Hz.

Aim of the present study is to propose a new algorithm for mECG estimation by introducing a segmented-beat modulation procedure. Differently from the other template-based approaches, this algorithm adjusts the template length to maternal physiological heart-rate variability (HRV) and reduces the level of noise. Eventually, the algorithm was tested using two 4-channel abdominal recordings to evaluate its ability to provide an improved estimation of the mECG signal.

2 Clinical Data and Methods

2.1 Clinical Data

SBMM was applied to two (ARec1 and ARec2) 4-channel abdominal recordings obtained from pregnant women between 38 and 41 weeks of gestation available at the "Abdominal and Direct Fetal Electrocardiogram Database" of Physionet (www.physionet.org). ARec1 and ARec2 were both 60 s long and characterized by similar heart rate (HR) and thus containing a similar number of beats (Nb) but different HRV (Table 1).

2.2 Maternal ECG Estimation

The new proposed algorithm to estimate the mECG from an abdominal recording is based on the segmented-beat modulation method (SBMM; see below) [7, 8], a technique which can be used to get a clean ECG estimation from a noisy recording [8]. The mECG was estimated by applying the SBMM to a channel of an abdominal recording, which was considered the noisy tracing. It works under the assumption of knowing the maternal R-peaks position and, in case of multi-channel abdominal recordings, the procedure had to be reiterated for each channel.

The segmented-beat modulation method. The SBMM is based on the practical observation that, in first approximation, the QRS-complex duration is independent on heart rate (HR), whereas duration of all other waves linearly varies with it. More specifically, if the duration of the QRS complex is independent from the preceding RR interval, the duration of the other waves is proportional to it. Consequently, each cardiac beat is divided in two segments: the QRS segment and the TUP segment, as represented in Fig. 1. Considering the beginning of the cardiac cycle (CC) in the PQ segment (i.e. the ECG segment between the P-wave offset and Q-wave onset) rather than at the beginning of the P wave, as usually done, the QRS segment is identified as $\pm\Delta t$ ms (for example $\Delta t = 40$ ms for adults) around the R peak, while the TUP segment is identified as the ECG segment within the time interval that begins Δt ms after the R peak and ends Δt ms before of the subsequent

Table 1 Characteristics of the two (Arec1 and Arec2) 4-channel abdominal recordings

	ARec1	ARec2
Number of channels	4	4
Length (s)	60 s	60 s
Nb	78	79
Mean RR (ms)	750	730
HR (bpm)	80	82
HRV (ms)	42	139

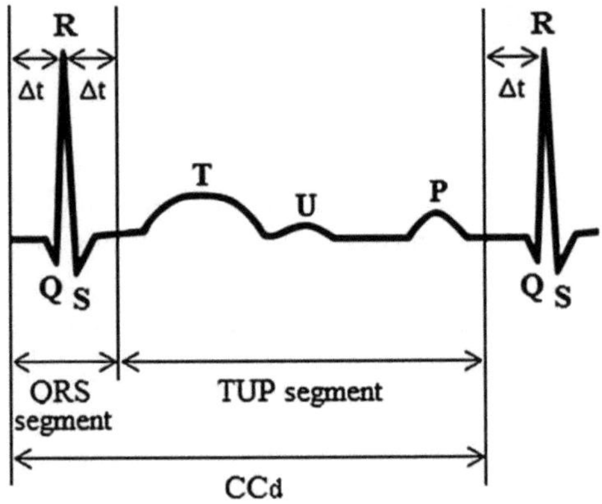

Fig. 1 Segmentation of the cardiac cycle (CC) in two segments: QRS and TUP (CCd: CC duration)

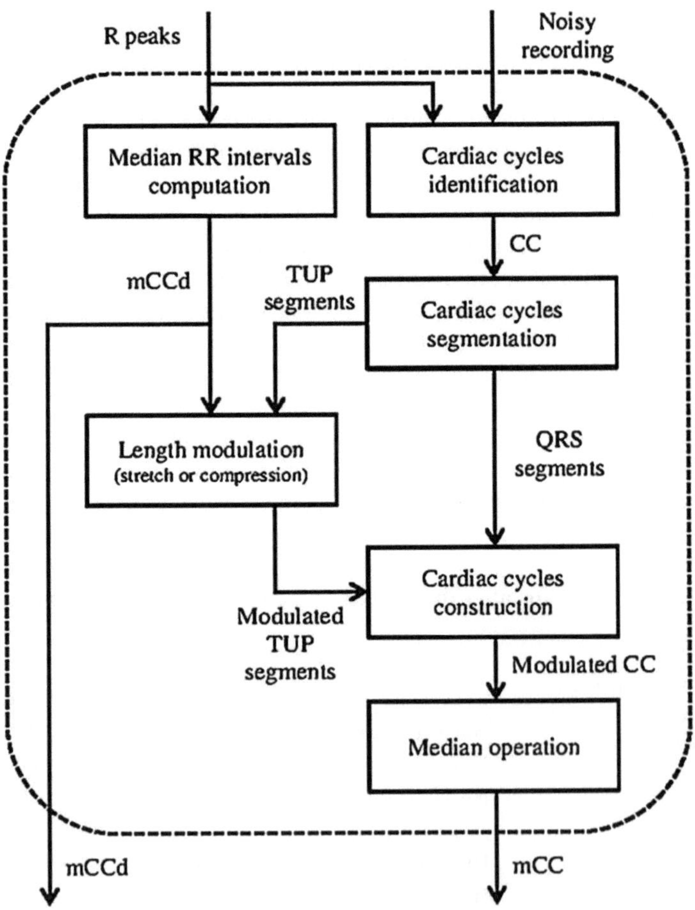

Fig. 2 Block diagram relative to the median cardiac cycle (mCC) computation

R peak. Then, the CC is the portion of ECG included between Δt ms before the R peak and Δt ms before the subsequent R peak. Consequently, each CC included in the ECG is characterized by its own duration (CCd), which may vary due to HRV. However, the duration of all the QRS segments is the same in all the CC and equal to 2 Δt, whereas the duration of the TUP segments is beat-dependent and equal to CCd-2 Δt.

After having segmented all the N beats of a noisy ECG tracing, the median CC duration (mCCd) is computed. The computation of the median CC (mCC; Fig. 2) requires all beats to have the same length. Thus, all the CC are modulated (stretched or compressed) in order to have their length to match mCCd. Since all the QRS segments have the same duration, the modulation involves only the TUP segments (Fig. 2).

Eventually, the mCC is obtained as the median of all cardiac cycles constructed using the original QRS segments and the modulated TUP segments. For the median operator properties, the noise level affecting the mCC is drastically reduced (Fig. 3).

The mCC represents the basic beat from which to derive the clean estimated ECG. Initially, an ECG tracing is obtained by an N-fold repetition of the mCC to make it to contain the same number of beats of the original one. Each mCC constituting this tracing is then segmented and the median TUP segment is demodulated (compressed or stretched) to match the length of the TUP segment of the corresponding beat in the original ECG tracing (Fig. 4). All beats are then reconstructed using the median QRS and the demodulated median TUP. Optimization processes, involving cross-correlation maximization and error minimization between the reconstructed beat and the corresponding one in the original signal, are performed in order to

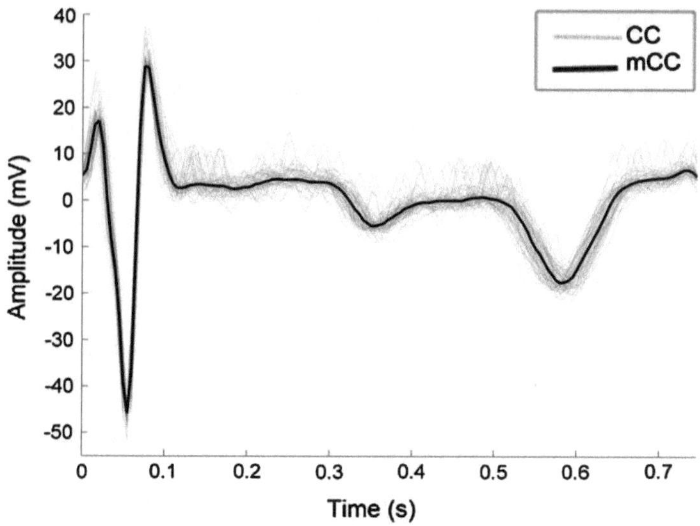

Fig. 3 Example of a median cardiac cycle (mCC) computation from all modulated cardiac cycles (CC)

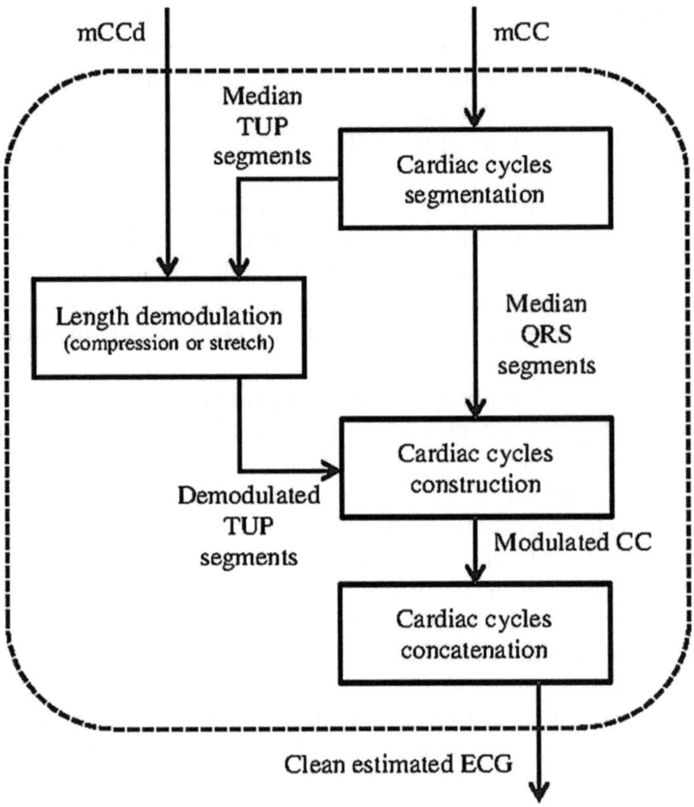

Fig. 4 Block diagram relative to the computation of the clean ECG estimation

compensate for possible small inter-beat, HR-independent variations of the CC waveforms. Eventually all reconstructed beats are concatenated to get the clean estimated ECG (Fig. 4).

2.3 Statistical Analysis

Goodness of the SBMM in the extraction of the fECG was evaluated independently for each channel of ARec1 and ARec2 calculating the error as defined in Eq. 1:

$$\varepsilon_{ch,seg}(i) = median|estimated\ mECG_{ch}(i) - abdominal\ signal_{ch}(i)|_{seg} \quad (1)$$

where ch = 1, 2, 3, 4, thus indicating the used channel of ARec1 and ARec2, i = 1, 2, …, Nm, thus indicating the maternal cardiac beat, and seg = QRS segment, TUP segment, thus indicating the ECG segment. Thus, $\varepsilon_{ch,seg}(i)$ indicates the error relative to the segment seg of the beat i in the channel ch, and was calculated as median of the absolute differences between the estimated mECG and the original abdominal signal. For each channel and each segment, $\varepsilon_{ch,seg}(i)$ were reported in terms of 50th (median) [25th–75th] percentiles. For each channel, the median errors relative to the QRS segment and TUP segment were compared using the Wilcoxon Rank-Sum test for non-normal distributions. Statistical significance level was set at 0.05.

3 Results

Figures 5 and 6 qualitatively show the results relative to the application of the SBMM to the first channel of ARec1. More specifically, the SBMM was first applied to the abdominal recording to estimate the mECG (Fig. 5), which could then be subtracted to get the fECG (Fig. 6). The quantitative results in terms of mean errors relative to all 4 channels of ARec1 and ARec2 were reported in Tables 2 and 3, respectively. Globally (when considering all leads together), median errors ranged from 0.87 and 2.37 μV. In ARec1 the median error in correspondence of the QRS segment was significantly greater than that of the TUP segment in all leads (Table 2). Vice versa, in ARec2 the median error in correspondence of the TUP segment was significantly greater than that of the QRS segment in all leads (Table 3). Moreover, the global median errors in the ARec2 were significantly higher than the corresponding ones in the ARec1, for both QRS and TUP segments.

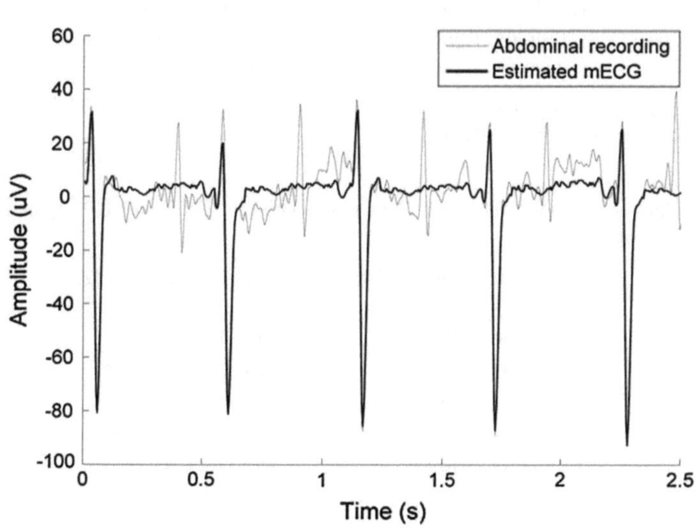

Fig. 5 Estimated mECG (*solid line*) overlapped to original abdominal tracing (*dotted line*) relative to first channel of ARec1

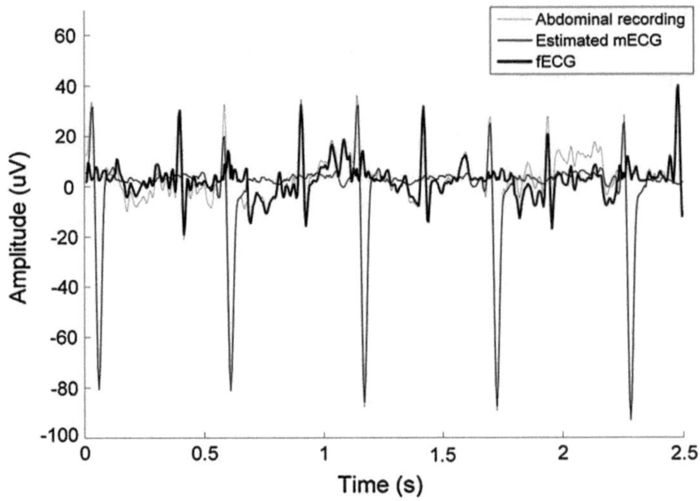

Fig. 6 fECG (*solid line*) overlapped to estimated mECG (*bold dotted line*) and original abdominal tracing (*dotted line*) relative to first channel of ARec1

Table 2 Median errors (μV; 50th [25th–75th]) associated to QRS and TUP segment calculated for each channel of ARec1

Channels of ARec1	QRS segment 50th [25th–75th]	TUP segment 50th [25th–75th]	*P*
1	1.54 [1.10–2.03]	1.06 [0.89–1.25]	$<10^{-7}$*
2	2.16 [1.56–3.24]	1.05 [0.78–1.26]	$<10^{-16}$*
3	1.70 [1.24–2.20]	0.39 [0.30–6.78]	$<10^{-23}$*
4	1.43 [0.95–1.86]	0.86 [0.64–1.12]	$<10^{-7}$*
Total	1.65 [1.17–2.25]	0.87 [0.59–1.14]	$<10^{-48}$*

*Statistically significant ($P < 0.05$) when comparing QRS segment versus TUP segment errors relative to the same channel of ARec1

Table 3 Median errors (μV; 50th [25th–75th]) associated to QRS and TUP segment calculated for each channel of ARec2

Channels of ARec2	QRS segment 50th [25th–75th]	TUP segment 50th [25th–75th]	*P*
1	1.90 [1.36–2.43]§§	2.15 [1.87–2.73]§§§	$<10^{-3}$*
2	2.95 [2.33–3.65]§§	3.87 [3.34–5.02]§§§	$<10^{-5}$*
3	1.36 [1.01–1.74]§	1.72 [1.44–2.08]§§§	$<10^{-4}$*
4	1.87 [1.48–2.86]§	2.43 [2.01–3.07]§§§	$<10^{-3}$*
Total	1.98 [1.35–2.79]§§	2.37 [1.83–3.35]§§§	$<10^{-9}$*

*$P < 0.05$ when comparing QRS segment versus TUP segment errors relative to the same channel of ARec2

§$P < 0.05$, §§$P < 10–3$, §§§$P < 10–24$ when comparing ARec1 versus ARec2 median errors relative to a specific channel and segment

4 Discussion

In the future, if able to provide reliable the fECG tracings, fetal monitoring algorithms could be integrated in medical devices [9, 10], either mobile or not-mobile, for real-time analysis. Such devices could be employed as recorders for signals that could be remotely analyzed to provide information when the only current surveillance strategy is maternal perception of reduced fetal movements [11, 12], and may promise an additional fetal monitoring to the traditional cardiotocography. At the present time, the fetal monitoring devices simply produce either the sound of the fetal heart beats, provide the instantaneous heart rate, or display the heart-rate signal on a screen [13]. Clearly, an accurate assessment of fetal well-being also requires other information linked to the ECG signal and its morphology. However, algorithms for the fECG monitoring have only occasionally been integrated on medical devices. Indeed, only rarely the noninvasive fetal electrocardiography has been able to reliably deliver an undistorted fECG, mainly because of a poor mECG estimation that causes the fECG to be characterized by a low-signal-to-noise ratio [5].

This study proposes a new algorithm for improving mECG estimation by means of a procedure based on the SBMM. As several other methods before, such technique is based on the construction of a template-beat [14], which is then concatenated to obtain the estimated mECG signal. However, only the SBMM is able to track variation in the maternal heart rhythm. Indeed, it involves the maternal CC segmentation into two segments, the QRS segment and the TUP segment, respectively independent and proportional to preceding RR interval [15]. The latter segment undergoes modulation to adjust for physiological maternal HRV. More specifically, the TUP-segment modulation (compression or stretch) is initially used to force all CC to have the same duration. Such condition is indeed required for computing the template-beat as the median beat. Then, the estimated mECG is obtained by concatenating the template-beat after the TUP-segment demodulation (stretch or compression) in order to obtain the reconstructed maternal beats whose duration matches that of the corresponding beats in the abdominal recording. The modulation/demodulation procedure performed on the TUP segments strongly improves accuracy of the estimated mECG, as results of this study demonstrate, and, consequently, the fECG extraction will also be improved. Figure 7 qualitatively shows an example of a mECG-beat estimation from an abdominal recording performed without (upper panel) and with (lower panel) the modulation/demodulation process. The figure clearly shows a higher error ($\varepsilon = 1.35$ μV) in the TUP segment associated to the former rather than the latter case ($\varepsilon = 1.10$ μV). A higher error implies a higher amplitude artifact in the fECG, since it is obtained by subtracting the estimated mECG from the original abdominal recording. The SBMM can be applied to the abdominal recording only if the R peaks relative to the maternal ECG signal are known. In most clinical cases, such condition is satisfied because the maternal ECG is the highest-amplitude component and standard algorithms for R peak detection (such as the Pan-Tompkins's one [16]) from the ECG can be applied. In less frequent cases in which the R peaks are not directly obtained from the abdominal

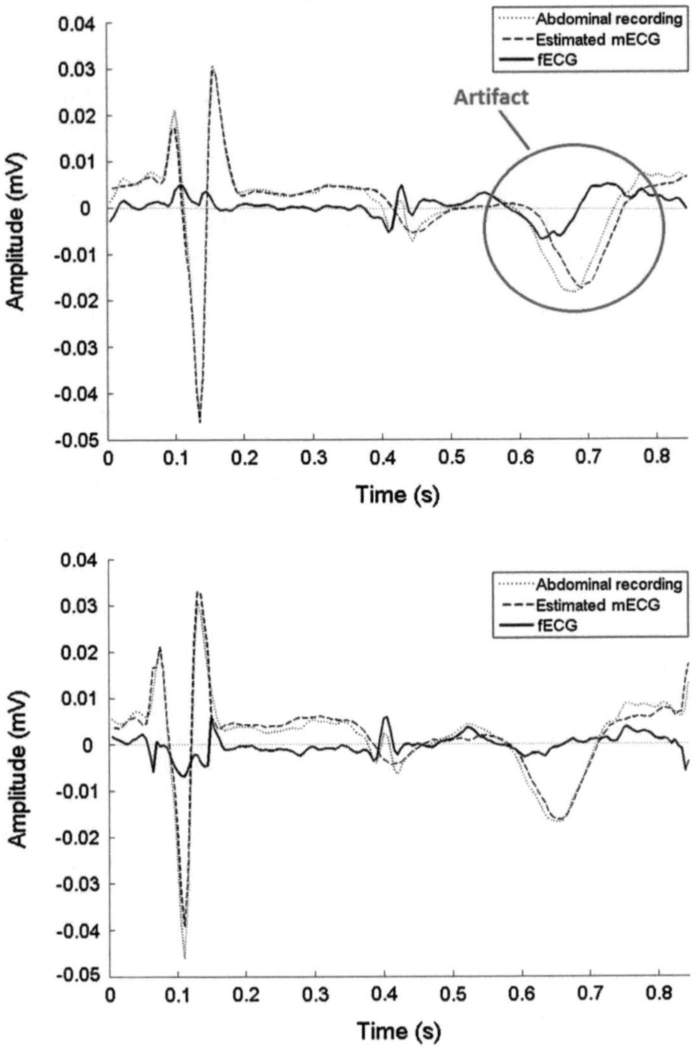

Fig. 7 Example of a beat of the fECG (*solid line*) overlapped to estimated mECG (*bold dotted line*) and original abdominal tracing (*dotted line*) obtained without (*upper panel*) and with (*lower panel*) the modulation/demodulation process. *Upper panel* shows an artefact that is largely reduced in the lower panel, thanks to the modulation/demodulation process

recording, the R peaks have to be indirectly obtained, for example using other ECG leads (for example the R-peaks position can be derived using thoracic leads) or from other signals before the SBMM can be applied.

In this study, the feature of the SBMM to adjust for physiological maternal HRV was tested in 60 s long ECG tracings. The duration of the recordings is a compromise between the need of obtaining a template that statistically represents the maternal beat (long recordings) and the need of having a stable signal in terms of HRV (short recordings). Quantitative results show that in ARec1 the errors relative to the QRS segment are significantly greater than those relative to the TUP segment, whereas in ARec2 the largest errors occur in correspondence of the TUP segment. Moreover, the errors relative to the two recordings are practically (not statistically) comparable in correspondence of the QRS segment (ARec1:1.65 μV; ARec2: 1.98 μV), whereas in correspondence of the TUP segments the median errors are almost 3 times larger in ARec2 (2.37 μV) than in ARec1 (0.87 μV). Such results can be interpreted on basis of the HRV. When HRV is low (i.e. when RR standard deviation is low) like in ARec1 (Table 1), both QRS and TUP segments of the estimated mECG are correctly aligned to the original ones in the abdominal recording. Consequently the larger errors occur in correspondence of the segment characterized by higher amplitude and higher frequency component since because even very small misalignments can cause relatively large errors during the subtraction step. Instead, if the HRV is high, as in the case of ARec2 (Table 1), the greatest errors occur in correspondence of the TUP segments, because in the estimated mECG they are likely less aligned than the QRS ones. These results also indicate that the hypothesis on which the SBMM relies, that is that the TUP duration is proportional to the previous RR is valid only in first approximation and that future optimizations are possible.

This study provides a preliminary validation of the SBMM algorithm by computing the errors relative to the QRS and the TUP segments. A greater number of tests are needed for a definite validation of the algorithm, and will be matter of future works. Imminent quantitative comparisons with other methods are also desirable. Tests performed in noisy recordings confirmed the SBMM ability to properly estimate the ECG [17].

5 Conclusion

In conclusion, the proposed algorithm based on the SBMM allows a clean mECG estimation from abdominal recordings thanks to a modulation procedure introduced to track physiological variation in the maternal heart rhythm.

References

1. Velayo C, Sato N, Ito T et al (2011) Understanding congenital heart defects through abdominal fetal electrocardiography: case reports and clinical. J Obstet Gynaecol 37:428–435
2. Hutter D, Kingdom J, Jaeggi E (2010) Causes and mechanisms of intrauterine hypoxia and its impact on the fetal cardiovascular system: a review. Int J Pediatr 2010:1–9
3. Regitz-Zagrosek V, Blomstrom-Lundqvist C, Borghi C et al (2001) ESC guidelines on the management of cardiovascular diseases during pregnancy. The task force on the management

of cardiovascular diseases during pregnancy of the European Society of Cardiology (ESC). Eur Heart J 32:3147–3197
4. Agostinelli A, Grillo M, Biagini A et al (2015) Noninvasive fetal electrocardiography: an overview of the signal electrophysiological meaning, recording procedures, and processing techniques. Ann Noninvasive Electrocardiol 20:303–313
5. Hasan M, Reaz M, Ibrahimy MI et al (2009) Detection and processing techniques of FECG signal for fetal monitoring. Biol Proced Online 11:263–295
6. Sameni R, Clifford G (2010) A review of fetal ECG signal processing; issues and promising directions. Electrophysiol Ther J 3:4–20
7. Agostinelli A, Giuliani C, Di Nardo F et al (2014) Extraction of the fetal ECG in noninvasive recordings using a new method based on segmented-beat modulation. Workshop mBiDa, pp 269–279
8. Agostinelli A, Giuliani C, Burattini L (2014) Extracting a clean ECG from a noisy recording: a new method based on segmented-beat modulation. Comput Cardiol 2014(41):49–52
9. Wolfber AJ (2012) The future of fetal monitoring. Rev Obstet Gynecol 5:132–136
10. Kriščiukaitis A, Lukoševičius M, Petrolis R et al (2013) Wearable system concept for monitoring of maternal and fetal heart activity. In: 17th international conference biomedical engineering, vol 17, pp 36–40
11. Hofmeyr GJ, Novikova N (2012) Management of reported decreased fetal movements for improving pregnancy outcomes. Cochrane Database Syst Rev 4:1–38
12. Mangesi L, Hofmeyr GJ, Smith V (2012) Fetal movement counting for assessment of fetal wellbeing (Review). Cochrane Libr 8:1–22
13. Goddard R (2001) Electronic fetal monitoring. Brit Med J 322:1436–1437
14. Kotas M, Jezewski J, Matonia A et al (2012) Towards noise immune detection of fetal QRS complexes. Comput Meth Prog Biomed 57:383–394
15. Malik M, Hnatkova K, Sisakova M et al (2008) Subject-specific heart rate dependency of electrocardiographic QT, PQ, and QRS intervals. J Electrocardiol 41:491–497
16. Pan J, Tompkins WJ (1985) A real-time QRS detection algorithm. IEEE Trans Bio-Med Eng 32:230–236
17. Agostinelli A, Giuliani C, Di Nardo F et al (2015) The segmented-beat modulation method for ECG estimation. IEEE Eng Med Biol Soc 2856–2859

Part V
Gait Analysis

Webcam and Smartphone for the Measure of Spatial-Temporal Parameters of Gait for Treadmill Use

Vinicio Barone, F. Verdini, F. Di Nardo, E. Maranesi, L. Burattini and S. Fioretti

Abstract A very low cost prototype has been developed for the spatial and temporal analysis of human movement using an integrated system of last generation smartphones and a high-definition webcam, controlled by a laptop. The system can be used to analyse mainly planar motions in non-structured environments. In this paper, the accelerometer signal captured by the 3D sensor embedded in one smartphone, and the position of coloured markers extracted from the analysis of the webcam frames, are used for the computation of spatial-temporal parameters of gait. The system has been tested on a treadmill at different gait speeds. Accuracy of results is compared with that obtainable by a gold-standard stereometric instrumentation. The system is characterised by a very low cost and a very high level of automation. It has been thought to be used by non-expert users in ambulatory settings.

1 Introduction

Treadmill training has recently emerged as an effective gait rehabilitation intervention. Furthermore, kinematic data collected during treadmill walking are commonly used for assessing gait performance also in pathological subjects [1–3]. In fact the use of a treadmill allows the simultaneous collection of spatial and temporal step kinematics for a large number of continuous steps. Several papers showed results obtained by different systems (opto-electronic system, instrumented walkway, foot-switches) related to spatial and temporal gait parameters on treadmill for healthy [4–9], and pathological subjects [1–3]. Generally, the results reported in literature show the variability of such parameters with respect to age of subjects, walking speed, pathological or experimental (over ground versus treadmill) conditions.

V. Barone (✉) · F. Verdini · F. Di Nardo · E. Maranesi · L. Burattini · S. Fioretti
Università Politecnica delle Marche, Ancona, Italy
e-mail: baronevinicio@libero.it

M. Conti et al. (eds.), *Mobile Networks for Biometric Data Analysis*,
Lecture Notes in Electrical Engineering 392, DOI 10.1007/978-3-319-39700-9_21

In papers [4, 5] from Owing et al., the focus was on the variability of step width and length with respect to age of subjects and walking speed. The results obtained in both these studies, at normal speed, showed that step width variability (measured by the standard deviation) of older adults was significantly larger than that of young adults. They found that step length and related variability were not age-dependent. In Stolze et al. study [6], the influence of treadmill on gait parameters with respect to the overground walking condition was examined between children and adults at their preferred speed. They found that step length significantly reduced in treadmill condition while step width significantly increased both for adults and children group. In Wearing et al. paper [7], temporal-spatial gait parameters were estimated in healthy adults while walking over an instrumented walkway (GAITRite) and instrumented treadmill system, at matched speed. They found that parameters for treadmill condition were not comparable with those obtained by the instrumented walkway because of different spatial resolution of both systems.

No study, to the best knowledge of the authors, was found related to the comparison between different acquisition systems on treadmill-gait parameters estimation.

In this paper a new prototype system based on a high definition (HD) webcam and a 3-D accelerometer, has been developed for the extraction of spatial-temporal gait parameters during treadmill walking. Its performance has been compared with respect to a 3-D gold-standard optoelectronic stereometric system.

3-D accelerometers and webcams are assuming an increasing interest for the ambulatory assessment of human motion, not only for the extraction of spatial-temporal parameters of gait [10–14], but also for a simplified analysis of motor tasks deemed clinically significant for the functional evaluation of motor disabilities. Literature reports many attempts to use accelerometers, often embedded in more complex measurement units, for the analysis of gait. They are able to give accurate estimates of the temporal gait parameters like, f.i. stride or step duration and walking cadence, but the estimation of the spatial parameters of gait are affected by a rather high incertitude [10]. The problem can be solved by the use of expensive commercial systems based on inertial measurement units. The use of webcams, on the contrary, has reduced timing accuracy, due to the low frame rate that characterizes this low cost instrumentation, but can give satisfactory accuracy in the estimate of spatial gait parameters. If a markerless approach is adopted, an acceptable level of accuracy can be reached, usually, at the expense of a higher computational effort and making particular care at the measurement environment (f.i. lighting conditions, use of suitable clothing, or background characteristics) [13, 14]. Much better results in terms of accuracy can be obtained by the use of webcams and passive markers. Of course, the gold standard for the kinematic analysis of movement is constituted by optoelectronic stereo-photogrammetric systems [15]. These are very accurate both in time and space but are very expensive and require structured environments (i.e. a laboratory setting and specialized personnel). The approach followed in this paper relies on the integration of both accelerometer and photogrammetric data in a unique system, very cheap and suitable to be used in non-structured environments like ambulatory settings by non-expert users like

General Practitioners. In particular the accelerometer data are captured by the inertial measurement units embedded in all last generation smartphones and is used mainly for timing accuracy [16]. The use of one webcam and colored markers allows the computation of marker position on a plane: in the present paper the plane is the rolling plane of a treadmill. The prototype described in this paper, is based on one smartphone, placed on the trunk at the pelvis level, and a webcam tracking six colored markers. The algorithm used in the present version of the prototype system is an extension of the algorithm described in [16], while using the same hardware architecture. With respect to [16], the new algorithm allows the use of the system both for treadmill use and for level walking. Substantial differences with this previous work will be illustrated in the Materials and Methods section.

2 Materials and Methods

The system is constituted by a laptop that controls up to 3 Android smartphones (Samsung S4), a webcam (Logitech HD Pro C920) and an optional wi-fi access point. For last generation smartphones this latter is not strictly necessary, as any smartphone can act as a router and as a sensor as well. The system architecture is shown in Fig. 1.

A suitable *app* collects all the inertial data from the internal sensors (3-axis accelerometer with gyroscope K330 ST Microelectronics, and YAS532 magnetic sensor Yamaha Corporation) and sends them, in wi-fi modality, to the laptop. The data transmission protocol is UDP. The acquisition frequency is about 50 Hz; the sampling frequency is variable because it depends on the real-time characteristics of Android operating system. In order to know exactly the time instant of the sensor data acquisition, the *app* associates to each data-packet the absolute system time before sending it to the laptop. The *app* needs to know only two parameters: the IP address of the laptop and the identification number of the smartphone (N = 0, 1, 2). The data acquired from each smartphone are shown graphically on the laptop in real

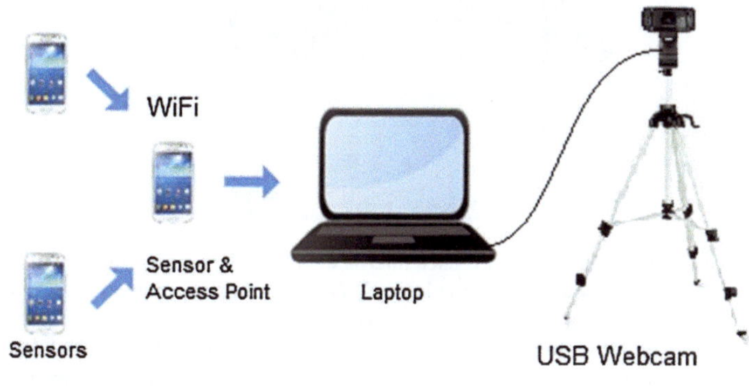

Fig. 1 System architecture

time. In this work, we used only one Samsung S4 smartphone positioned on the dorsal side of the trunk at the pelvis level and held by bandages firmly strapped around the pelvis.

Six healthy young subjects (3 males and 3 females, 21 ± 2 years old) were tested, walking on the treadmill and wearing their regular shoes and their usual clothing, at two different treadmill speeds: 3 and 5 km/h, respectively. All subjects gave their informed consent.

The characteristics of the webcam are: maximum capture rate of 30 fps, and maximum resolution 1920 × 1080 pixels. The camera was positioned orthogonally to the motion of the subject pointing downwards at an angle of approximately 40° from the horizontal plane of the treadmill.

Passive reflective coloured markers were employed. They were flat circles of 2 cm diameter coloured in red and yellow.

The measurement field was 2 m wide. For a better identification of the coloured markers, low light conditions of the environment were adopted. Attention was paid at the colour of the background that has to be different from that of the markers. A led lamp was also used in order to highlight the retroreflective markers with respect to the rest of the scene.

At the end of the acquisition period, the system automatically splits the video into frames, synchronizes the first frame with the starting time instant of the inertial sensor data acquisition, and reads on the video file the number of video frames per second (Fig. 2, L label) in order to maintain synchronization of the video with the acceleration tracks. The red markers were put on the left shoe while the yellow ones were placed on the right shoe. In particular they were placed on the shoes in

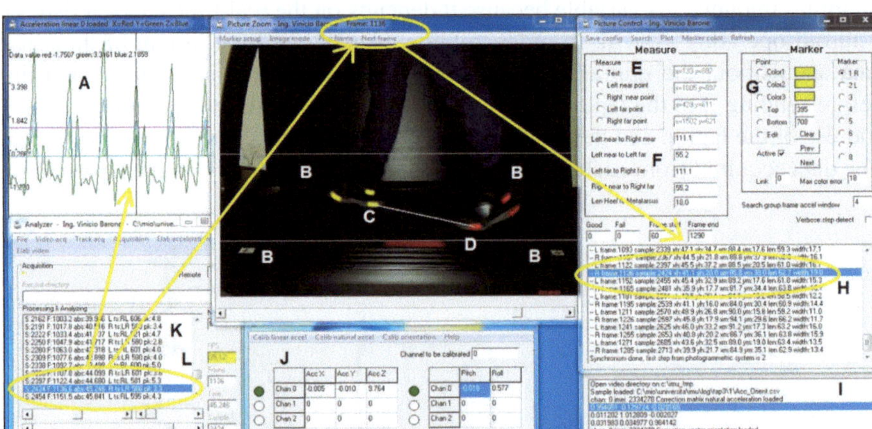

Fig. 2 Some windows of the software user interface; *A* Accelerometer tracks; *B* Four reference markers for the homographic transformation; *C*, *D* retroreflective markers; *E* Coordinates of reference markers on the frame *F* distances between reference markers; *G* color sample and max color tolerance *H* results from frame elaboration; *I* accelerometer calibration matrix; *J* accelerometer test calibration windows; *K* results from accelerometric elaboration *L* frame per second (FPS). The *yellow arrows* show the link between the accelerometric and the photogrammetric results

correspondence of the heels, of the 1st or the 5th metatarsal head (for the right and left foot, respectively), and of the malleoli (Fig. 2, C and D labels). The aim is to track these markers and to derive the gait spatial parameters, i.e. right and left step length and width. Of course, it is not possible to acquire 3D data by one camera, but it is possible to compute the position of a point belonging to a plane by a suitable homographic transformation.

2.1 Accelerometer Data Elaboration

The purpose of the acceleration measurement is to know when heel strikes occur: these events give rise to well defined peaks of the accelerometer signal.

The processing of the accelerometer trace is much faster than video frames processing; the knowledge of the time instant in which the heel impacts the floor allows to speed up the successive video analysis because it is necessary to analyze only few video frames per step. Thus, assuming a mean step duration of 0.6 s, and a capture rate of 30 fps, the analysis of only four frames in the neighborhood of the peak vertical acceleration, corresponding to the heel strike, allows to speed-up more than 4 times the video processing time. Moreover, the use of an accelerometer to identify the heel strike instants allows to obtain a more precise measure of the time: in fact the acquisition frequency of the accelerometer signal is 50 sample/s while for the webcam it is only 30 frame/s.

In order to identify the peaks on the accelerometer signal, the vertical acceleration components is analyzed by a double threshold algorithm. Figure 2, (A-label) shows in blue the vertical acceleration signal. The elaboration of the blue signal by a double threshold algorithm produces the green signal (Fig. 2, A-label) where the peaks identify the heel strike events in correspondence of the peaks of the blue signal.

2.1.1 Accelerometer Calibration

For the static calibration of the acceleration sensor, the system guides the operator through three measures:

- smartphone lying on its left side: the ideal acceleration vector should be $\mathbf{a}_1 = \begin{bmatrix} g & 0 & 0 \end{bmatrix}^T$ while the real measure are $\mathbf{a}_1^r = \begin{bmatrix} a_{1x}^r & a_{1y}^r & a_{1z}^r \end{bmatrix}^T$
- smartphone held vertical: the ideal acceleration vector should be $\mathbf{a}_2 = \begin{bmatrix} 0 & g & 0 \end{bmatrix}^T$
  while the real measure are $\mathbf{a}_2^r = \begin{bmatrix} a_{2x}^r & a_{2y}^r & a_{2z}^r \end{bmatrix}^T$
- smartphone held horizontal with the screen upwards: $\mathbf{a}_3 = \begin{bmatrix} 0 & 0 & g \end{bmatrix}^T$ while the real measure are $\mathbf{a}_3^r = \begin{bmatrix} a_{3x}^r & a_{3y}^r & a_{3z}^r \end{bmatrix}^T$

These steps are facilitated by a dedicated support equipped with a bubble level.

On the basis of the data acquired in these steps the system creates a transformation matrix **T** based on the assumption that in each position the corresponding acceleration axis should sense an acceleration equal to g (the gravity one).

The transformation matrix is given by

$$\mathbf{T} = g\begin{bmatrix} a_{1x}^r & a_{1y}^r & a_{1z}^r \\ a_{2x}^r & a_{2y}^r & a_{2z}^r \\ a_{3x}^r & a_{3y}^r & a_{3z}^r \end{bmatrix}^{-1}$$

Indicating with $\mathbf{a}_{\mathrm{m}}$ the generic acceleration measure, the corrected one is computed as: $\mathbf{a}_{corr} = \mathbf{T}\mathbf{a}_m$

Although the calibration procedure is very fast, it is not always necessary; in fact during the calibration phase the system associates the calibration matrix to the unique IMEI code of each smartphone, so that, after the first calibration, the system is able to automatically recognize the sensor and the associated matrix. It is possible to verify the correctness of the measure in the calibration window (Fig. 2, J label).

2.2 Photogrammetric Data Elaboration

The algorithm starts from the extraction of all frames from the movie, and then searches within each frame all pixels that belong to a previously defined set of colours. In particular, the colours are identified by an appropriate colour sampling previously done by the user: from three colour samples, chosen for each marker, in three different frames, and taking into account a user defined tolerance for the maximum allowed colour error (Fig. 2, G label), the system determines a range for the RGB values within which each marker has to be found. The next step consists in building a 3D map where x and y are the frame coordinates of a pixel and z is the density of points that match the colour set (Fig. 3). In this surface the system searches the three highest peaks and identifies the heel, the metatarsus and the malleolus starting from the assumption that the person is walking from right to left and, consequently, the marker closest to the left side of the frame is on the metatarsus, the marker closest to the right bottom corner is on the heel and the remainder is on the malleolus. This latter is only used for future expansions of the procedure.

To find the step parameters the algorithm searches the frame where the heel marker of one foot and the metatarsal marker of the contralateral foot reach the minimum value of Y-coordinate in the frame reference system; this means that both markers on the heel and on the metatarsal head belong to the treadmill walking plane. In order to speed up this minimum search, only the frames just before or soon after the heel strike instant detected by the accelerometric signal are analysed. At this instant, the homographic transformation allows to convert the image frame

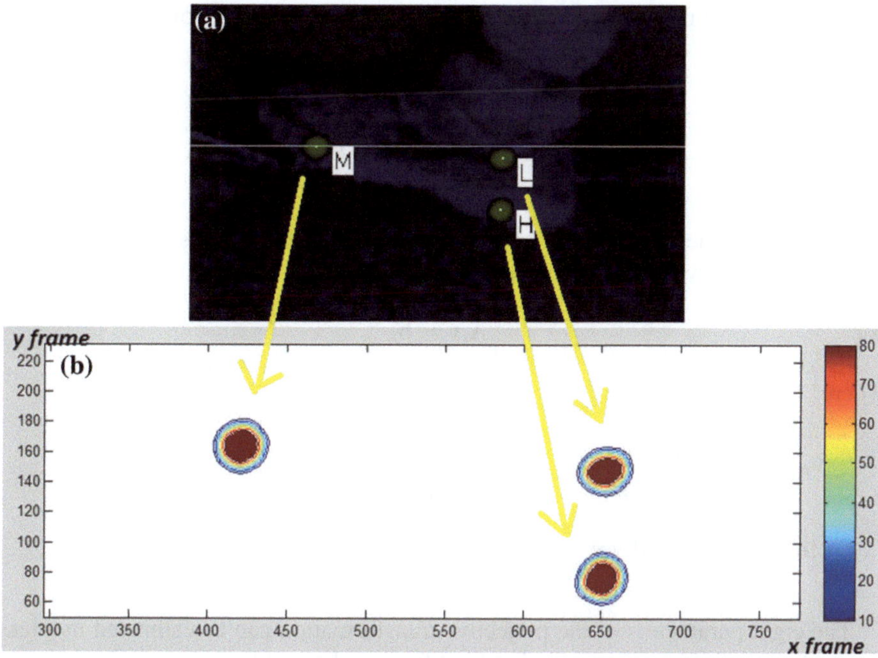

Fig. 3 *a* Particular of a frame after elaboration: once processed the markers appear coloured in *green*. The centre of each circular spot is marked in *yellow*. The *green* intensity decreases from the centre towards the marker contour. Furthermore, the software determines the label to be associated to each marker: *H* Heel, *M* Metatarsal head, *L* Malleolus. *b* 3D contour map: it shows the density of points that match the colour set. The *white colour* of the map indicates z-values lower than 10

coordinates of heel and metatarsal contact points to the walking plane coordinates. To calculate the length of a step, it is necessary to know the distance between the heel and the metatarsal markers placed on the same shoe; this distance has to be added to the distance between the heel and the contralateral metatarsal marker.

2.2.1 Webcam Calibration

Webcam calibration: The correspondence between the image plane coordinates $\mathbf{p}_i = (x_i, y_i)$ and the walking plane coordinates of point $\mathbf{P}_i = (X_i, Y_i)$ as seen by the webcam, is a projective homographic transformation that, in homogeneous coordinates, can be modeled as:

$$\begin{bmatrix} \lambda x_i \\ \lambda y_i \\ \lambda \end{bmatrix} = \begin{bmatrix} p_{11} & p_{12} & p_{13} \\ p_{21} & p_{22} & p_{23} \\ p_{31} & p_{32} & 1 \end{bmatrix} \begin{bmatrix} X_i \\ Y_i \\ 1 \end{bmatrix}$$

Dividing the first and second equations by the third one, it follows:

$$x_i = \frac{p_{11}X_i + p_{12}Y_i + p_{13}}{p_{31}X_i + p_{32}Y_i + 1}$$
$$y_i = \frac{p_{21}X_i + p_{22}Y_i + p_{23}}{p_{31}X_i + p_{32}Y_i + 1}$$

Multiplying the two members of the above equations by the same denominator and writing the two resulting equations in matrix form, one obtains:

$$\mathbf{A}_i \mathbf{v} = \mathbf{b}_i$$

where:

$$\mathbf{A}_i = \begin{bmatrix} X_i & Y_i & 1 & 0 & 0 & 0 & -x_i X_i & -x_i Y_i \\ 0 & 0 & 0 & X_i & Y_i & 1 & -y_i X_i & -y_i Y_i \end{bmatrix}$$
$$\mathbf{b}_i = \begin{bmatrix} x_i \\ y_i \end{bmatrix} \quad \text{and} \quad \mathbf{v} = \begin{bmatrix} p_{11} & p_{12} & p_{13} & p_{31} & p_{32} & p_{21} & p_{22} & p_{23} \end{bmatrix}^T$$

The eight parameters of the projective transformation can be estimated in a least squared method if one disposes of at least four markers of known coordinates in the walking plane. The four reference markers are placed at the vertices of a rectangle as shown in Fig. 2 (B label); Fig. 2 (E label) shows the pixel coordinates of the four vertices, while Fig. 2 (F-label) shows the lengths of the rectangle sides. The estimate of the unknown vector **v** is given by:

$$\mathbf{v} = (\mathbf{A}^{\mathrm{T}}\mathbf{A})^{-1}\mathbf{A}^{\mathrm{T}}\mathbf{B}$$

where:

$$\mathbf{A} = \begin{bmatrix} \mathbf{A}_1 \\ \mathbf{A}_2 \\ \vdots \\ \mathbf{A}_n \end{bmatrix}, \mathbf{B} = \begin{bmatrix} \mathbf{b}_1 \\ \mathbf{b}_2 \\ \vdots \\ \mathbf{b}_n \end{bmatrix} \quad \text{and} \quad n \geq 4$$

Once estimated the eight projective parameters, the coordinates of point $\mathbf{P}_{\mathrm{i}}$ in the walking plane can be computed solving the following equation system:

$$\begin{bmatrix} X_i \\ Y_i \end{bmatrix} = \begin{bmatrix} p_{11} - p_{22}x_i & p_{12} - p_{23}x_i \\ p_{31} - p_{22}y_i & p_{32} - p_{23}y_i \end{bmatrix}^{-1} \begin{bmatrix} x_i - p_{13} \\ y_i - p_{21} \end{bmatrix}$$

2.2.2 Differences of This Approach with the Previous Work [16]

With respect to a similar previous work [16], here the differences lie in the determination of step length and width on a moving surface with unknown and variable speed. In the previous work the system was thought to analyse the gait on a fixed and rigid walkpath looking for two different frames corresponding to right and left heel strikes, respectively. Of course the position of the right and left heel markers is different with respect to the walking plane reference system so that the step distance can be computed simply as the difference between the two positions. On the contrary, in the case of a treadmill, the treadmill reference system, rigid with the moving surface of the treadmill, moves at an unknown and variable speed with respect to the external fixed reference system, i.e. with the one used for the webcam calibration by the homographic transformation. Consequently, for treadmill use, it is necessary to develop an algorithm that uses only one specific frame for step length calculation. To this purpose, the frame corresponding to the weight transfer from one foot to the contralateral one is looked for. Consequently, two markers are necessary for each feet: one on the heel and the other one on the fifth or first metatarsal heads. So, in this paper. step length is computed as the sum of the distance between the metatarsal marker of one foot, and the heel marker of the contralateral foot, plus the constant distance between the metatarsal and the heel markers placed on the same foot.

A further difference of the present algorithm with respect to the one proposed in [16], lies on the fact that the previous algorithm was able to identify markers of the same colour only if they were placed sufficiently far one with respect to the other: consequently it was impossible to place more than one marker on each foot. On the contrary, the present algorithm allows the determination of an a-priori known number of markers of the same colour placed on different anatomical positions on each foot.

Moreover, in this new approach the smartphone accelerometric data is used to identify the heel strike instants. In this manner it was possible to minimize the search of the frame where to compute the step length. In fact only a very limited number of frames have to be analysed, two before and two soon after each heel strike, thus reducing the computation time and allowing to analyse a large number of steps as usual in the treadmill use.

3 Results

The study analysed six subjects while walking on the treadmill at two different speeds, for 1 min per trial. The spatial-temporal parameters (step length, width and duration) obtained by the system have been validated with respect to those obtained by an optoelectronic stereo-photogrammetric system (6-camera Elite BTS, 100 Hz). The markers used for both systems were the same coloured markers needed for the prototype system.

Table 1 Mean step length, width, and duration (±SD), measured by the prototype system

	3 km/h	5 km/h
Left step length (cm)	53.2 (±1.3)	70.5 (±2.1)
Right step length (cm)	55.1 (±1.2)	72.9 (±2.4)
Left step width (cm)	7.2 (±1.3)	9.1 (±1.0)
Right step width (cm)	9.4 (±1.2)	12.1 (±1.8)
Left step duration (s)	0.76 (±0.02)	0.52 (±0.02)
Right step duration (s)	0.75 (±0.02)	0.51 (±0.02)

Table 2 Mean error of spatial temporal parameters estimated by the prototype system versus the gold-standard stereometric walking on the treadmill (High Power Miami 2012), at 3 and 5 km/h

	3 km/h	5 km/h
Left step length (cm)	0.15	0.32
Right step length (cm)	0.25	0.61
Left step width (cm)	0.50	0.80
Right step width (cm)	0.55	0.90
Left step duration (s)	0.02	0.02
Right step duration (s)	0.02	0.02

Table 1 shows mean step length, width and duration, measured at the two different treadmill speeds for the left and right steps, respectively, obtained by the proposed system.

Table 2 shows for the two different speeds, the mean error for the right and left step length, width and duration computed as the difference between the values obtained by the proposed system and those obtained by the reference stereophotogrammetric one.

4 Discussion and Conclusions

This paper describes the realization of a prototype system based on smartphones and a webcam for the estimation of spatial-temporal parameters of gait during treadmill walking. The system is characterized by a very low cost and by a very high level of automation. It has been thought to be used by non-expert users for common walkpath gait as well as for treadmill use. The inertial unit embedded in the Android smartphone is exclusively used for the timing of gait events because accelerometer data are very accurate to this purpose; step length and width are determined by the photogrammetric procedures associated to the use of the webcam.

Treadmill training has recently emerged as an effective gait rehabilitation intervention. Furthermore, kinematic data collected during treadmill walking are commonly used for assessing gait performance also in pathological subjects [1–3]. Consequently a low cost system able to automatically provide a reliable and repeatable evaluation of some gait events and parameters during treadmill walking, appears relevant from a clinical point of view.

Results of this study showed that a high accuracy in the spatial gait parameters estimation can be reached, comparable with those obtainable through a sophisticated and expensive system like the optoelectronic stereo-photogrammetric instrumentation.

The heel strike events are automatically detected by the system and the heel strike recognition for the right and left foot was reached without errors for each tested treadmill speed, in all the tested cases.

Table 2 shows mean errors between the system proposed and the gold standard one. Mean errors resulted lower for the left step length determination than for the right one. This is due to the lower distance from the webcam of the left leg than the right one. Indeed, the errors in the determination of the position of a point lying in the treadmill plane depend in a direct manner on its distance from the camera.

Table 2 shows that errors in the direction perpendicular to the optical axis of the webcam are lower than errors in the direction of the optical axis (depth direction). The entity of this latter error depends, mainly, on the height of the webcam with respect to the treadmill plane.

In terms of percentage errors, relative to step length and width computed with respect to the average value of step length and width shown in Table 1, errors affecting step length with respect to step width resulted lower because step length is much higher than step width.

As the speed increases, the determination of the position of the heel strike contact point on the treadmill plane, is affected by a higher error. With respect to walking on a firm surface, the treadmill moves the supporting foot with respect to the camera so that the position of the heel contact point translates with the same speed of the treadmill mat. This causes motion blur that affects the webcam images (see red and yellow markers shown in Fig. 2). In fact, the limited sensibility of the CMOS sensor, and the high frame rate for the webcam (30 FPS) impose a higher open shutter time interval with consequent blurry images.

Results showed in the paper indicate that the system, validated in this study, can be usefully adopted for the estimation of some spatial gait parameter during treadmill walking. The low cost of the instrumentation and the implemented software allow the automatic recognition of the heel strike event and the calculation of step length and width. Associated with a not-instrumented treadmill it can record many steps for both the legs and it can provide accurate measures for spatial and temporal gait parameters. This aspect can constitute a relevant characteristic for the use of the system in a clinical context.

References

1. Kesar T, Binder-Macleod SA, Hicks GE, Reisman DS (2011) Minimal detectable change for gait variables collected during treadmill walking in individuals post-stroke. Gait Posture 33: 314–317
2. Savin DN, Morton SM, Whitall J (2014) Generalization of improved step length symmetry from treadmill to overground walking in persons with stroke and hemiparesis. Clin Neurophysiol 125:1012–1020
3. Kalron A, Achiron A (2014) The relationship between fear of falling to spatiotemporal gait parameters measured by an instrumented treadmill in people with multiple sclerosis. Gait Posture 39(2):739–744
4. Owings TM, Grabiner MD (2004) Variability of step kinematics in young and older adults. Gait Posture 20(1):26–29
5. Owings TM, Grabiner MD (2004) Step width variability, but not step length variability or step time variability, discriminates gait of healthy young and older adults during treadmill locomotion. J Biomech 37(6):935–938
6. Stolze H et al (1997) Gait analysis during treadmill and overground locomotion in children and adults. Electroencephalogr Clin Neurophysiol/Electromyogr Motor Control 105(6): 490–497
7. Wearing SC, Reed LF, Urry SR (2013) Agreement between temporal and spatial gait parameters from an instrumented walkway and treadmill system at matched walking speed. Gait Posture 38:380–384
8. Di Nardo F, Fioretti S (2014) Emg-based analysis of treadmill and ground walking in distal leg muscles. IFMBE Proc 41:611–614. doi:10.1007/978-3-319-00846-2_151
9. Maranesi E, Di Nardo F, Ghetti G et al (2014) A goniometer-based method for the assessment of gait parameters. In: 10th international conference on mechatronic and embedded systems and applications (MESA), IEEE/ASME. Publisher IEEE. doi:10.1109/MESA.2014.6935539
10. Zijlstra W, Hof AL (2003) Assessment of spatio-temporal gait parameters from trunk accelerations during human walking. Gait Posture 18:1–10
11. Henriksen M, Lunda H, Moe-Nilssen R, Bliddal H, Danneskiod-Samsøe B (2004) Test–retest reliability of trunk accelerometric gait analysis. Gait Posture 288–297
12. Martin E (2011) Novel method for stride length estimation with body area network accelerometers. In: Biomedical wireless technologies—networks, and sensing systems (BioWireleSS), pp 79–82
13. Surer E, Cereatti A, Grosso E, Della Croce U (2011) A markerless estimation of the ankle–foot complex 2D kinematics during stance. Gait Posture 532–537
14. Poppe R (2007) Vision-based human motion analysis: an overview. Comput Vis Image Underst 108:4–18
15. Whittle MW (1996) Gait analysis: an introduction. Butterworth-Heinemann, Oxford
16. Barone V, Maranesi E, Fioretti S (2014) Integration of smartphones and webcam for the measure of spatio-temporal gait parameters. In: IEEE EMBC conference, Chicago, IL; pp 5948–5951

The Wireless Surface Electromyographic Technique for Supporting the Assessment of Fall Risk in Diabetic Neuropathic Patients

E. Maranesi, S. Fioretti, V. Barone, G.G. Ghetti, O. Mercante, R.A. Rabini, L. Burattini and F. Di Nardo

Abstract The Functional Reach (FR) Test is commonly used in the clinical practice to identify elderly subjects at risk of recurrent falls. This study was designed to provide a useful tool for supporting the assessment of fall risk in diabetic neuropathic (DN) subjects: a non-invasive analysis of muscle activation patterns involved in the FR test using wireless surface electromyography (sEMG). Kinematic analysis was also performed to support results achieved by sEMG technique. A population of diabetic non-neuropathic subjects was considered as control group (CTRL). Kinematics was acquired by a 6-camera Elite optoelectronic system and passive markers were placed according to Davis protocol, with an additional marker placed on dominant wrist. sEMG analysis of Sternocleidomastoideus, Rectus Abdominis, Erectores Spinae at L4 level (L4), Rectus Femoris, Hamstrings (Ham), Tibialis Anterior (TA) and Soleus (Sol) was performed. No significant differences (*p*-value for evaluation of statistical significance was <0.05) were detected between groups in the computed kinematics parameters. sEMG analysis shows that in CTRL subjects, TA was identified as the first motor, since it activates before the start of FR movement; its action can be attributed to the anticipatory muscular activity necessary to initiate the body forward displacement. Sol and Ham, followed by L4, act mainly as tonic muscles, opposing the movement and preventing falls. Compared to CTRL group, DN subjects show an anticipatory recruitment of TA. The earlier

E. Maranesi (✉) · S. Fioretti · V. Barone · L. Burattini · F. Di Nardo
Department of Information Engineering, Università Politecnica delle Marche,
Via Brecce Bianche, 60131 Ancona, Italy
e-mail: e.maranesi@univpm.it

G.G. Ghetti · O. Mercante
Posture and Movement Analysis Laboratory, INRCA Geriatric Hospital,
Via della Montagnola, 60125 Ancona, Italy

R.A. Rabini
Diabetology Department, INRCA Geriatric Hospital, Via della Montagnola,
60125 Ancona, Italy

M. Conti et al. (eds.), *Mobile Networks for Biometric Data Analysis*,
Lecture Notes in Electrical Engineering 392, DOI 10.1007/978-3-319-39700-9_22

activation of TA in DN group suggests that this anticipation could be a predictor of a greater propensity of the DN subjects to fall compared to non-neuropathic diabetic subjects.

1 Introduction

Falls are a leading cause of disability, injury and death in older and pathological people and represent a major public health problem with substantial medical and economic consequences. The possibility of quantify fall risk by means of non-invasive approaches, which allow patients to maintain their mobility, is a relevant issue for health maintenance in aging societies. Diabetic peripheral neuropathy is associated with decreased stability while standing [1, 2] as well as during dynamic conditions [3, 4]. Thus, peripheral neuropathic patients exhibit a very high fall risk. The Functional Reach Test developed by Duncan et al. [5] has been proposed as a measure of balance able to identify elderly subjects at risk of recurrent falls. Functional Reach (FR) has been defined as the maximum distance one can reach forward beyond arm's length while maintaining a fixed base of support in the standing position. It is based on the idea that to investigate limits of stability in the absence of external perturbations, the maximum, voluntary, inclined posture can be used. In fact, limits of stability, quantified by the maximum, intentional displacement of the body in a given direction without losing balance, are influenced by body biomechanics as well as by subjective perception, and internal postural control abilities [6]. Therefore, a greater reach distance indicates a larger limit of stability and hence a better ability to maintain stability while leaning forward [7].

Such tools to quantify fall risk typically consist of a rating or scoring system designed to reflect the cumulative effect of known risk factors. Recently, instrumental tools were developed to help the clinical practice for fall risk assessment.

Due to decreased stability in standing as well as during dynamic conditions, the motor skills of the diabetic neuropathic subjects should be analyzed using non-invasive techniques, allowing to maintain the natural mobility of the subject. Generally, the detailed analysis of the FR test is studied by classic methodologies by means of sophisticated equipment, such as 6-camera stereophotogrammetric systems, and dynamometric platforms [8]. Recently, methods based on surface electromyography are getting increasing attention, not only for motion and gait analysis over ground [9–11] or different surfaces [12], but also for the fall-risk assessment [13–15].

Thus, aim of the present study is to provide a further tool for supporting the assessment of the fall risk in diabetic neuropathic subjects by means of a non-invasive analysis of muscle activation patterns involved in the FR test using wireless surface electromyography. Kinematics has been used to verify the results obtained by the surface electromyographic technique. A population of diabetic non-neuropathic subjects was also tested, and considered as control group.

2 Materials and Methods

2.1 Subjects and Experimental Set-up

The FR test was applied to 10 diabetic patients affected by type-2-diabetes mellitus: 5 diabetic non-neuropathic subjects (CTRL) and 5 diabetic neuropathic patients (DN). Their clinical data, together with the mean FR distance reached by each subjects normalized to the subject's height (FR_H), are shown in Table 1. FR distance was measured using a yardstick secured to the wall at the height of the shoulder.

The measurement protocol consisted in standing barefoot on a dynamometric platform (Kistler, 9281 type) sampled at 480 Hz. The dominant arm was extended and kept perpendicular to the trunk. The test consisted in moving the dominant arm as far forward as possible, and immediately backward again maintaining the wrist above a yardstick positioned at shoulder height (Fig. 1). The test was performed at the maximum possible speed in order to have more repeatability [16]. The FR test was repeated three times by each subject.

Kinematics was acquired by a 6-camera Elite optoelectronic system (BTS SMART D, sample rate 120 Hz). Passive markers were placed according to the protocol described in [17] with an additional marker placed on the dominant wrist.

Electromyographic signals were acquired using a wireless 8-channel surface electromyography system (BTS FREE EMG 300, sample rate 1000 Hz). Seven EMG active probes were attached unilaterally (dominant side) over:

1. Sternocleidomastoideus (Scm)
2. Erectores Spinae at L4 level (L4)
3. Rectus Abdominis (RAbd)
4. Rectus Femoris (RF)

Table 1 Clinical data and FR_H value of the analyzed subjects

A. CTRL	Subject A	Subject B	Subject C	Subject D	Subject E
Age (years)	74	66	67	77	79
Height (cm)	165	166	182	152	150
Foot length (cm)	27	24.5	27	24	24
Body mass index (kg•m^{-2})	34.3	28.3	27.1	29.2	32
HbA1c (%)	7.2	6.9	7.4	7.0	6.8
FR_H	0.14 ± 0.01	0.16 ± 0.01	0.15 ± 0.01	0.09 ± 0.01	0.13 ± 0.01
B. DN	Subject A_n	Subject B_n	Subject C_n	Subject D_n	Subject E_n
Age (years)	74	68	72	71	63
Height (cm)	166	163	173	169	174
Foot length (cm)	26	25	26	26.5	27
Body mass index (kg•m^{-2})	29.3	27.1	25.3	33.6	24.1
HbA1c (%)	7.5	7.8	7.9	7.5	7.7
FR_H	0.12 ± 0.01	0.17 ± 0.01	0.11 ± 0.01	0.07 ± 0.01	0.07 ± 0.01

CTRL Diabetic non-neuropathic subjects; *DN* diabetic neuropathic patients; *HbA1c* glycosylated hemoglobin

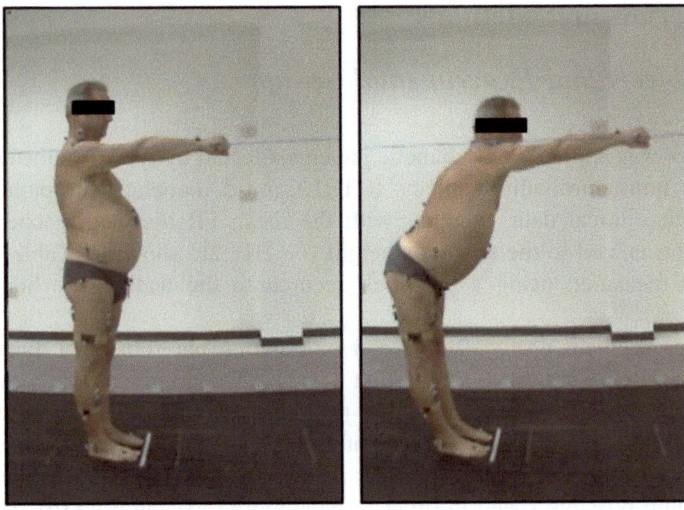

Fig. 1 Functional reach test

5. Hamstrings (Ham)
6. Tibialis Anterior (TA)
7. Soleus (Sol)

following the SENIAM recommendations [18, 19] (Fig. 2).

2.2 Data Analysis

FR-start instant was defined as the point in time immediately preceding the backward displacement of the centre of pressure (COP) excursion measured by means of the dyanamometric plaform (Kistler 9281 C). FR-end instant was defined as the instant of the maximum anterior displacement of the COP trajectory. The time interval was normalized with respect to the FR-period.

All kinematic data were filtered by a 4-th order low-pass Butterworth filter at a 5 Hz cut-off frequency. FR distance was defined as the difference between the point of maximum forward extension of the wrist from its initial starting position and was normalized to the subject's height (FR_H). Table 2 reports the kinematic parameters computed as described in [8] and in particular, the kinematic parameters computed are:

1. Hip flexion: angle between the femur axis and pelvis;
2. Ankle plantarflexion: angle between the tibia axis and foot;
3. Trunk rotation: maximum thoracic rotation computed as the maximal internal angle between shoulder line segment and the line segment at the 12th Thoracic Vertebra level;
4. Initial Angle: initial orientation of the lower leg relative to the vertical axis.

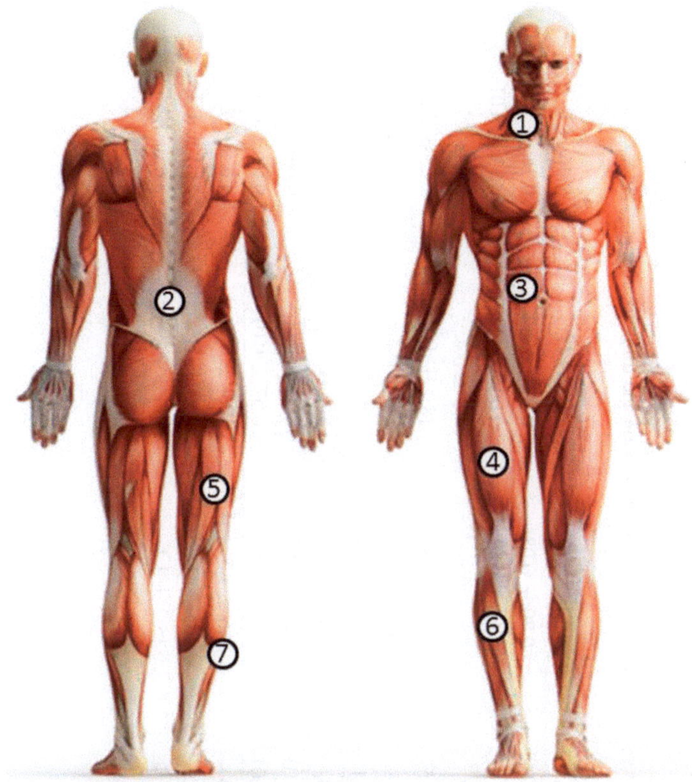

Fig. 2 EMG active probe position

The EMG signals were full-wave rectified and band-pass filtered (35–500 Hz). The root mean square value of the surface electromyographic signals was computed and then processed by a double-threshold detector that provides the onset and offset time instants of muscle activity. This technique [20] consists of selecting a first threshold for the signal and identifying the sample when the signal exceeds the first threshold and remains above it for at least 50 ms (the value of this interval represents the second threshold); this sample is acknowledged as the ON-instant of the muscle activity. In the same way, the first sample after the ON-instant, from which the signal remains below the first threshold for at least 50 ms, is acknowledged as the OFF-instant. The setting of the first threshold is based on the assessment of the background noise level, acquired in non-contracted and isometric muscle conditions, for 5 s before the FR actual start. For each subject the ON–OFF time instants relative to each muscle were averaged over the three trials performed.

Table 2 Mean value ± standard deviation of reach time, FR_H, initial angle, hip flexion, ankle plantarflexion and trunk rotation angles for CTRL (panel A) and DN (panel B) groups

A. CTRL	Subject A	Subject B	Subject C	Subject D	Subject E
Reach time (s)	0.94 ± 0.09	0.76 ± 0.08	0.70 ± 0.08	0.87 ± 0.02	0.76 ± 0.10
FR_H	0.14 ± 0.01	0.16 ± 0.01	0.15 ± 0.01	0.09 ± 0.01	0.13 ± 0.01
Initial angle (°)	1.3 ± 0.3	1.2 ± 0.2	3.2 ± 0.6	4.8 ± 0.4	2.9 ± 0.7
Hip flexion (°)	18.2 ± 1.1	36.9 ± 1.5	31.9 ± 1.8	18.1 ± 1.2	29.5 ± 2.5
Ankleplantarflexion (°)	10.0 ± 0.3	26.3 ± 8.9	20.0 ± 2.0	5.6 ± 0.3	10.1 ± 1.1
Trunk rotation (°)	15.1 ± 3.2	11.6 ± 4.1	12.8 ± 3.4	3.7 ± 2.4	3.9 ± 1.6
B. DN	Subject A_n	Subject B_n	Subject C_n	Subject D_n	Subject E_n
Reach time (s)	1.13 ± 0.03	0.74 ± 0.05	0.69 ± 0.05	0.53 ± 0.005	0.64 ± 0.07
FR_H	0.12 ± 0.01	0.17 ± 0.01	0.11 ± 0.01	0.07 ± 0.01	0.07 ± 0.01
Initial angle (°)	−0.4 ± 0.3	4.1 ± 0.4	2.6 ± 1.5	10.7 ± 0.2	4.6 ± 0.2
Hip flexion (°)	17.5 ± 3.2	44.2 ± 1.9	26.9 ± 1.5	28.7 ± 2.8	21.1 ± 1.2
Ankle plantarflexion (°)	14.2 ± 0.3	15.9 ± 0.5	15.2 ± 1.1	8.6 ± 2.2	8.6 ± 1.3
Trunk rotation (°)	11.7 ± 1.5	11.8 ± 2.8	8.6 ± 1.8	6.0 ± 1.1	7.4 ± 1.4

Positive values for initial angle indicate a forward tibial inclination

2.3 *Statistics*

The Lilliefors test was used to evaluate the hypothesis that each data vector had a normal distribution. The Wilcoxon Rank Sum Test was applied to test the presence of statistically significant differences between the two groups of analyzed subjects. Statistical significance was set at 5 %, i.e. p-value <0.05.

3 Results

Table 2 shows the mean value of reach time, FR_H, initial angle, hip flexion, ankle plantarflexion and trunk rotation angles for each subject in both categories (CTRL: panel A, and DN: panel B). In particular, the mean FR_H distance for CTRL and DN groups was found to be 0.13 ± 0.02 and 0.11 ± 0.04, respectively and did not result to be statistically significant.

Mean values (mv) and standard deviations (SD) of the instant of muscle activations in CTRL and DN subjects expressed in percentage of FR task duration are shown in Table 3.

Figure 3 shows the muscle activation intervals of the CTRL subjects. Generally, the anterior muscles of the body (Scm, RAbd, RF, TA) are active before the beginning of the movement. In each subject the first motor is TA that contracts

Table 3 Mean values (mv) and standard deviations (SD) of the instant of muscle activations in CTRL and DN subjects expressed in percentage of FR task

	Instant of muscle activation (% of FR period) mv (SD)				
A. CTRL	Subject A	Subject B	Subject C	Subject D	Subject E
Scm	20.8 (33.8)	3.2 (11.8)	−14.5 (9.1)	18.2 (27.8)	−10.8 (4.8)
L4	29.7 (5.7)	19.8 (2.9)	29.4 (4.8)	45.3 (26.8)	19.4 (3.6)
RAbd	−3.3 (3.1)	−9.2 (2.2)	−12.4 (5.9)	–	15.9 (9.9)
RF	−16.3(2.4)	−24.9 (5.8)	−30.6 (7.9)	−24.4 (1.7)	−33.7 (16.0)
Ham	22.1 (11.4)	−7.2 (16.0)	28.7 (11.2)	12.1 (4.8)	−2.9 (13.0)
TA	−20.5 (0.4)	−26.8 (5.2)	−27.1 (6.3)	−24.6 (1.3)	−30.7 (4.6)
Sol	25.7 (7.7)	11.9 (8.5)	11.0 (6.7)	14.29 (5.8)	22.9 (12.5)
	Instant of muscle activation (% of FR period) mv (SD)				
B. DN	Subject A_n	Subject B_n	Subject C_n	Subject D_n	Subject E_n
Scm	18.8 (5.1)	−10.4 (6.9)	24.3 (13.2)	−17.1 (8.1)	−5.3 (10.6)
L4	37.4 (0.5)	12.7 (7.3)	22.5 (4.3)	11.6(3.6)	–
RAbd	38.5 (11.2)	7.7 (27.3)	12.4 (3.1)	−4.0 (31.6)	−16.7 (3.3)
RF	–	−24.8 (5.9)	2.7 (7.8)	−31.5 (2.6)	−28.5 (3.8)
Ham	24.3 (34.4)	−5.0 (20.3)	56.2 (11.2)	29.4 (14.6)	8.5 (12.0)
TA	1.2 (9.3)	−28.9 (4.2)	−25.3 (14.8)	−37.2 (5.8)	−31.3 (2.4)
Sol	21.1 (7.9)	7.4 (9.6)	17.8 (27.9)	7.9 (22.8)	9.0 (0.6)

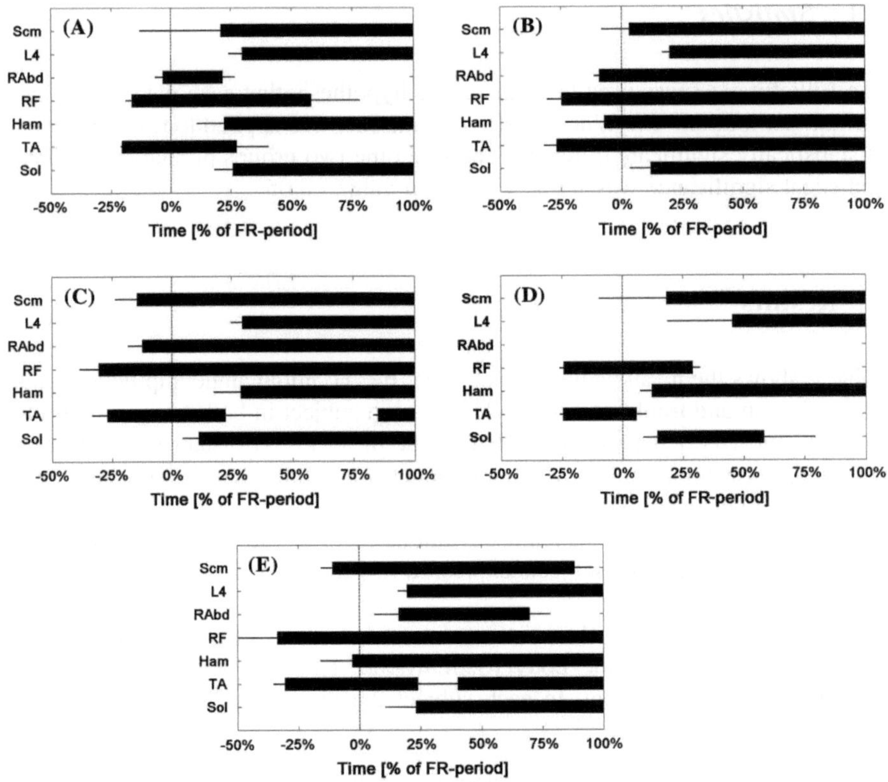

Fig. 3 Muscle activation intervals in the five CTRL subjects (**a–e**). The diagram shows mean value and standard deviation over the three repeated tests. Time axis is normalized between −50 and 100 %. The 0 % is FR-start instant (*dashed line*); the 100 % is the instant of maximum reach

at −25 % of the FR-period. RF contracts at the same time as TA and seems to be synchronous with this muscle. According to the number (one or two) and modality of TA and RF activations, the CTRL subjects can be divided in two groups: in the first group (subjects A and D), TA is active before the start and turns off within the 25 % of the FR-period; the second group (subjects B, C, E) is characterized by the same activity detected for the first group and by a further TA activity in the second half of the FR-period. The activation modality in this second group can occur as a single prolonged activity for the whole duration of the motor task (subject B) or as a double activation (subjects C and E), around the beginning and near the end of the FR-period.

L4 and Sol behave in similar manner, showing a simultaneous activation, approximately at 20 % of the FR-period and a high repeatability among subjects. Also Ham presents a similar trend; however, in two subjects (B and E) it acts before the movement start, about −5 % of the FR-period.

Figure 4 shows the muscle activation intervals of the DN subjects.

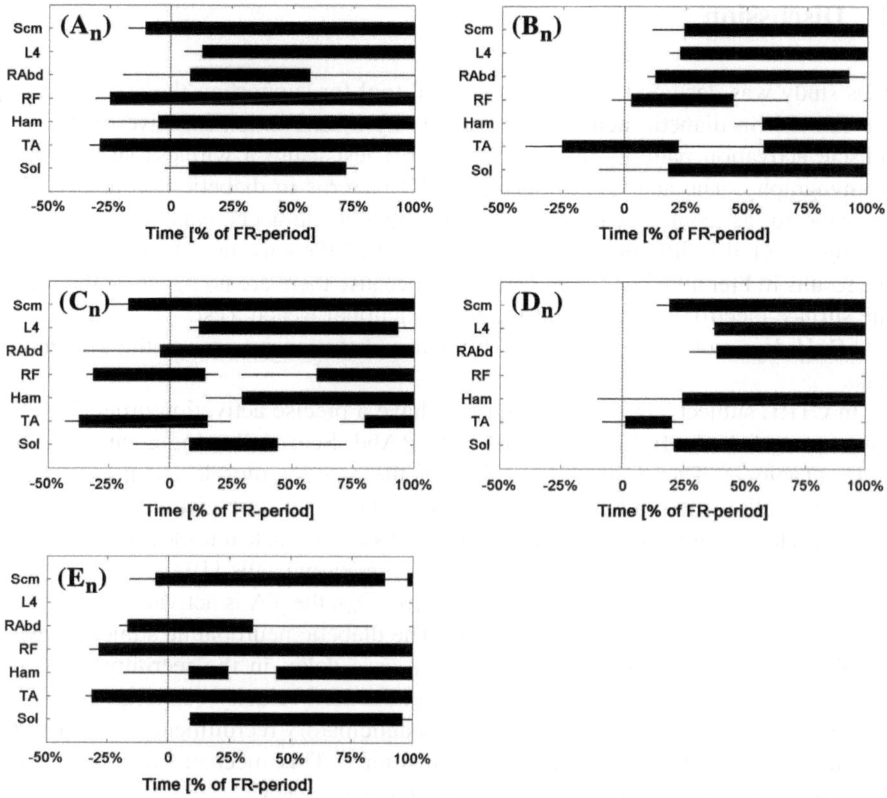

Fig. 4 Muscle activation intervals in the five DN subjects (An-En). The diagram shows mean value and standard deviation over the three repeated tests. Time axis is normalized between −50 and 100 %. The 0 % is the FR-start instant (*dashed line*); the 100 % is the instant of maximum reach

Compared to the CTRL group, a larger inter-subject variability is detected. In particular, in three DN subjects (An, Cn and En) there is a clear activation of Scm and RAbd before the FR-start. Subjects Bn and Dn, on the other hand, show a delayed activation of the above two muscles.

Also the activation and deactivation modalities of posterior muscles are more variable. The synchronous activity detected in the CTRL subjects between Sol and L4 is not present, especially in the OFF-phase, which can occur earlier for Sol than for L4 (subjects An and Cn). Moreover, Ham confirms its large variability in the ON-phase as already seen in CTRL group. Otherwise, similarly to the CTRL group, the anterior muscles of the lower limb (RF and TA) showed a myoelectric activity before the beginning of the movement. In four subjects (An, Bn, Cn, En), TA is still the first motor: its activation occurs at around −30 % of the movement and generally acts synchronously with RF. During the test, TA can show a double or a prolonged activation. Subject Dn shows a different picture of muscle activation, characterized by a generalized delay of the ON instants.

4 Discussion

This study was designed to provide a useful tool for supporting the assessment of the fall risk in diabetic neuropathic patients by means a non-invasive analysis of muscle activation patterns involved in the FR test using a wireless surface electromyography. The analysis of surface EMG patterns in diabetic non-neuropathic (considered as control subjects) and neuropathic subjects was able to reveal analogies but also differences in the sequence of EMG activation. Comparison with the results in literature cannot be performed because there are no paper on the use of the surface electromyography during the Functional Reach Test.

FR_H does not show significant differences between groups, as the kinematics data.

In CTRL subjects, the anterior muscles have a precise activation order from the lower part of the body to the top (TA, RF, RAbd, Scm) following a caudo-cranial order (Table 3). Therefore, a synergism of the anterior muscle chain exists: this synergism allows the imbalance of the body. There is a synergistic tonic action of Sol and Ham followed by L4 activation. In fact, the back muscles act mainly as tonic muscles that oppose to the movement, preventing falls [19].

In DN subjects, except for a subject (subject D_n), the TA is activated earlier than its activation in non-neuropathic subjects. The diabetic neuropathic subjects have a problem of proprioception which manifests as a delay in the recruitment of the motor units and as a reduced nervous conduction velocity; consequently it is probable that DN subjects put into action an anticipatory recruitment to compensate for the delay and to adjust the movement timing. This mechanism describes the early muscle predisposition to perform the movement [21].

In all CTRL and DN subjects, the TA can be recognized as the first muscle to contract and its action can be attributed to the anticipatory muscular activity necessary to initiate the movement creating useful conditions for forward displacement of the body. Its activation, preceding the movement starts, is necessary to move the COP backward and to give rise to the ankle dorsi-flexor moment, thus unbalancing the body forward. RF is activated before or close to the movement start: in fact it works as a hip flexor, and consequently it contributes to bend the trunk forward and it acts when acceleration is required, namely when the movement has to begin. When TA is inhibited, there is a synergistic tonic action of L4, Ham and Sol that opposes to the forward body fall (Figs. 3 and 4). At the end of the movement, the tonic action of the posterior muscles is present. This muscular activation is common to both groups of subjects and it is necessary to prevent the forward fall of the body. This action confirms the important role of these muscles that, though they do not participate in the movement generation, limit the excessive forward unbalance of the body.

Although there is no statistically significant difference in the kinematic variables examined (initial angle, hip flexion, ankle plantarflexion and trunk rotation angles), a difference in the TA activation was detected. So, electromyography can provide useful information about how the FR test is performed. In particular, results suggest that the anticipatory muscular activity exerted during the FR task not only serve to control the final body posture or to counterbalance the forthcoming movements perturbations, but is responsible of creating the proper center of pressure displacement within the base of support necessary to create initial destabilizing moments and thus to promote the forward movement of the body. In particular, the earlier activation of TA in the diabetic neuropathic subjects could be the predictor of a greater propensity of this groups to fall compared to diabetic subjects without neuropathy.

Thus, the surface electromyographic analysis is suggested as a non-invasive technique useful to help in assessing the fall risk in diabetic neuropathic subjects and to provide additional information besides the classical clinical FR measure.

5 Limitations of the Study

The relatively small size of our population samples might appear as a limitation of this study and one may argue that some variables could have not shown any statistical differences between groups possibly because of a type II error. On the other hand, this study exercised special care in recruiting subjects, in order to avoid age, height, weight and body mass index (BMI) as confounders. Our restrictive selection reduced the number of patients eligible for the study but strengthened the reliability of results in terms of performance of the test.

References

1. Simoneau GG, Ulbrecht JS, Derr JA et al (1994) Postural instability in patients with diabetic sensory neuropathy. Diab Care 17:1411–1421
2. Fioretti S, Scocco M, Ladislao L et al (2010) Identification of peripheral neuropathy in type-2 diabetic subjects by static posturography and linear discriminant analysis. Gait Posture 32:317–320
3. Bloem BR, Allum JH, Carpenter MG et al (2000) Is lower leg proprioception essential for triggering human automatic postural responses? Exp Brain Res 130:375–391
4. Inglis JT, Horak FB, Shupert CL et al (1994) The importance of somatosensory information in triggering and scaling automatic postural responses in humans. Exp Brain Res 101:159–164
5. Duncan PW, Weiner DK, Chandler J et al (1990) Functional reach: a new clinical measure of balance. J Gerontol 45:M192–M197
6. Duncan PW, Studenski S, Chandler J et al (1992) Functional reach: predictive validity in a sample of elderly male veterans. J Gerontol 47:M93–M98
7. http://www.who.int/diabetes/en/index.html
8. Maranesi E, Ghetti G, Rabini RA et al (2014) Functional reach test: movement strategies in diabetic subjects. Gait Posture 39:501–505
9. Di Nardo F, Mengarelli A, Maranesi E et al (2015) Assessment of the ankle muscle co-contraction during normal gait: A surface electromyography study, J Electromyogr Kinesiol 25(2):347–354

10. Di Nardo F, Ghetti G, Fioretti S (2013) Assessment of the activation modalities of gastrocnemius lateralis and tibialis anterior during gait: a statistical analysis. J Electromyogr Kinesiol 23(6):1428–1433
11. Di Nardo F, Fioretti S (2013) Statistical analysis of surface electromyographic signal for the assessment of rectus femoris modalities of activation during gait. J Electromyogr Kinesiol 23 (1):56–61
12. Di Nardo F, Fioretti S (2014) Emg-based analysis of treadmill and ground walking in distal leg muscles. IFMBE Proc 41:611–614. doi:10.1007/978-3-319-00846-2_151
13. Hubble RP, Naughton GA, Silburn PA (2014) Trunk muscle exercises as a means of improving postural stability in people with Parkinson's disease: a protocol for a randomised controlled trial. BMJ Open 4
14. Lee D, Lee S, Park J et al (2013) The effect of fixed ankle and knee joints on postural stability and muscle activity. J Phys Ther Sci 25:33–36
15. Zhou H, Hu H (2008) Human motion tracking for rehabilitation—A survey. Biomed Signal Process Control 3:1–18
16. Kozak K, Ashton-Miller JA, Alexander NB (2003) The effect of age and movement speed on maximum forward reach from an elevated surface: a study in healthy woman. Clin Biomech 18(3):190–196
17. Davis RB, Ounpuu S, Dennis T et al (1991) A gait analysis data collection and reduction technique. Hum Mov Sci 10:575–587
18. Freriks B, Hermens HJ, Disselhorst-Klug C (1999) The recommendations for sensors and sensor placement procedures for surface electromyography. In: Hermens HJ, Freriks B, Merletti R, Stegeman D, Bok J, Rau G et al (eds) European recommendations for surface electromyography. Results of the SENIAM project. Enschede: Roessingh Research and Development
19. Maranesi E, Fioretti S, Ghetti GG et al (2016) The surface electromyographic evaluation of the functional reach in elderly subjects. J Electromyogr Kinesiol 26:102–110
20. De Luca CJ (1997) The use of electromyography in biomechanics. J Appl Biomech 13 (2):135–163
21. Maranesi E, Di Nardo F, Rabini RA et al (2016) Muscle activation patterns related to diabetic neuropathy in elderly subjects: a functional reach test study. Clin Biomech 32:236–240

A Method for Computing the Functional Ambulation Profile Score by Stereophotogrammetric Data

E. Maranesi, M. Petrini, G.G. Ghetti, O. Mercante, V. Barone, L. Burattini, F. Di Nardo and S. Fioretti

Abstract The Functional Ambulation Profile (FAP) Score is a performance index based on the assessment of selected time and distance parameters, during standard gait test. FAP is considered a reliable global parameter for gait evaluation. The first and original FAP score is provided in an automatic way by the GAITRite System, a portable electronic mat embedded with pressure-activated sensors. The present study was designed to achieve a FAP score computation which could be independent from its native environment, and could be implemented by means of classic gait analysis systems, as stereophotogrammetry. This goal was pursued computing the parameters, which FAP definition is based on, by means of processing of stereophotogrammetric data. The reliability of FAP computation was tested by direct comparison with the original GAITRite FAP score, in a single subject, during different walking trials (comfortable gait, very slow gait; very fast gait) and simulating several pathological gaits (gait with small step; "foot-drop" gait; "festinating" gait; "spastic" gait). No relevant differences detected between stereophotogrammetry-based and GAITRite-based estimates indicate that the proposed method is able to provide a reliable assessment of FAP in different kind of natural/pathological walking. The capability of FAP index to identify alterations of walking is also preserved. In conclusion, this study proposes the stereophotogrammetry-based computation of FAP as a valid alternative to the original GAITRite FAP. The convenience of an easy integration with classic stereophotogrammetry-based gait analysis parameters, supports the usefulness of stereophotogrammetry-based FAP in providing a complete picture of subject walking, without needing a further measurement system.

E. Maranesi (✉) · M. Petrini · V. Barone · L. Burattini · F. Di Nardo · S. Fioretti
Department of Information Engineering, Università Politecnica delle Marche,
Via Brecce Bianche, 60131 Ancona, Italy
e-mail: e.maranesi@univpm.it

G.G. Ghetti · O. Mercante
Posture and Movement Analysis Laboratory, INRCA Geriatric Hospital,
Via della Montagnola, 60125 Ancona, Italy

M. Conti et al. (eds.), *Mobile Networks for Biometric Data Analysis*,
Lecture Notes in Electrical Engineering 392, DOI 10.1007/978-3-319-39700-9_23

1 Introduction

The gait is a complex motor act, resulting from a balance and posture control system, which instantly adjusts the interactions between internal and external body forces by the action of the muscles. Moreover, it is a regular act whose functional unit, called Gait Cycle, consists of well-defined stages, characterized by specific functional goals. The need for reliable methods to assess spatio-temporal variables of the gait cycle is particularly important in the gait evaluation in healthy and pathological subjects. From this point of view, many efforts have been made both using classical measure systems [1, 2] and/or simplified techniques [3, 4].

In order to have a performance index of gait as one single global parameter, the Functional Ambulation Profile (FAP) Score has been developed. The FAP scoring system which was originally developed by Nelson [5], integrates selected time and distance parameters, obtained during a standard gait test (linear walking of 5 m) at self-selected natural speed. The score provides a quantitative means of assessing gait without the subjective qualification that most rating scales require [6]. The FAP score is based on the linear relationship of step length/leg length ratio to step time when the velocity is normalized to leg length in healthy adults. The FAP score ranges from 95 to 100 points in the healthy adult population [5, 7–10]. The FAP score is a valid, reliable and objective method of measuring various gait parameters [8]. FAP is a reliable tool for the evaluation of gait performance and it has been largely used in literature to evaluate gait and to estimate the risk of falling among elderly people [11] and to assess gait variations in patients affected by Down syndrome [7], multiple sclerosis [9], Parkinson's disease [8], and chronic stroke [12]. It was also used to estimate the progress of patients walking further to a rehabilitation program [13].

The first and unique FAP obtained in an automatic way is that provided by GAITRite Walkway System. The GAITRite system consists of a portable electronic mat embedded with pressure-activated sensors. The electronic mat detects the timing of sensor activation and feeds this information into application software that computes spatial and temporal gait parameters [14, 15]. The reliability of the GAITRite system of FAP scoring is reported in several studies [16–18].

The aim of this work was to obtain the FAP score from data acquired by a classic stereophotogrammetric system, "the stereophotogrammetric FAP", and to compare it with the score provided by GAITRite. In this way, the FAP score could be computed directly with classic instrumentation without the need of using one further measurement system as GAITRite, with the consequent benefit of having a diagnostic index for any gait disturbances, as well as feedback on the effectiveness of therapeutic approaches.

2 Materials and Methods

The two methods used to compute the FAP score obtained by GAITRite (GaitRite FAP Score) and by the stereophotogrammetric system (Stereo FAP Score) are shown below.

2.1 GAITRite FAP Parameters

The FAP score is based on the following parameters [14, 15]:

1. Step/Extremity Ratio: ratio between Step Length and Leg Length of the subject; the leg length is measured under load, from the greater trochanter to the ground, intersecting the lateral malleolus;
2. Mean Normalized Velocity: ratio between Speed (Step Length/Step Time) and Leg Length;
3. Step Time;
4. Asymmetry of Step Length: absolute value of the difference between right and left Step/Extremity Ratio;
5. Dynamic Base of Support: distance between the heel center and the progression line of the opposite foot. In Fig. 1, the Dynamic Base of Support is the segment CD.

Except for Leg Length, which is manually measured by an operator, the other parameters are automatically calculated by the system.

While the subject is walking on the electronic mat at a natural self-selected speed, the FAP Score is obtained by subtracting from the maximum score of 100, the penalties, expressed as percentage, identified as follows:

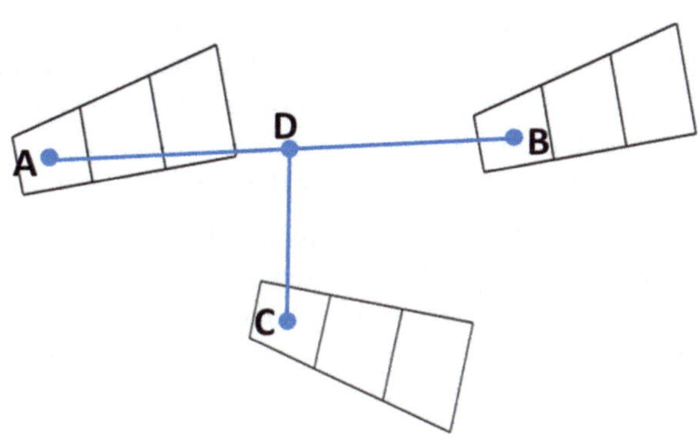

Fig. 1 Graphic representation of the dynamic base of support (CD segment). **A** is left heel center at the first step; **B** is left heel center at the second step; **AB** is the progression line of the left foot; **C** is right heel center at the first step; **D** is the orthogonal projection of the point **C** on the line AB

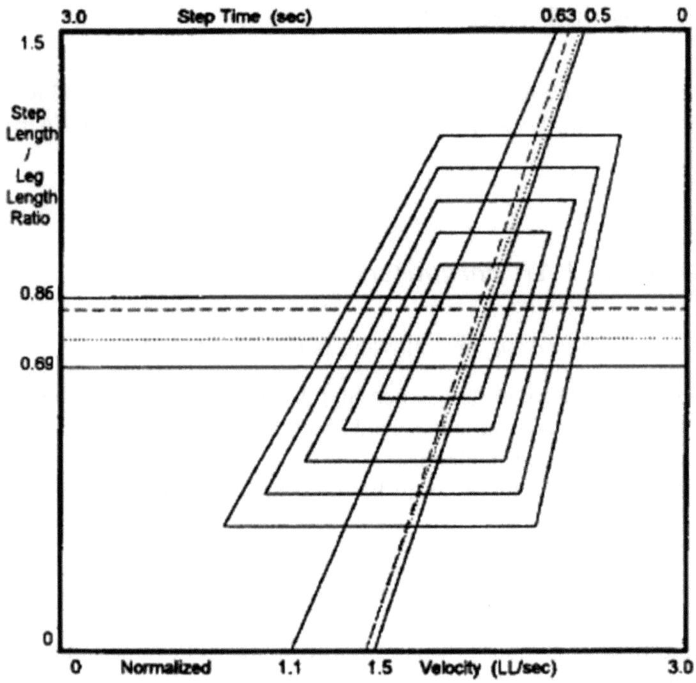

Fig. 2 The area of normality of mean normalized velocity, step/extremity ratio and step time

1. Step Function (0–44 %). For the parameters Mean Normalized Velocity, Step/Extremity Ratio and Step Time, the normal range in the database of the software are shown in a graph defining an area of normality (Fig. 2).
2. Differential Step/Extremity Ratio (0–8 %). If Asymmetry of Step Length parameter is less than 0.05 no penalty is applied; otherwise, the increase of its value, corresponds to a decrease of score up to a maximum of 8.
3. Dynamic Base of Support (0–8 %). The average between right and left Dynamic Base of Support is computed. This value is compared with the respective normal value; if it is too wide or too narrow, points are subtracted by the score up to a maximum of 8.
4. Mobility aids and assistive devices (0–10 %).

Figure 3 shows the flow chart of the FAP computation algorithm. In this Figure, StepPtL and StepPtR are the scores for right and left step function, DiffStepRatio is the score for Step/Extremity Ratio, BaseSupp is the score for Dynamic Base of Support, WalkDev is the score for the use of assistance devices and WalkAid is the score for the use of walking aids.

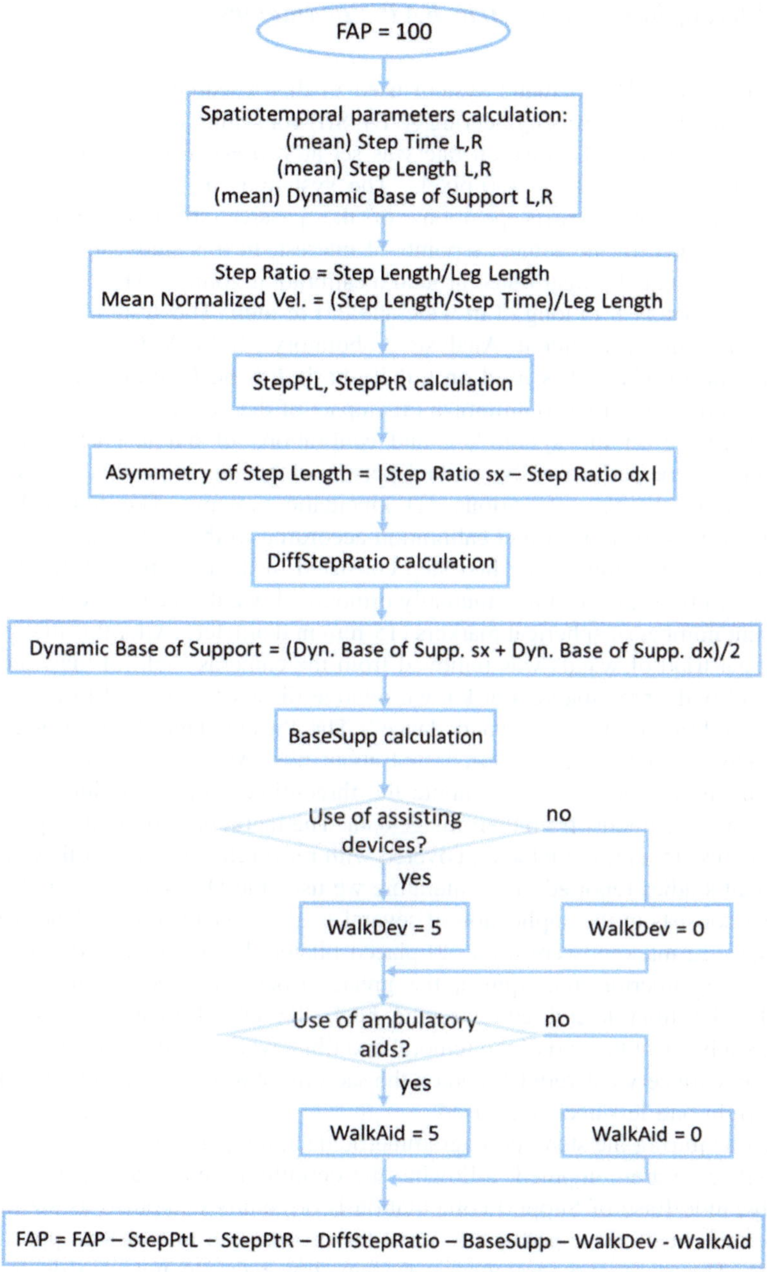

Fig. 3 Flow chart of the FAP computation algorithm

2.2 Stereophotogrammetric FAP Parameters

The commercial 3D kinematic system used in this research study was Smart-D, manufactured by BTS Bioengineering (BTS, Srl) consisting of six infrared cameras and passive markers. Acquisition rate was set at 120 Hz. Resolution of the CCD digital cameras was 640 × 480 pixels. The system is an optoelectronic system which actually reconstructs positions of the passive spherical retro-reflective markers, attached to the subject's points of interest. In this study, the six cameras involved were equally distributed around a calibration volume. The calibrated field was approximately 8 m long, 3 m wide and 2.0 m high. The system was set up in the Posture and Movement Analysis Laboratory, INRCA Geriatric Hospital, (Ancona, Italy) where it is used on a daily basis for the functional evaluation of motor disorders. In order to minimize the impact of the human factor and the effect of inexperience on the system use and calibration, all our test and calibration procedures were performed by trained personnel having more than ten years of experience in the use of various 3D kinematic systems. The first calibration experiment was made by wand calibration according to the system manufacturer's instructions. Accordingly, the first step consisted in defining the laboratory reference system by means of three mutually orthogonal wands each one endowed with a different number of spherical markers (15 mm in diameter). After this first step the orthogonal triad of wands was removed from the cameras' field of view and only one wand with three markers at known relative distance was used to perform the second calibration step—a "wand dance". The "wand dance", according to the manufacturer's recommendations, lasted about 60s. At the end of the calibration procedure it was possible to compute the three-dimensional position of markers fixed to body segments from their image data. The markers were lightweight plastic hemispheres, 15 mm in diameter, covered with retro-reflective tape. Likewise most of the gait studies reported in the literature we used the Davis's protocol [19]. This protocol consists in the application of several markers on anatomical landmarks. In particular, 22 markers were used: 20 placed bilaterally on the shoulder acromion, the anterior superior iliac spinae, the greater trochanter, the femur lateral epicondyle, the tibia lateral condyle, the heel, the lateral malleolus, the second metatarsal head, at the end of the femoral and tibial wands, and 2 placed unilaterally on the seventh cervical vertebra and on the sacrum. In this study, additional markers on the right and left heel were used.

After acquiring the stereophotogrammetric data, the parameters used to compute the FAP score are computed, following the definitions reported in paragraph 2.1. The Dynamic Base of Support computation is originally computed as the distance between the heel center position (C) and the progression line of the opposite foot (Fig. 1). The stereophotogrammetric system, however, do not provide the footprints, so the heel center position (C) is not available. Therefore, the Dynamic Base of Support has been computed as the distance (C′D′) between the heel points (C′) and the progression line of the opposite foot (A′B′), as shown in detail in Fig. 4.

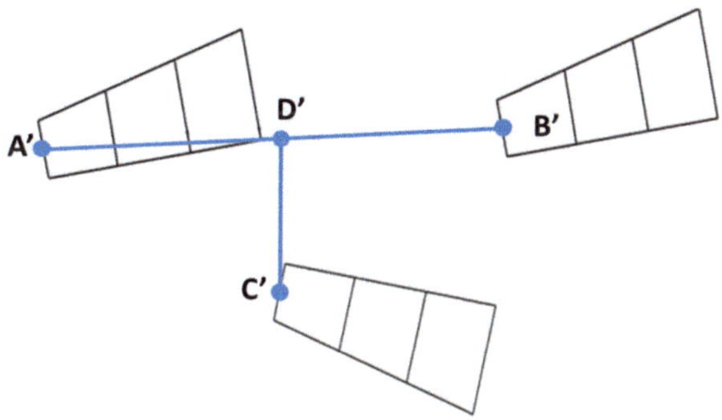

Fig. 4 Graphic representation of the dynamic base of support computed using GAITRite (segment CD) and stereophotogrammetric data (C′D′)

2.3 Comparison of the FAP Score Computed by GAITRite and by the Stereophotigrammetric System

One healthy female adult volunteer, (22 years old, 55 kg, 165 cm and 20 kg*m^{-2}), was selected for the validation procedure. The subject did not suffer neurological pathology, orthopaedic surgery and acute or chronic knee and ankle pain or pathology. The FAP score computation was performed using data acquired from a stereophotogrammetric system during a gait test. Matlab software was used for this purpose. Data from GAITRite and stereophotogrammetric system were acquired at the same time while the subject walked on the GAITRite electronic walkway.

The GAITRite system consists of a portable electronic walkway embedded with pressure-activated sensors. The electronic mat detects the timing of sensor activation as well as the relative distances between the activated sensors, and feeds this information into application software that computes spatial and temporal gait parameters for individual footfalls as well as an overall average for each parameter. The GAITRite system active area is 61 cm wide and 732 cm long. The sensors are placed 1.27 cm apart (total of 27,648 sensors). The sampling rate of the system is 80 Hz [14, 15].

The subject was instrumented with passive markers as described above. Data from GAITRite and stereophotogrammetric system were acquired contemporaneously while the subject walked on the GAITRite's electronic walkway. Data were acquired at the Posture and Movement Analysis Laboratory at the INRCA Hospital (Ancona).

The subject walked naturally and simulating several pathological gaits:

1. Very slow gait;
2. Very fast gait;
3. Gait with small step, typical of a an elderly person who has a very limited ability to flex the hip and knee;
4. "Foot-drop" gait, due to paralysis of the pretibial and peroneal muscles that cause the fall of the forefoot;

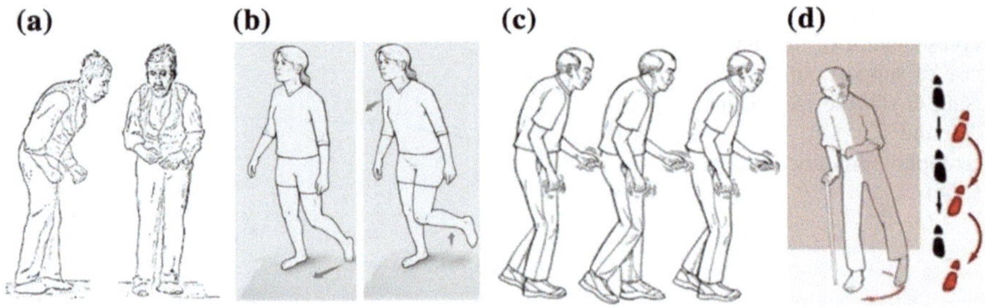

Fig. 5 Pathological gait. **a** Gait with small steps; **b** "Foot-drop" gait; **c** "Festinating" gait; **d** "Spastic" gait

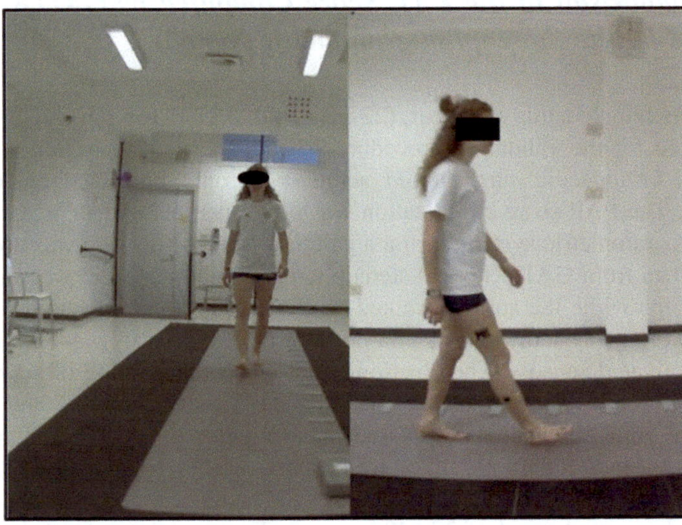

Fig. 6 The experimental set-up

5. "Festinating" gait, typical of Parkinson's disease in an advanced stage;
6. "Spastic" gait, characteristic of a hemiplegic patient.

A schematization of these pathological gaits are shown in Fig. 5.
The experimental set-up is shown in Fig. 6.

3 Results

Tables 1, 2, 3, 4, 5, 6 and 7 show the values of parameters computed for the FAP Score assessment and the FAP score, computed using stereophotogrammetric system (Stereo FAP Score) and provided by GAITRite (GaitRite FAP Score) during

Table 1 Values of parameters computed for the FAP Score assessment and the FAP score, computed using stereophotogrammetric system and provided by GAITRite (GaitRite FAP Score), during natural walking

	Natural						
	GAITRite		Stereo		Difference (%)		Range of normality
	L	R	L	R	L	R	
Step length (cm)	60.4	59.6	60.8	59.2	0.54	0.70	
Step time (s)	0.5	0.5	0.5	0.5	0	0	0.5–0.63
Mean normalized velocity	1.4	1.4	1.4	1.4	0	0	1.1–1.5
Step/extremity ratio	0.7	0.7	0.7	0.7	0	0	0.69–0.86
Dynamic base of support (cm)	9.9	10	9.0	8.9	8.60	11.38	1.5–13
Asymmetry of step length	0		0.01				<0.05
FAP score	100		100		0		95–100

The difference in the parameters computed by the two systems are also reported, together with the range of normality for each parameter

Table 2 Values of parameters computed for the FAP Score assessment and the FAP score, computed using stereophotogrammetric system and provided by GAITRite (GaitRite FAP Score), during very slow walking. The difference in the parameters computed by the two systems are also reported, together with the range of normality for each parameter

	Slow						
	GAITRite		Stereo		Difference (%)		Range of normality
	L	R	L	R	L	R	
Step length (cm)	48.1	47.1	48.1	47.3	0.02	0.40	
Step time (s)	0.8	0.8	0.8	0.8	0	0	0.5–0.63
Mean normalized velocity	0.7	0.7	0.7	0.7	0	0	1.1–1.5
Step/extremity ratio	0.6	0.6	0.6	0.6	0	0	0.69–0.86
Dynamic base of support (cm)	7.8	8.2	5.6	6.8	28.35	16.82	1.5–13
Asymmetry of step length	0.005		0.002				<0.05
FAP score	80		79		1		95–100

each task performed. In particular, Table 1 describes the results for the trial at natural speed; Table 2 for the trial at slow speed; Table 3 for the trial at fast speed; Table 4 for small step trial; Table 5 for the foot-drop trial; Table 6 for the fastinating walking; Table 7 for the spastic walking.

Table 3 Values of parameters computed for the FAP Score assessment and the FAP score, computed using stereophotogrammetric system and provided by GAITRite (GaitRite FAP Score), during very fast walking

	Fast						
	GAITRite		Stereo		Difference (%)		Range of normality
	L	R	L	R	L	R	
Step length (cm)	67.8	66.5	66.3	67.9	2.24	1.99	
Step time (s)	0.4	0.4	0.4	0.4	0	0	0.5–0.63
Mean normalized velocity	1.9	1.8	1.9	1.8	0	0	1.1–1.5
Step/extremity ratio	0.8	0.8	0.8	0.8	0	0	0.69–0.86
Dynamic base of support (cm)	9.4	7.7	8.0	7.2	14.27	6.51	1.5–13
Asymmetry of step length	0.006		0.029				<0.05
FAP score	87		86		1		95–100

The difference in the parameters computed by the two systems are also reported, together with the range of normality for each parameter

Table 4 Values of parameters computed for the FAP Score assessment and the FAP score, computed using stereophotogrammetric system and provided by GAITRite (GaitRite FAP Score), during gait with small step

	Small step						
	GAITRite		Stereo		Difference (%)		Range of normality
	L	R	L	R	L	R	
Step length (cm)	20.1	18.2	20.5	17.9	1.68	1.91	
Step time (s)	0.9	0.9	0.9	0.9	0	0	0.5–0.63
Mean normalized velocity	0.3	0.2	0.3	0.2	0	0	1.1–1.5
Step/extremity ratio	0.2	0.2	0.2	0.2	0	0	0.69–0.86
Dynamic base of support (cm)	25.7	25.8	23.8	23.9	7.27	7.17	1.5–13
Asymmetry of step length	0.02		0.03				<0.05
FAP score	50		50		0		95–100

The difference in the parameters computed by the two systems are also reported, together with the range of normality for each parameter

The most significant difference in the FAP scores provided by the two systems is recorded during the festinating (5 %) and spastic (6 %) gaits. These differences depend on the constant difference reported in the Dynamic Base of Support computation obtained from the two systems. This parameter shows the greater difference as compared by the two.

Table 5 Values of parameters computed for the FAP Score assessment and the FAP score, computed using stereophotogrammetric system and provided by GAITRite (GaitRite FAP Score), during foot-drop gait

	Foot-drop						
	GAITRite		Stereo		Difference (%)		Range of normality
	L	R	L	R	L	R	
Step length (cm)	43.0	30.4	42.6	30.6	0.86	0.52	
Step time (s)	0.7	0.7	0.7	0.7	0	0	0.5–0.63
Mean normalized velocity	0.7	0.5	0.7	0.5	0	0	1.1–1.5
Step/extremity ratio	0.5	0.4	0.5	0.4	0	0	0.69–0.86
Dynamic base of support (cm)	11.4	11.5	10.1	10.3	11.30	11.1	1.5–13
Asymmetry of step length	0.160		0.144				<0.05
FAP score	63		64		1		95–100

The difference in the parameters computed by the two systems are also reported, together with the range of normality for each parameter

Table 6 Values of parameters computed for the FAP Score assessment and the FAP score, computed using stereophotogrammetric system and provided by GAITRite (GaitRite FAP Score), during festinanting gait

	Festinating						
	GAITRite		Stereo		Difference (%)		Range of normality
	L	R	L	R	L	R	
Step length (cm)	31.9	32.2	34.9	31.8	9.49	1.15	
Step time (s)	0.4	0.4	0.4	0.4	0	0	0.5–0.63
Mean normalized velocity	0.9	0.9	0.9	0.9	0	0	1.1–1.5
Step/extremity ratio	0.4	0.4	0.4	0.4	0	0	0.69–0.86
Dynamic base of support (cm)	19.9	19.6	14.7	14.4	25.73	26.59	1.5–13
Asymmetry of step length	0.008		0.034				<0.05
FAP score	67		72		5		95–100

The difference in the parameters computed by the two systems are also reported, together with the range of normality for each parameter

Table 7 Values of parameters computed for the FAP Score assessment and the FAP score, computed using stereophotogrammetric system and provided by GAITRite (GaitRite FAP Score), during spastic gait

	Spastic						
	GAITRite		Stereo		Difference (%)		Range of normality
	L	R	L	R	L	R	
Step length (cm)	33.9	40.7	33.9	41.4	0	1.69	
Step time (s)	0.7	0.9	0.7	0.9	0	0	0.5–0.63
Mean normalized velocity	0.6	0.5	0.6	0.5	0	0	1.1–1.5
Step/extremity ratio	0.4	0.5	0.4	0.5	0	0	0.69–0.86
Dynamic base of support (cm)	19.1	18.8	17.9	17.7	6.13	5.69	1.5–13
Asymmetry of step length	0.1		0.099				<0.05
FAP score	57		63		6		95–100

The difference in the parameters computed by the two systems are also reported, together with the range of normality for each parameter

4 Discussion

The present study was designed to achieve a FAP score computation which can be integrated in the classic gait analysis performed by a stereophotogrammetric system. This goal was pursued computing all FAP parameters by means of the processing of stereophotogrammetric data and the comparison with the original GAITRite FAP score. The reliability of the FAP score computation has been tested on a single subject, during natural walking and simulating several pathological gaits.

The fundamental parameters to compute the FAP score are Step Length, Step Time, Mean Normalized Velocity, Step/Extremity Ratio and Dynamic Base of Support. Comparison with GAITRite-based estimates (Tables 1, 2, 3, 4 and 5) shows that the method based on the stereophotogrammetric data is able to provide a reliable assessment of the first four parameters (Step Length, Step Time, Mean Normalized Velocity, Step/Extremity Ratio) in all walking trial performed, both in normal and pathological gaits. Small differences (mainly during very slow and festinating walking) have been detected in Dynamic Base of Support (Tables 1, 2, 3, 4, 5, 6 and 7). This is due to the fact that the stereophotogrammetric system do not provide the footprints, so the heel center position (C) is not available and the Dynamic Base of Support needs to be computed as the distance between the heel points and the progression line of the opposite foot (Fig. 4). No significant differences between the two methods were detected in the assessment of FAP Scores during the normal walking test, (value of 100/100 for both GAITRite and Matlab software), and for the slow, fast, small steps and foot-drop walking (maximum differences between the values of FAP obtained by the two systems of a one percentage point). On the other hand, the slight differences detected in Dynamic Base of Support partially affect FAP values assessed in spastic and festinating walking trials. Despite these small percentage differences, however, the ability of the FAP index to classify the walking trial as normal or affected by clinical disorders is not compromised. Nevertheless, a possible improvement of the methodology could be the definition of a new protocol, adding several markers on the foot, in order to define, in more detail, the foot.

In conclusion, results of the present study suggest that the stereophotogrammetric system could represent a valid alternative to the GAITRite system in providing the FAP score in normal and pathological walking. Indeed, since the stereophotogrammetric FAP could be integrated with the classic gait analysis parameters assessed automatically by the stereophotogrammetric system, it is able to provide a complete picture of the subject walking, without the need of a further measurement system. In this way, the "stereophotogrammetric FAP" could be used in pathological walking, for monitoring the effectiveness of the rehabilitation treatments index, also in classic movement analysis laboratories, usually equipped with stereometric instrumentation but not endowed with the GAITRite system [11]. Moreover, the stereophotogrammetry-based computation of the FAP score may be particularly useful for the assessment of the step quality in studies that analyze a high (hundreds) number of steps [11, 20–26].

The present methodology was tested on several types of normal and pathological walking, even if only a single subject were analysed. Thus, further studies with a wider population of healthy and pathological subjects are evoked to confirm these findings.

References

1. Zijlstra W (2004) Assessment of spatio-temporal parameters during unconstrained walking. Eur J Appl Physiol 92:39–44
2. Oberg t, Karsznia A, Oberg K (1993) Basic gait parameters: reference data for normal subjects, 10–79 years of age. J Rehabil Res Dev 30(2):210–223
3. Maranesi E, Di Nardo F, Ghetti G et al (2014) A goniometer-based method for the assessment of gait parameters. In: 10th international conference on mechatronic and embedded systems and applications (MESA), IEEE/ASME. Publisher IEEE. doi:10.1109/MESA.2014.6935539
4. Maranesi E, Barone V, Fioretti S (2014) Assessment of walking speed by a goniometer-based method. In: 36th annual international conference of engineering in medicine and biology society (EMBC). ISBN 978-1-4244-7929-0/14; pp. 1202–1205; doi:10.1109/EMBC.2014.6943812
5. Nelson AJ (1974) Functional ambulation profile. Phys Ther 54:1059–1065
6. Gouelle A (2014) Use of functional ambulation performance score as measurement of gait ability: review. J Rehab Res Dev 51(5):665–674
7. Gretz HR (1998) Functional ambulation performance testing of adults with Down syndrome. NeuroRehabilitation 11:211–225
8. Nelson AJ (2002) The validity of the GaitRite and the functional ambulation performance scoring system in the analysis of Parkinson gait. NeuroRehabilitation 17:255–262

9. Givon U (2009) Gait analysis in multiple sclerosis: characterization of temporal-spatial parameters using GaitRite functional ambulation system. Gait Posture 29:138–142
10. Gouelle A (2011) Validity of functional ambulation performance score for the evaluation of spatiotemporal parameters of children's gait. J Mot Behav 43:93–98
11. Nelson AJ (1999) The functional ambulation performance of elderly fallers and non-fallers walking at their preferred velocity. NeuroRehabilitation 13:141–146
12. Peurala SH, Titianova EB, Mateev P et al (2005) Gait characteristics after gait-oriented rehabilitation in chronic stroke. Restorative Neurol Neurosci 23:57–65
13. Freedland RL, Festa C, Sealy M et al (2002) The effects of pulsed auditory stimulation on various gait measurements in persons with Parkinson's disease. NeuroRehabilitation 17:81–87
14. GaitRite Electronic Walkway (2013) Technical Reference
15. GaitRite Operating Manual (2013) Appendix A- FAP score
16. McDonough AL, Batavia M, Chen FC et al (2001) The validity and reliability of the GaitRite System's measurements: a preliminary evaluation. Arch Phy Med Rehabil 82(3):419–425
17. Bilney B, Morris M, Webster K (2003) Concurrent related validity of the GAITRite walkway system for quantification of the spatial and temporal parameters of gait. Gait Posture 17(1):68–74
18. Webster KE, Wittwer JE, Feller JA (2005) Validity of the GAITRite walkway system for the measurement of averaged and individual step parameters of gait. Gait Posture 22(4):317–321
19. Davis RB (1991) A gait analysis data collection and reduction technique. Hum Mov Sci 10:575–587
20. Di Nardo F, Mengarelli A, Maranesi E et al (2015) Assessment of the ankle muscle co-contraction during normal gait: A surface electromyography study, J Electromyogr Kinesiol 25(2):347-54
21. Di Nardo F, Ghetti G, Fioretti S (2013) Assessment of the activation modalities of gastrocnemius lateralis and tibialis anterior during gait: a statistical analysis. J Electromyogr Kinesiol 23(6):1428–1433
22. Di Nardo F, Fioretti S (2013) Statistical analysis of surface electromyographic signal for the assessment of rectus femoris modalities of activation during gait. J Electromyogr Kinesiol 23 (1):56–61
23. Di Nardo F, Mengarelli A, Maranesi E, Burattini L, Fioretti S (2014) Gender differences in the myoelectric activity of lower limb muscles in young healthy subjects during walking. Biomed Sig Process Control 19:14–22
24. Di Nardo F, Maranesi E, Mengarelli A, Ghetti G, Burattini L, Fioretti S (2015) Assessment of the variability of vastii myoelectric activity in young healthy females during walking: a statistical gait analysis. J Electromyogr Kinesiol 25(5):800–807
25. Di Nardo F, Mengarelli A, Maranesi E, Burattini L, Fioretti S (2015) Assessment of the ankle muscle co-contraction during normal gait: A surface electromyography study. J Electromyogr Kinesiol 25(2):347–354
26. Di Nardo F, Fioretti S (2014) Emg-based analysis of treadmill and ground walking in distal leg muscles. IFMBE Proc 41:611–614. doi:10.1007/978-3-319-00846-2_151

Fall Detection Using Wearable Accelerometers and Smartphone

Luca Basili, Giuseppe DeMaso-Gentile, Cristiano Scavongelli, Simone Orcioni, Stefano Pirani and Massimo Conti

Abstract The governments are investing in research on solutions for independent living, active ageing, at home health monitoring, with the objective of a significant prolongation of personal autonomy of older people. Fall avoidance and fall detection are important aspects of health care of ageing people. The proposed fall detection system consists in a wireless network with a smartphone and a board with another 3-axis accelerometer. The algorithm for fall detection is a part of a user friendly application developed for the smartphone. A comparison with existing fall detection systems and algorithms is reported.

1 Introduction

The risk of degradation of quality of life and the exclusion to the active participation in the society is high for ageing people. The increment of the number of ageing people in industrialized countries makes this risk a social problem. Furthermore the cost for maintaining services for the health of ageing people that the government or the single person must pay is increasing.

Therefore the governments are investing in research on solutions for independent living, active ageing, at home health monitoring, leading to a significant prolongation of personal autonomy.

The falls, in particular, are among the most damaging events for the health of an ageing people [1]. The risk of falling increases with age and in many cases fall happens at home, for a person living alone. The ultimate goal of the fall caring is instantly detect the fall, estimate the degree of gravity and help quickly and efficiently the person.

L. Basili · G. DeMaso-Gentile · C. Scavongelli · S. Orcioni · S. Pirani · M. Conti (✉)
Dipartimento di Ingegneria dell'Informazione, Università Politecnica delle Marche, via Brecce Bianche, 12, 60131 Ancona, Italy
e-mail: m.conti@univpm.it

M. Conti et al. (eds.), *Mobile Networks for Biometric Data Analysis*,
Lecture Notes in Electrical Engineering 392, DOI 10.1007/978-3-319-39700-9_24

Many research works have been developed for the design of devices and/or algorithm for fall detection [1–20].

In [2] the authors investigated the feasibility of a portable preimpact fall detector in detecting falls. It was hypothesized that a single sensor with the appropriate kinematics measurements and detection algorithms, located near the body center of gravity, would be able to distinguish an in-progress and unrecoverable fall from nonfalling activities.

In [3] a complex system consisting of several battery powered wireless sensor boards with a 3-axial accelerometer and a biaxial gyroscope. The data are elaborated offline with a complex signal processing developed in MATLAB to identify the movements of the monitored person.

In [4] an 802.15.4 wireless network with six nodes posed in different places of the body of an elderly has been developed. The data are collected by a node connected via USB to a PC that elaborates the signals of the six triaxial accelerometers to identify a possible fall. In [5] an application for fall detection on a smartphone using only the triaxial accelerometer of the smartphone is presented. In [6] system for fall detection consists in a single board with accelerometer wireless connected to a PC with the algorithm for fall detection.

2 Description of the System

2.1 Hardware Components

The proposed system is composed of battery supplied wearable board, a smartphone, and a smartphone application. The wearable board consists of a 3-axis accelerometer, a microcontroller and a Bluetooth transceiver. The board, shown in Fig. 1 and described in detail in [7], is based on the Freescale Freedom KL25Z demoboard with a low-power 3-axis accelerometer (MMA8451Q). A Bluetooth shield has been connected to the board via the UART port. At start up the

Fig. 1 Wearable board with freescale freedom KL25Z demoboard and the Bluetooth shield

microcontroller creates the connection with the Bluetooth module and the 3-axis accelerometer. Then the Bluetooth module establishes a Bluetooth connection with the smartphone (the board is slave and the smartphone is master). Finally the microcontroller continuously collects data from the accelerometer and sends data through the Bluetooth module to the smartphone.

The power consumption of the module has been measured, with the current probe of a high performance oscilloscope. On the basis of the analysis reported in [7], the average current during continuous transmission is about 36.9 mA. As a consequence the life of the device that is continuously monitors the person can be estimated as 3 days using a 2500 mAh battery.

2.2 *Fall Detection Algorithm*

In order to test the device and to acquire preliminary data useful for the definition of the fall detection algorithm some measurements have been carried out.

The smartphone and the device have been placed in different positions of the body to identify the best solution for the fall detection. The smartphone has been placed in the waist and in the shirt pocket and the device has been placed in the foot, ankle and arm as shown in Fig. 2.

The accelerometers measure the 3 components of the acceleration A_x, A_y, A_z. If the person does not move the acceleration magnitude is the gravity acceleration. Therefore, measuring the 3 components of the acceleration of the smartphone and of the device, and knowing the position of smartphone and device in the body, it is possible to estimate the orientation of the smartphone and of the device with respect to the vertical position. This information can be used to estimate the position of the person (sitting, standing or lying). The idea is summarized in Fig. 3, in the case in which the smartphone in a pocket in the waist and the device is in the ankle.

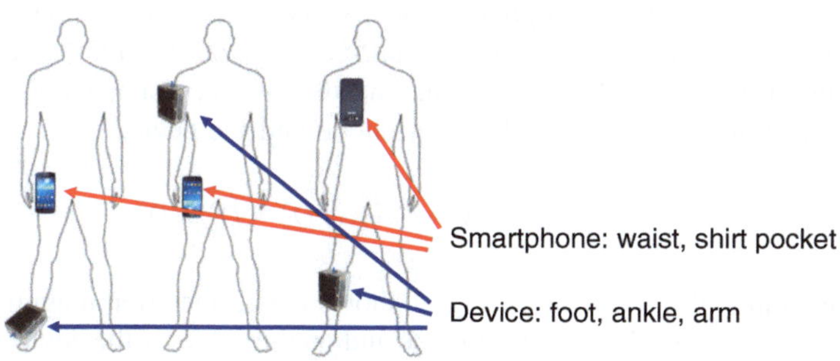

Fig. 2 Different test with smartphone and device placed in different positions

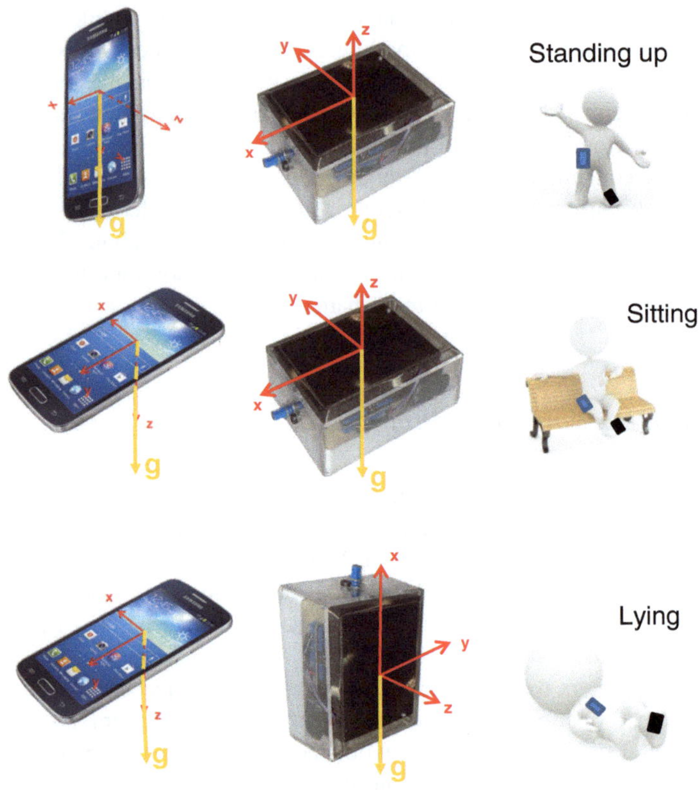

Fig. 3 Estimation of the position of the person using the 3-dimensional accelerations of the smartphone and of the device

Preliminary test measurements evidenced that the acceleration measured on the ankle and foot is high when the person is moving (walking, going upstairs or down stairs) and comparable to the acceleration when he is falling. On the other hand the acceleration of the waist or shirt pocket is low while the person is moving and high while sitting, rising or falling. Let us define S_x, S_y, S_z, D_x, D_y, D_z the acceleration components measured by the smartphone and the device, respectively, and S and D the acceleration magnitude by the smartphone and the device

$$S = \sqrt{S_x^2 + S_y^2 + S_x^2} \quad D = \sqrt{D_x^2 + D_y^2 + D_x^2} \tag{1}$$

As an example of the test cases, Fig. 4 shows the acceleration magnitude S and D measured on the waist by the smartphone and on the ankle by the device while the user is going upstairs for the first 28 seconds and falling at about the 30-th second. In the ankle the positive and negative variation of the acceleration with respect to the gravity acceleration is high going and falling. Conversely, the variation is high only during a fall for the acceleration measured by the smartphone in the waist.

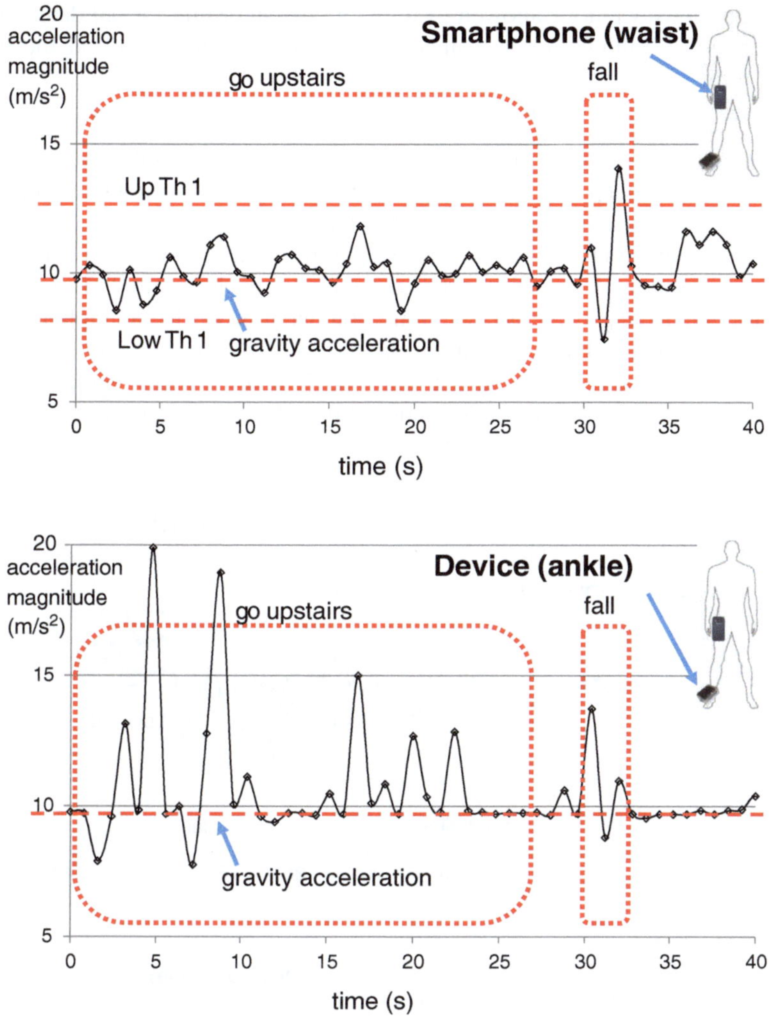

Fig. 4 Acceleration magnitude of waist and ankle going upstairs and falling

Figure 5 reports for the same test the acceleration components defined as in Fig. 4 measured on the waist by the smartphone and on the ankle by the device. Before the fall, the position of the smartphone in the waist is vertical (see Fig. 4), that is the S_y component is approximately the gravity acceleration. Few seconds after a fall the position of the person is lying and the S_y component is lower.

On the other hand, before the fall the position of the device is horizontal, that is the D_z component is approximately negative and equal to the gravity acceleration. After a fall the position is lying and the D_z component is approximately zero.

Previous considerations allowed us to define the algorithm for fall detection has been defined as follows, for the smartphone placed in the waist and the device placed in the ankle.

The accelerations are monitored and displayed on the smartphone.

(1) When the acceleration magnitude S of the smartphone (on the waist) is higher than a threshold value (Low_Th_1 in Fig. 4) or lower than another threshold value (Up_Th_1 in Fig. 4), a possible fall is considered.
(2) After a defined number (n) of seconds the position of the person is estimated.
(3) If the position is "lying" the fall detection event is asserted. The person is considered lying if the acceleration component S_y of the smartphone (in the waist) is higher than a threshold value (Low_Th_2 in Fig. 5) or lower than another threshold value (Up_Th_2 in Fig. 5) and the acceleration component D_z of the device (in the ankle) than a threshold value (Low_Th_3 in Fig. 5).

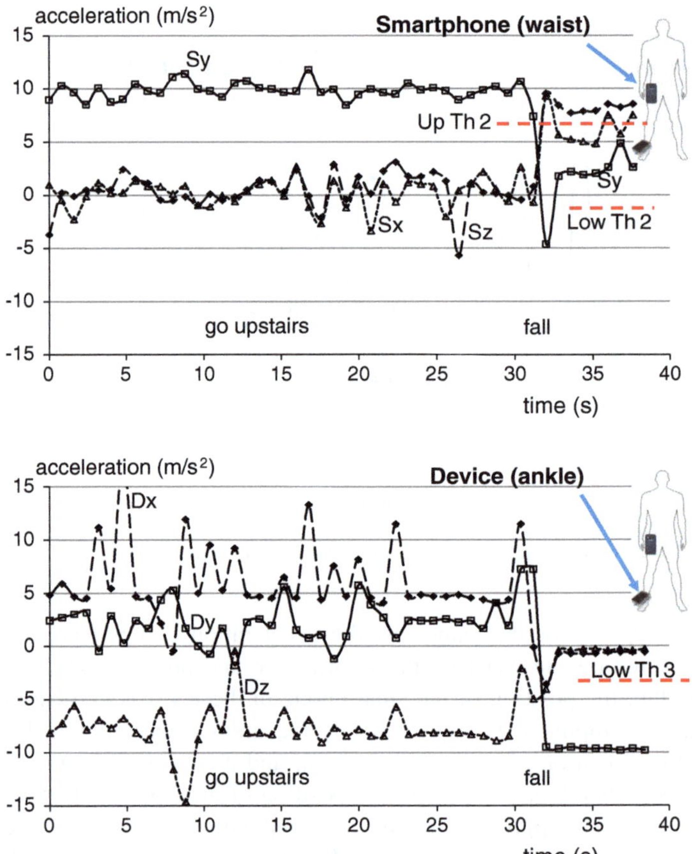

Fig. 5 Acceleration components of waist and ankle going upstairs and falling

3 Smartphone Application

An application for Android smartphone has been developed for the fall detection applying the algorithm described above. The application can be started by the user or approaching the smartphone to the device using the NFC feature. The user selects the emergency phone numbers that will be called in event of fall and he must activate the GPS in order to send the actual position of the user to the emergency assistance.

Pressing the start button the Bluetooth connection with the device is activated and the fall monitoring starts. When monitoring is active, the menu (4) in Fig. 6 is displayed, with the actual data of the accelerometers, the GPS coordinates, date and time. If requested, the data are continuously stored in the smartphone memory. The user can pause monitoring or send an alarm request manually.

When the fall event is detected, the menu (5) in Fig. 6 is displayed, an alarm is activated and sms messages or phone calls are scheduled to be sent after few seconds. The user can deactivate the alarm and avoid the automatic calls pressing a button in (5) of Fig. 6.

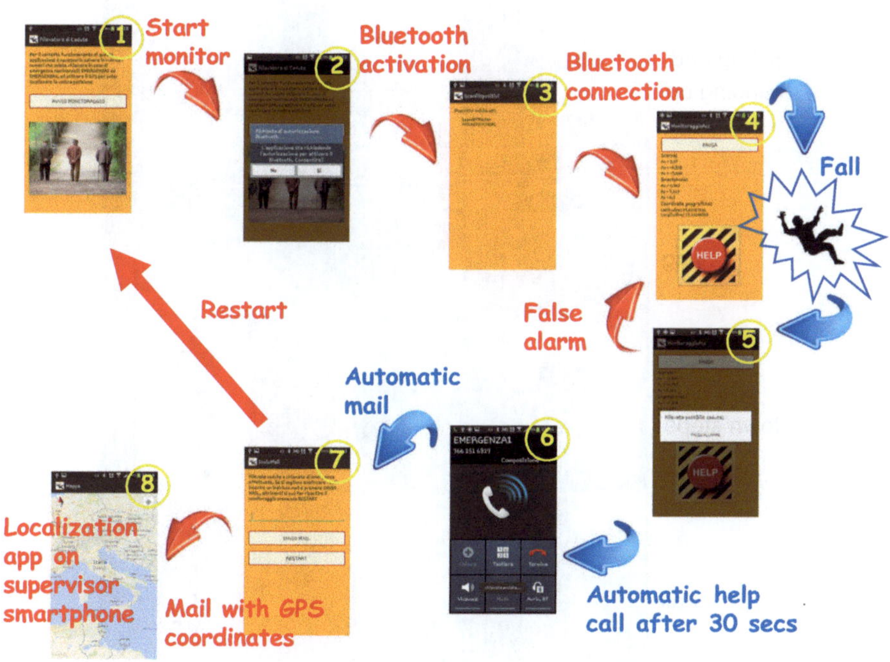

Fig. 6 Flow diagram and screenshots of the application developed

4 Test of the System

The complete system (hardware, algorithm and application) has been primarily tested on a reduced number of persons. Different test situations have been carried out: placing the board in different positions (foot, arm and ankle) and the smartphone placed in the pocket (waist), while the person was sitting (20 cases), lying (20 cases), standing up (20 cases), going upstairs and downstairs (20 cases) and falling (41 cases), for a total of 111 tests. The test conditions are summarized in Fig. 7. We used the data stored in the smartphone and applied the proposed algorithm using different parameter settings and we applied the algorithms reported in [4–6] for comparison.

To verify and compare the goodness of the algorithms, we used the standard formulae commonly accepted in the literature for Sensitivity, Specificity, Accuracy and False Negative Rate (FNR). The parameter FNR indicates the percentage of not detected falls with respect to the total number of falls. In a fall detection system it can be more important to detect all the falls accepting possible false alarms, that the user can correct using the smartphone application developed.

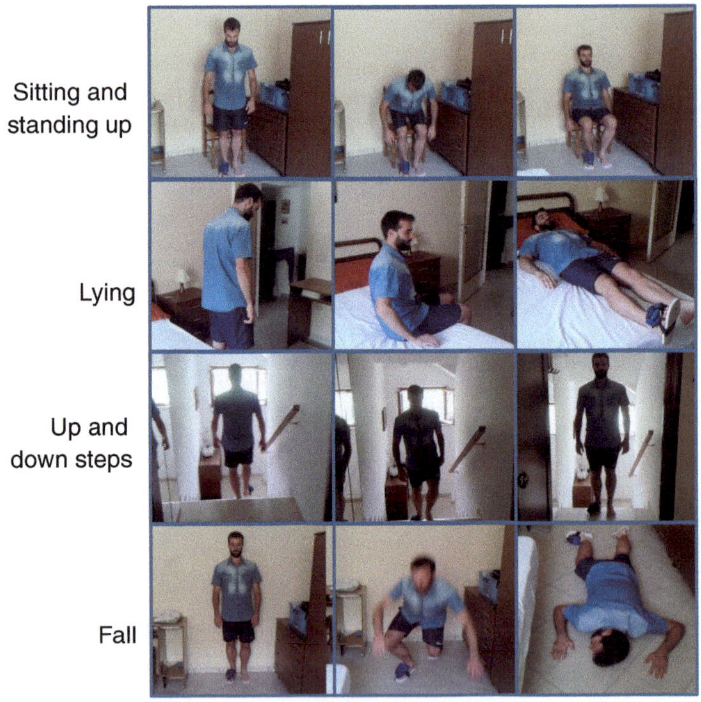

Fig. 7 Test of the complete system and fall detection algorithm

	Fall	Not fall
Fall detected	TP (true positive)	FP (false positive)
Fall not detected	FN (false negative)	TN (true negative)

$$\text{Sensitivity} = \frac{\text{True Positive}}{\text{True Positive} + \text{False Negative}} = \frac{\text{fall correctly identified}}{\text{fall}}$$

$$\text{Specificity} = \frac{\text{True Negative}}{\text{True Negative} + \text{False Positive}} = \frac{\text{not fall correctly identified}}{\text{not fall}}$$

$$\text{Accuracy} = \frac{\text{True Positive} + \text{True Negative}}{\text{all cases}} = \frac{\text{correct responses}}{\text{all cases}}$$

$$\text{FNR} = \frac{\text{False Negative}}{\text{True Positive} + \text{False Negative}} * 100 = \frac{\text{fall not detected}}{\text{fall}} * 100$$

The values of the five parameters (Up_Th_1, Low_Th_1, Up_Th_2, Low_Th_2, Low_Th_3) of the algorithms affect the performance of the system. The performance of three setting have been reported in this works. The values of the settings are reported in Table 1. Setting A has been used as a reference example, setting B has been chosen to optimize the specificity, white setting C allows the best value of the sum of sensitivity, specificity and accuracy, this setting allows the best value of FNR too.

Table 2 reports the results for different algorithms, different settings and different test data. The test data column indicates the results obtained applying the algorithm to the complete data set of 111 cases (total), the 61 cases in which the device is placed in the foot (foot), 25 cases in the arm (arm) and 25 cases in the ankle (ankle).

The row "reference results" indicates the results reported in the reference. The algorithm reported in [4] has been developed for a smartphone and up to six additional devices, for our test case it has been applied to the data of the accelerometer and of the device. The algorithms reported in [5, 6] use only the acceleration of the smartphone, therefore the data of device are not used. The parameters of the algorithm of the papers [4–6] have been chosen to optimize the sum of sensitivity, specificity and accuracy.

In general the results obtained, with a parameter optimization, are better with our algorithm with respect to the algorithms [4–6].

Better results are obtained if the device is placed in the foot or in the ankle with respect to the placement in the arm. Changing the parameters settings of our algorithms we can improve sensitivity at the cost of reducing specificity or vice versa. Figure 8 reports the performances as a function of the values of the parameters, in particular the parameter Low_Th_1.

Table 1 Values of the 5 parameters of the proposed algorithm for three settings

Setting	Up_Th_1 (m/s^2)	Low_Th_1 (m/s^2)	Up_Th_2 (m/s^2)	Low_Th_2 (m/s^2)	Low_Th_3 (m/s^2)
A	10.5	9.0	6.0	−6.0	−8.0
B	11.3	7.0	6.0	−3.0	−6.4
C	10.7	9.3	6.1	−4.4	−6.4

Table 2 Test results of the detection accuracy for different values of the parameters of the algorithm and for different algorithms in the literature

Algorithm	Setting	Test data	TP	FP	TN	FN	Sensitivity (%)	Specificity (%)	Accuracy (%)	FNR (%)
Proposed algorithm	A	Total	30	37	43	1	96.8	53.8	65.8	3.2
	B	Total	23	0	80	8	74.2	100	92.8	25.8
	C	Total	31	15	65	0	100	81.3	86.5	0
	C	Foot	21	4	36	0	100	90	93.4	0
	C	Arm	5	7	13	0	100	65	72	0
	C	Ankle	5	4	16	0	100	80	84	0
[4]	Reference results		–	–	–	–	–	–	99.55	–
	D	Total	19	3	77	12	61.3	96.3	86.5	38.7
	D	Foot	15	0	40	6	71.4	100	90.2	28.6
	D	Arm	1	1	19	4	20	95	80	80
	D	Ankle	3	2	18	2	60	90	84	40
[5]	E	Total	19	36	44	12	61.3	55	56.8	38.7
[6]	Reference results		–	–	–	–	97.7	94.8	–	–
	F	Total	20	28	52	11	64.5	65	64.9	35.5

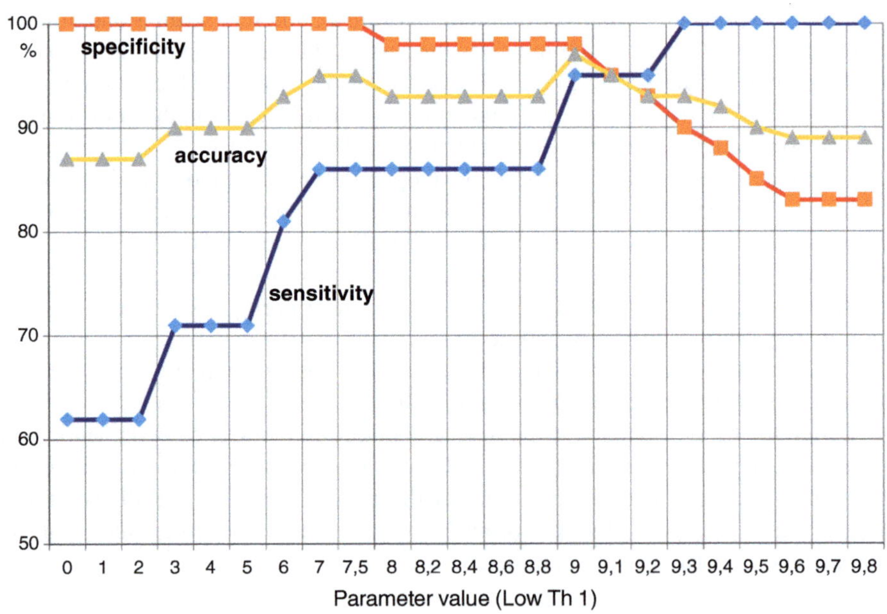

Fig. 8 Performances as a function of the values of the parameter Low_Th_1

Figures 9, 10 and 11 report the number of detected and not detected falls for the different test cases: sitting (20 cases), lying (20 cases), standing up (20 cases), going upstairs and downstairs (20 cases) and falling (41 cases). Setting A, B and C has been used in Figs. 9, 10 and 11 respectively. Figure 8 shows that setting B allows no erroneous falls have been detected but 8 real falls have not detected. Conversely, setting C detects all the real falls at the cost of some False positive, in particular laying is sometimes confused with a fall.

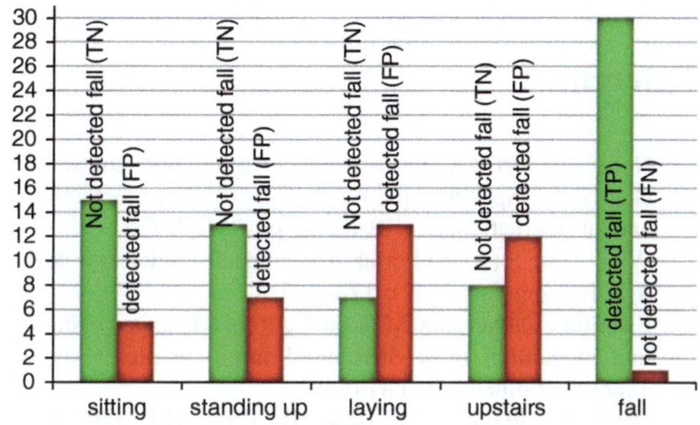

Fig. 9 Number of detected and not detected falls for the different test cases for setting A

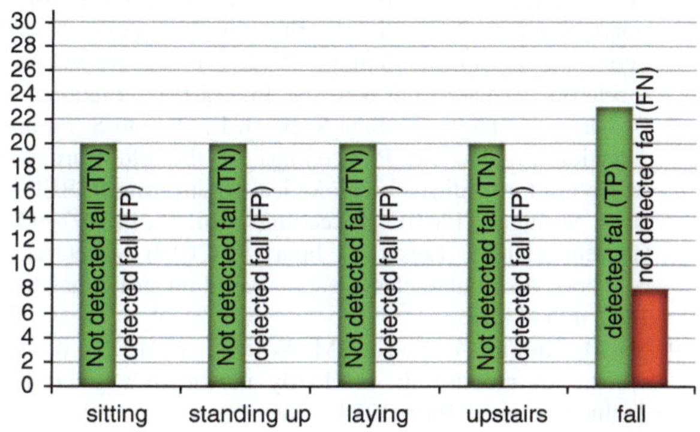

Fig. 10 Number of detected and not detected falls for the different test cases for setting B

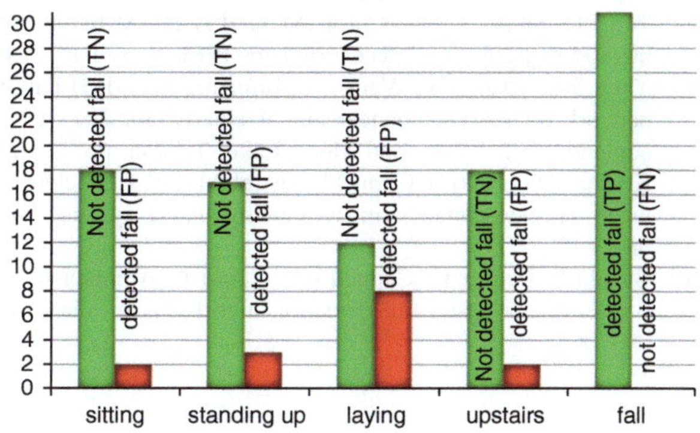

Fig. 11 Number of detected and not detected falls for the different test cases for setting C

5 Conclusions

The algorithm for fall detection presented in this work uses the measures of the 3-axis accelerometers of a smartphone and of a Bluetooth device designed specifically for this purpose. A user friendly application has been developed for the smartphone. The system has been tested in many different situations and the results compared with other algorithm reported in literature.

References

1. Noury N (2002) A smart sensor for the remote follow up of activity and fall detection of the elderly. In: Special topic conference on microtechnologies in medicine & biology 2nd annual international IEEE-EMB, pp 314–317
2. Wu G, Xue S (2008) Portable preimpact fall detector with inertial sensors. IEEE Trans. Neural Syst Rehabil Eng 16(2): 178–183
3. Estudillo-Valderrama M, Roa L, Reina-Tosina J, Naranjo-Hernandez D (2009) Design and implementation of a distributed fall detection system—personal server. IEEE Trans Inf Technol Biomed 13(6):874–881
4. Lai C-F, Huang Y-M, Park JH, Chao H-C (2010) Adaptive body posture analysis for elderly-falling detection with multisensors. IEEE Intell Syst 2010:20–30
5. Thammasat E, Chaicharn J (2012) A simply fall-detection algorithm using accelerometers on a smartphone. In: Biomedical engineering international conference, 5–7 Dec 2012
6. Leone A, Rescio G, Siciliano P (2013) Supervised machine learning scheme for tri-axial accelerometer-based fall detector. In: IEEE international conference sensors
7. De Maso-Gentile G, Malizia S, Basili L, Orcioni S, Pirani S, Conti M (2014) Low power fall detection system. In: Proceedings of the 6th European conference of the international federation for medical and biological engineering, MBEC2014, Dubrovnik, Croatia, 7–11 Sept 2014. Springer, IFMBE proceedings, vol. 45, pp 897–901
8. Ghasemzadeh H, Loseu V, Jafari R (2010) Structural action recognition in body sensor networks: distributed classification based on string matching. IEEE Trans Inf Technol Biomed 14(2):425–435
9. Yu M, Rhuma A, Naqvi SM, Wang L (2012) A posture recognition-based fall detection system for monitoring an elderly person in a smart home environment. IEEE Trans Inf Technol Biomed 16(6):1274–1286
10. Mariani B, Rochat S, Bula CJ, Aminian K (2012) Heel and toe clearance estimation for gait analysis using wireless inertial sensors. IEEE Trans Biomed Eng 59(11):3162–3168
11. Tong L, Song Q, Ge Y, Liu M (2013) HMM-based human fall detection and prediction method using tri-axial accelerometer. IEEE Sens J 13(5):1849–1856
12. Yang S-H, Zhang W, Wang Y, Tomizuka M (2013) Fall-prediction algorithm using a neural network for safety enhancement of elderly. In: 2013 CACS international automatic control conference, 2–4 Dec, Sun Moon Lake, Taiwan, pp 245–249
13. Kozina S, Gjoreski H, Gams M, Lustrek M (2013) Efficient activity recognition and fall detection using accelerometers. In: Evaluating AAL systems through competitive benchmarking, pp 13–23
14. Bagalà F, Becker C, Cappello A, Chiari L, Aminian K, Hausdorff JM (2012) Evaluation of accelerometer-based fall detection algorithms on real-world falls. doi:10.1371/journal.pone.0037062

15. Rakhman AZ, Nugroho LE, Kurnianignih K (2014) Fall detection system using combination accelerometer and gyroscope. In: Conference: 2014 1st international conference on information technology, Computer and Electrical Engineering (ICITACEE), 2014
16. Wibisono W, Arifin DN, Pratomo BA, Ahmad T, Ijtihadie RM (2013) Falls detection and notification system using tri-axial accelerometer and gyroscope sensors of a smartphone. In: Conference on technologies and applications of artificial intelligence (TAAI), 2013, pp 382–385
17. Bourke AK, Van de Ven P, Gamble M, O'Connor R, Murphy K et al (2010) Evaluation of waist-mounted tri-axial accelerometer based fall-detection algorithms during scripted and continuous unscripted activities. J Biomech 43:3051–3057
18. Luo J, Zhong B, Lv D (2015) Fall monitoring device for old people based on triaxial accelerometer. (IJACSA) Int J Adv Comput Sci Appl 6(5)
19. Lan C-C, Hsueh Y-H, Hu R-Y (2012) Real-time fall detecting system using a tri-axial accelerometer for home care. In: International conference on biomedical engineering and biotechnology (iCBEB), 2012, Macao, pp 1077–1080
20. Purwar A, Jeong DU, Chung WY (2007) Activity monitoring from real-time triaxial accelerometer data using sensor network. In: International conference on control, automation and systems 2007, pp 2402–2406

[illegible] ... Fall detection system ... [illegible]
[illegible] ... (ICHI [illegible]), 2015
[illegible] ... IEEE ... [illegible]
[illegible]
[illegible]
[illegible]
[illegible]

Step Length Estimation for Freezing of Gait Monitoring in Parkinsonian Patients

Lucia Pepa, Marco Rossini, Luca Spalazzi and Federica Verdini

Abstract Step length (SL) is a fundamental parameter for the characterization of normal and pathologic gait. In particular, Freezing of Gait (FOG) in Parkinson's Disease is a motor disorder associated with a markedly reduction of SL, hence this parameter has the potential to be used for FOG detection and monitoring. In this paper, we present a non-obtrusive and non-invasive architecture for the estimation of SL from trunk acceleration. We compared the reliability of our methodology with a stereophotogrammetric system, which is the gold standard for gait analysis. Preliminary experimental results on three healthy subjects are reported. Results show that the architecture could be employed to detect SL reduction prior to a FOG event.

1 Introduction

Parkinson's Disease (PD) is a neurodegenerative pathology affecting the basal ganglia [3], which have a primary role in motor learning and in the execution of automatic movement. As a consequence this kind of patients is affected by motor disorders like Freezing of Gait (FOG), typical of the late stages of pathology.

L. Pepa (✉) · M. Rossini · L. Spalazzi · F. Verdini
Dipartimento di Ingegneria dell'Informazione, UNIVPM,
via Brecce Bianche, 60131 Ancona, Italy
e-mail: l.pepa@univpm.it

M. Rossini
e-mail: mrc.rossini@yahoo.it

L. Spalazzi
e-mail: l.spalazzi@univpm.it

F. Verdini
e-mail: f.verdini@univpm.it

M. Conti et al. (eds.), *Mobile Networks for Biometric Data Analysis*,
Lecture Notes in Electrical Engineering 392, DOI 10.1007/978-3-319-39700-9_25

The freezing of gait (FOG) is a common and highly distressing motor symptom of patients with Parkinsons Disease (PD) ranging between 21 and 27 % of patients in the early stages and 60–80 % in the later stages [23]. FOG is defined as a 'brief, episodic absence or marked reduction of forward progression of the feet despite having the intention to walk' [5]. FOG induces an increased risk for falls, disease progression and mortality. It is still considered mysterious by clinicians because of its episodic, heterogeneous and erratic nature [21]. Clinicians try to elicit FOG in laboratory settings in order to have better insight on the mechanisms underlying FOG and on its characteristic features. In particular two FOG characterizations were found [4]: one approach is based on frequency analysis of leg movements (a leg trembling in the band 3–8 Hz was found during FOG) [7], while the other approach concentrates on spatiotemporal kinematic parameters of gait. For what concerns the last characterization, several studies found that FOG is preceded by a strong increase in step cadence (SC), decrease in step length (SL), linear velocity and angular excursion of leg joints [15, 16]. This information can be used to develop a system for FOG detection, with the possibility of monitoring FOG episodes during daily life, improving medical knowledge and clinical management of this disabling and mysterious symptom.

Another critical problem concerning FOG is its resistance to standard pharmacological treatments [17]. To overcome this drug resistance, clinicians found alternative approaches [8], such as the use of external rhythmic stimuli, said cues, that help the patient to overcome FOG. In the literature there is good agreement that auditory cues (e.g. walking at the rhythm of music or metronome) are the most efficient in enhancing gait [24]. Therefore a system able to detect FOG can be also used to deliver suitable cues at FOG occurrences, in order to help patients to resume walk. A number of wireless body sensor networks have been proposed in the literature to detect FOG [2, 10, 13, 14]. Such systems are composed of wearable inertial sensors attached to patient's legs. The acquired signals are sent to a computing unit for processing purpose. The major part of the algorithms used in the literature uses frequency analysis of acceleration signal for FOG detection, but some limitations were found in system performance. In particular this kind of approach showed a high variability of performance among patients, which means that it works well only for a group of patients. For this reason we want to investigate FOG detection by means of spatiotemporal parameters of gait, since in the literature FOG was related to the disruption of spatiotemporal characteristics of gait such as SC and SL, as discussed above [15, 16]. We have already used SC for FOG detection in previous works [18], but the addiction of SL information will improve detection accuracy. Prior to apply SL computation to parkinsonian patients we must address the question of SL estimation in healthy subjects by means of inertial sensors, such as accelerometers.

In this work we present a system that estimates SL in a non-obtrusive and non-invasive way, based on acceleration signal from trunk, and we report preliminary results on 3 healthy subjects. The architecture uses a smartphone for both sensing and processing functions. If such estimation will reveal sufficiently accurate to detect SL decrease before FOG, the proposed system could be used for FOG detection. The paper is structured as follows. In Sect. 2 we will summarize previous works in the literature with focus on gait analysis through accelerometers. In Sect. 3 we will explain the hardware used and the algorithm adopted to extract SL. Section 4 describes the experimental set up to validate the proposed methodology and reports the results. Finally in Sect. 5 we will draw conclusions.

2 Related Works

The gold standard for gait analysis, and hence SL computation, is the use of stereophotogrammetric system, which is inadequate to apply in a daily living condition. In recent years, researchers investigate alternative methods such as the use of accelerometers [1, 6, 29], given the advances in miniaturization and costs of inertial sensors. Several studies focused the attention on acceleration signal from trunk, because it is considered suitable position for an every-day usage. A central issue for SL estimate from trunk acceleration is the identification of heel strike, the time instant of initial contact. Zijstra and Hof [29] found a relationship between the basic pattern of trunk accelerations and gait cycle phases on the basis of an inverted pendulum model. Accordingly to their findings, heel strike coincides with the peak in antero-posterior acceleration before the change of sign (from positive to negative). For what concerns SL, it was estimated using vertical movements, because on the basis of inverted pendulum model changes in the height of centre of mass depend on SL [28]. Thus SL can be estimated according to the following formula [28]:

$$SL = 2\sqrt{2lh - h^2} \quad (1)$$

where l is the length of subject's leg, while h is the difference between maximum and minimum height reached by the centre of mass during gait cycle. In [11], Menz et al. described acceleration pattern of pelvis during normal walking focusing the attention on the location of initial contact. They also evaluate the effect of irregular surface on gait parameters. However, they evaluate SL in an approximative way as the ratio between the covered distance and the number of steps. Auvinet et al. [1] conducted a study about the influence of gender on some gait parameters such as stride frequency, gait symmetry and regularity. They used a system composed of one accelerometer and an electronic watch to measure walking speed. The accelerometer is placed over L3-L4 inter-vertebral space and blocked through a semi-elastic belt. Stride frequency was computed as the first harmonic of the Fast Fourier Transform. Additionally they also computed a mean value for SL over the entire trial as the ratio between mean speed and stride frequency. Gonzalez et al. [6] proposed an approach for real-time identification of heel strike and toe off time instants. For the detection of heel strike they used the findings of Zijstra and Hof [29], while for the toe off they developed an algorithm on the basis of conclusions reported by Auvinet et al. [1] and Menz et al. [11]. Moore et al. [12] used a geometric formula to estimate SL from angular velocity of the shank. Their aim was the monitoring of PD patients. They use an inertial measurement unit (with both accelerometer and gyroscope) attached to the shank and a wearable computer attached to the waist. Their architecture needs a complex and long calibration process and the methodology gives best results for short steps.

In these works the primary aim was a reliable identification of significative point of gait cycle such as heel strike or toe off. As a consequence, the gold standard used for comparison by authors used force platforms to evaluate the confidence of the detection of such points. There is a lack in the literature about comparison of SL estimation with the gold standard, e.g. gait analysis by infrared cameras and markers.

Moreover even if the systems proposed are wearable and low cost, they used inertial sensor and processing units to extract information. In an ambient assisted living contest, the users of these systems are elderly people, that generally do not have a background of technical or engineering education. It is highly probable that this kind of users will not feel confident with such technology, which is probably unknown for them. We tried to overcome this problem having particular attention to user's needs during the design of our architecture.

3 Materials and Methods

The objective of the paper was the development of an architecture for SL estimation. In this section we will explain hardware and software parts of the architecture.

3.1 Hardware

In order to find a solution apt to be used in a daily living context, we conducted the design in a user centered way, taking into account acceptability and usability requirements.

According to studies about acceptability in healthcare [25, 26] subjective technical confidence, technical expertise and technical attitude are user's characteristics required for a wide acceptance of assistive devices. Furthermore the same group of study pointed out that medical technology refers to 'taboo related' areas [26], because are associated with disease and illness. Since the patient do not want to be labeled as ill, an integrated device (which means a medical device integrated in every-day tools) is expected to be used for medical monitoring.

Usability is the other important requirement for monitoring or assistive systems, because the patient should be able to use the system alone during his daily activities. According to [22] usability is 'the extent to which a product can be used by specified users to achieve specified goals with effectiveness, efficiency and satisfaction in a specified context of use'. Hence the system should permit an easy, intuitive and efficient interaction.

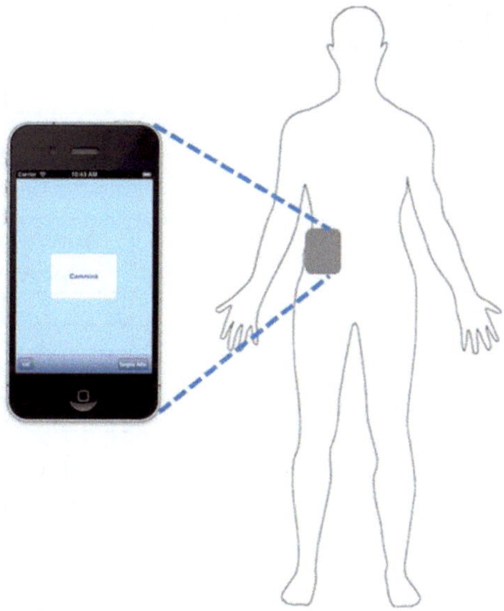

Fig. 1 Smartphone for step length estimation

We used a smartphone as sensing and processing unit, because of its popularity and hence social acceptability, ease of use and growing computational capabilities. Furthermore smartphones offer the possibility to customize user interface, in order to increment perceived ease of use. In Fig. 1 there is a schematic representation of the architecture proposed.

3.2 *Software*

The algorithm used in this study for SL estimate was implemented on the smartphone for the sensing part and on Matlab for the processing part, since in this research phase we are investigating the possibility of a reliable SL estimate through the smartphone accelerometer. In Fig. 2 we reported the algorithm flowchart.

The first step is the acquisition of acceleration signal at a frequency of 100 Hz, hence we built an application to carry out this function through the onboard accelerometer. The application saved all acquired data on a text file.

Acceleration data were low pass filtered by a fourth order Butterworth filter with a cut off frequency of 20 Hz. The third step was foot contact detection on the antero-posterior signal, according to [6, 29], the peak before the zero-crossing was selected as the heel strike. Figure 3 shows the shape of antero-posterior acceleration and peaks selected as heel strikes. Comparing this signal with previous works, we found that trunk acceleration obtained from the smartphone was in agreement with the typical shape found in the literature [29].

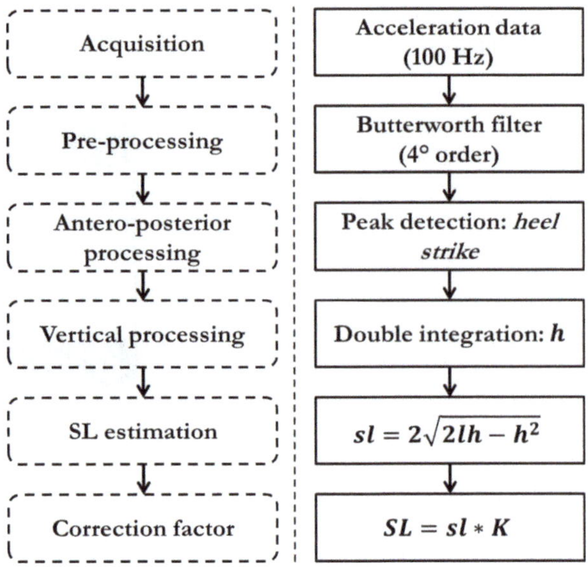

Fig. 2 Block scheme of the SL estimation algorithm

To estimate SL we must consider also the vertical component of the acceleration. In fact, initial conditions for linear velocity must be known to obtain SL from double integration of acceleration signal of antero-posterior direction, but unluckily we can not assume any valid hypothesis on the value of antero-posterior velocity at heel strike. Instead for the vertical velocity, we can assume that its value at the heel strike is close to zero, as confirmed by previous studies [9, 27]. The fourth step performs the double integration of vertical acceleration to find the height change of body centre of mass during a step (parameter h in the flowchart).

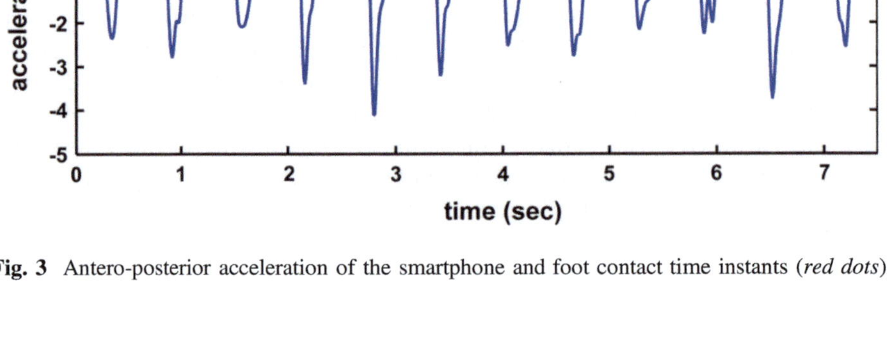

Fig. 3 Antero-posterior acceleration of the smartphone and foot contact time instants (*red dots*)

Finally, SL was evaluated according to the formula (1) of the inverted pendulum model. Since this formula underestimate SL, a corrective factor *K* is applied. The value of *K* was set to 1.25 in accordance with the literature [27].

Hence our algorithm uses both the antero-posterior and vertical components of acceleration for SL estimation. The antero-posterior one is used to find the heel strike time instants, e.g. the extremes of integration. The vertical one is used to calculate the height change of centre of mass during the gait cycle. Unlike the algorithm proposed by Lan and Shih [9], we decided to use the antero-posterior component to detect heel strike because in the literature there is good agreement on the relation between heel strike and the peak before zero-crossing in the antero-posterior acceleration. While for the vertical component there are different opinion about the heel strike location [1, 9].

4 Tests and Results

The reliability of our method for SL computation was evaluated by comparison with a stereophotogrammetric system, which is the gold standard for gait analysis. In this paper, we report preliminary results of the tests on 8 healthy subjects (7 males), aged between 22 and 30 years. Each subject performed 6 walk trials, 3 at their normal speed, and 3 at higher speed. Nine reflective markers were placed on pelvis and feet. The pelvis was identified by the three markers: two over the left and right anterior superior iliac spines and the third on the middle point between the posterior superior iliac spines. Each foot was identified by three markers placed on: heel, first and fifth metatarsal head. Markers' position is shown in Fig. 4.

During the trials, simultaneous acceleration signal was acquired from the smartphone placed at subject's lower back. The smartphone was fixed through an elastic belt and an appropriate socket at the level of L3-L4. The belt and the socket guaranteed the alignment of the accelerometer axis with the antero-posterior, medio-lateral and vertical axes of the subject. The algorithm described in the previous was applied offline on the acquired data to compute SL.

We examined a total of 455 steps, each value for SL estimated from the smartphone was compared to the corresponding SL computed from the gait analysis system. Results are summerized in Table 1. We obtain an error of 10.60 $\pm$ 7.11 % (mean $\pm$ standard deviation) for normal velocity and of $11.01\% \pm 6.96\%$ (mean $\pm$ standard deviation) for high velocity. The error manifested polarization (Fig. 5), that may indicate the need of a calibration for the correction factor K.

We also evaluated the difference among adjacent steps for the smartphone and the stereophotogrammetric system. We computed the difference between each step and the previous one and we found that 69.17 % of differences computed from the smartphone were in accordance with the gold standard. In Fig. 6 there is a graphical explanation of the way we computed this score.

Fig. 4 Markers' placement during tests

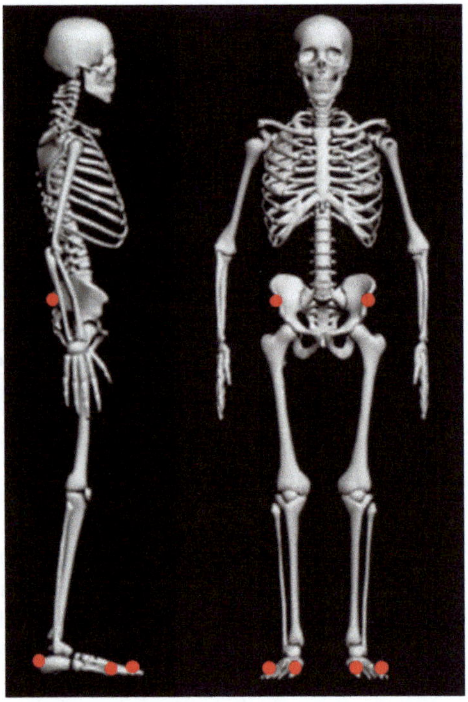

Table 1 Estimation error with respect to SL measurements from stereophotogrammetric system

Subject	Normal velocity	High velocity
	(mean ± std) (%)	(mean ± std) (%)
1	10.27 ± 5.97	16.03 ± 6.10
2	11.98 ± 9.39	12.80 ± 8.54
3	2.44 ± 1.79	3.46 ± 9.32
4	8.32 ± 6.93	7.16 ± 5.31
5	12.66 ± 8.16	15.50 ± 10.90
6	9.89 ± 4.70	11.08 ± 6.84
7	14.04 ± 8.42	11.59 ± 6.16
8	15.46 ± 8.75	9.98 ± 8.32
Mean	10.63 ± 7.11	11.01 ± 6.96

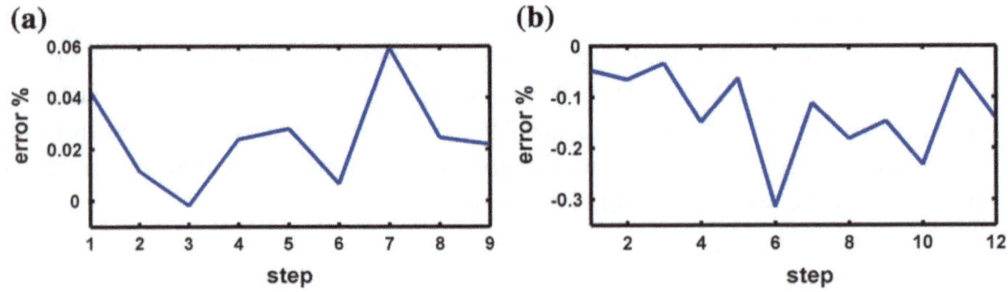

Fig. 5 Error trend for subject 1 (**a**) and subject 3 (**b**)

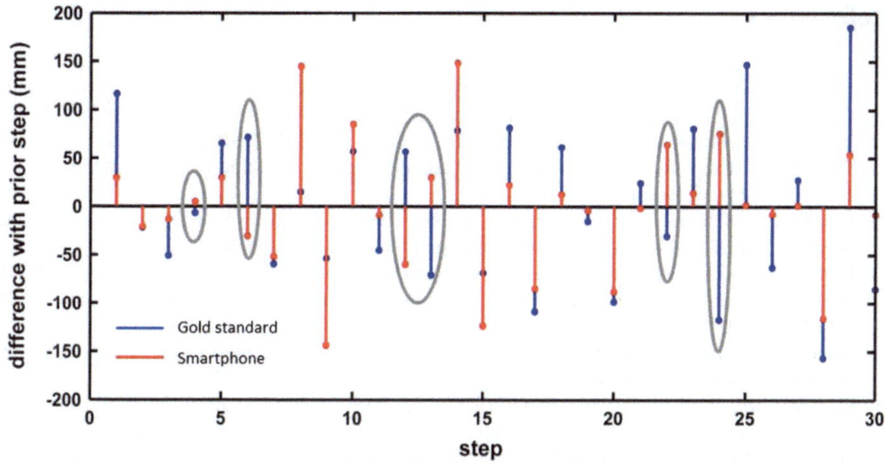

Fig. 6 Differences among adjacent steps: *blue stems* for the gold standard and *red stems* for the smartphone. *Grey circles* indicate a discordance of the smartphone difference with respect to the corresponding difference computed from stereophogrammetric system. The figure shows the differences for subject 1 steps

5 Conclusion and Future Works

In this paper we proposed an architecture for the estimation of SL, built on the basis of acceptability and usability requirements. For this reason we chose a popular daily living tool, the smartphone. We tested the system on 3 healthy subjects, but we aim to apply it for FOG monitoring in PD patients. We obtained a mean error of 10.60 % for normal velocity and 11.01 % for high velocity. The error polarization suggests the need of a customized correction factor.

The question that motivated the paper was whether the proposed methodology was able to reach a sufficient accuracy to detect FOG. Since previuos work in the literature concerning FOG characterization [15, 16] reported a mean reduction of 71.89 % in SL before a FOG episode, we can conclude that our accuracy is sufficient for FOG detection or even prediction. Furthermore in 69.17 % of cases the smartphone was able to detect differences among steps. It must be considered that subjects were instructed to walk regularly, and so little differences among steps were observed during the experiments.

The next step will be the implementation of the algorithm on the smartphone, for a real-time estimation of SL in order to improve the reliability of the application we previously used for FOG detection [19, 20]. This architecture can be easily applied in a daily living context to help parkinsonian patients to unfreeze gait or more generally to monitor the gait of patients with motor disturbances.

References

1. Auvinet B, Berrut G, Touzard C, Moutel L, Collet N, Chaleil D, Barrey E (2002) Reference data for normal subjects obtained with an accelerometric device. Gait Posture 16(2):124–134
2. Bächlin M, Plotnik M, Roggen D, Maidan I, Hausdorff JM, Giladi N, Tröster G (2010) Wearable assistant for Parkinson's disease patients with the freezing of gait symptom. IEEE Trans Inf Technol Biomed 14(2):436–446
3. Braak H, Ghebremedhin E, Rüb U, Bratzke H, Del Tredici K (2004) Stages in the development of Parkinson's disease related pathology. Cell Tissue Res 318(1):121–134
4. Delval A, Snijders A, Weerdesteyn V, Duysens J, Defebvre L, Giladi N, Bloem B (2010) Objective detection of subtle freezing of gait episodes in parkinson's disease. Mov Disord 25 (11):1684–1693
5. Giladi N, Nieuwboer A (2008) Understanding and treating freezing of gait in parkinsonism, proposed working definition, and setting the stage. Mov Disord 23(2):s423–s425
6. Gonzalez RC, Lopez AM, Rodriguez-Uria J, Alvarez D, Alvarez JC (2010) Real-time gait event detection for normal subjects from lower trunk accelerations. Gait Posture 31:322–325
7. Hausdorff JM, Balash Y, Giladi N (2003) Statistical mechanics and its applications: time series analysis of leg movements during freezing of gait in Parkinson's disease: akinesia, rhyme or reason? Phys A 321:565–570
8. Keus SH, Munneke M, Nijkrake MJ, Kwakkel G, Bloem BR (2009) Physical therapy in Parkinson's disease: evolution and future challenges. Mov Disord 24(1):1–14
9. Lan KC, Shih WY (2012) Using simple harmonic motion to estimate walking distance for waist-mounted PDR. In: IEEE wireless communications and networking conference: mobile and wireless networks
10. Mazilu S, Hardegger M, Zhu Z, Roggen D, Troster G, Plotnik M, Hausdorff JM (2012) Online detection of freezing of gait with smartphones and machine learning techniques. In: 6th international conference on pervasive computing technologies for healthcare (PervasiveHealth) and workshops
11. Menz HB, Lord SR, Fitzpatrick RC (2003) Acceleration patterns of the head and pelvis when walking on level and irregular surfaces. Gait Posture 18:35–46
12. Moore ST, MacDougall HG, Gracies JM, Cohen HS, Ondo WG (2007) Long-term monitoring of gait in Parkinson's disease. Gait Posture 26(2):200–207
13. Moore ST, MacDougall HG, Ondo WG (2008) Ambulatory monitoring of freezing of gait in Parkinson's disease. J Neurosci Methods 167(2):340–348
14. Moore ST, Yungher DA, Morris TR, Dilda V, MacDougall HG, Shine JM, Naismith SL, Lewis SJ (2013) Autonomous identification of freezing of gait in Parkinson's disease from lower-body segmental accelerometry. J Neuroeng Rehabil 10:19
15. Nieuwboer A, Chavret F, Willems AM, Desloovere K (2007) Does freezing in Parkinson's disease change limb coordination? A kinematic analysis. J Neurol 254:1268–1277
16. Nieuwboer A, Dom R, De Weerdt W, Desloovere K, Fieuws S, Broens-Kaucsik E (2001) Abnormalities of the spatiotemporal characteristics of gait at the onset of freezing in Parkinson's disease. Mov Disord 16(6):1066–1075
17. Nieuwboer A, Giladi N (2013) Characterizing freezing of gait in Parkinson's disease: models of an episodic phenomenon. Mov Disord 28(11):1509–1519
18. Pepa L, Ciabattoni L, Verdini F, Capecci M, Ceravolo MG (2014) Smartphone based fuzzy logic freezing of gait detection in Parkinson's disease. In: 10th IEEE/ASME international conference on mechatronics and embedded systems and applications (MESA)
19. Pepa L, Verdini F, Capecci M, Ceravolo MG (2015) Smartphone based freezing of gait detection for parkinsonian patients. In: IEEE international conference on consumer electronics (ICCE)
20. Pepa L, Verdini F, Capecci M, Ceravolo MG, Leo T (2014) An architecture for reducing the freezing of gait during the daily life of patients with Parkinson's disease. Gait and Posture 40: S2

21. Schaafsma JD, Balash Y, Gurevich T, Bartels AL, Hausdorff JM, Giladi N (2003) Characterization of freezing of gait subtypes and the response of each to levodopa in Parkinson's disease. Eur J Neurol 10(4):391–398
22. Schumacher RM, Lowry SZ (2010) Nist guide to the processes approach for improving the usability of electronic health records. National institute of standards and technology, NISTIR 7741
23. Tan DM, McGinley JL, Danoudis ME, Iansek R, Morris ME (2011) Freezing of gait and activity limitations in people with Parkinson's disease. Arch Phys Med Rehabil 92:1159–1165
24. Wittwer JE, Webster KE, Hill K (2013) Rhythmic auditory cueing to improve walking in patients with neurological conditions other than Parkinson's disease-what is the evidence? Disabil Rehabil 35(2):164–176
25. Ziefle M, Schaar AK (2010) Technical expertise and its influence on the acceptance of future medical technologies: what is influencing what to which extent? In: HCI in work and learning, life and leisure lecture notes in computer science, vol 6389, pp 513—529
26. Ziefle M, Wilkowska W (2010) Technology acceptability for medical assistance. In: Proceedings of the ICST conference on pervasive computing technologies for healthcare, IEEE
27. Zijstra W (2004) Assessment of spatio-temporal parameters during unconstrained walking. Eur J Appl Physiol 92:39–44
28. Zijstra W, Hof AL (1997) Displacement of the pelvis during human walking: experimental data and model predictions. Gait Posture 6:249–267
29. Zijstra W, Hof AL (2003) Assessment of spatio-temporal gait parameters from trunk accelerations during human walking. Gait Posture 18:1–10

MIX
Papier aus verantwortungsvollen Quellen
Paper from responsible sources
FSC® C105338

If you have any concerns about our products,
you can contact us on
ProductSafety@springernature.com

In case Publisher is established outside the EU,
the EU authorized representative is:
Springer Nature Customer Service Center GmbH
Europaplatz 3, 69115 Heidelberg, Germany

Printed by Libri Plureos GmbH
in Hamburg, Germany